Environmental Impacts of Road Vehicles
Past, Present and Future

ISSUES IN ENVIRONMENTAL SCIENCE AND TECHNOLOGY

SERIES EDITORS:

R. E. Hester, University of York, UK
R. M. Harrison, University of Birmingham, UK

EDITORIAL ADVISORY BOARD:

P. Crutzen, Max-Planck-Institut für Chemie, Germany, S. J. de Mora, Plymouth Marine Laboratory, UK, G. Eduljee, SITA, UK, L. Heathwaite, Lancaster University, UK, S. Holgate, University of Southampton, UK, P. K. Hopke, Clarkson University, USA, P. Leinster, Cranfield University, UK, P. S. Liss, School of Environmental Sciences, University of East Anglia, UK, D. Mackay, Trent University, Canada, Professor A. Proctor, Food Science Department, University of Arkansas, USA, Xavier Querol, Consejo Superior de Investigaciones Científicas, Spain, D. Taylor, WCA Environmental Ltd, UK.

TITLES IN THE SERIES:

1: Mining and its Environmental Impact
2: Waste Incineration and the Environment
3: Waste Treatment and Disposal
4: Volatile Organic Compounds in the Atmosphere
5: Agricultural Chemicals and the Environment
6: Chlorinated Organic Micropollutants
7: Contaminated Land and its Reclamation
8: Air Quality Management
9: Risk Assessment and Risk Management
10: Air Pollution and Health
11: Environmental Impact of Power Generation
12: Endocrine Disrupting Chemicals
13: Chemistry in the Marine Environment
14: Causes and Environmental Implications of Increased UV-B Radiation
15: Food Safety and Food Quality
16: Assessment and Reclamation of Contaminated Land
17: Global Environmental Change
18: Environmental and Health Impact of Solid Waste Management Activities
19: Sustainability and Environmental Impact of Renewable Energy Sources
20: Transport and the Environment
21: Sustainability in Agriculture
22: Chemicals in the Environment: Assessing and Managing Risk
23: Alternatives to Animal Testing
24: Nanotechnology
25: Biodiversity Under Threat
26: Environmental Forensics
27: Electronic Waste Management
28: Air Quality in Urban Environments
29: Carbon Capture
30: Ecosystem Services
31: Sustainable Water
32: Nuclear Power and the Environment
33: Marine Pollution and Human Health
34: Environmental Impacts of Modern Agriculture
35: Soils and Food Security
36: Chemical Alternatives Assessments
37: Waste as a Resource
38: Geoengineering of the Climate System
39: Fracking
40: Still Only One Earth: Progress in the 40 Years Since the First UN Conference on the Environment
41: Pharmaceuticals in the Environment
42: Airborne Particulate Matter
43: Agricultural Chemicals and the Environment: Issues and Potential Solutions, 2nd Edition
44: Environmental Impacts of Road Vehicles: Past, Present and Future

How to obtain future titles on publication:

A subscription is available for this series. This will bring delivery of each new volume immediately on publication and also provide you with online access to each title *via* the Internet. For further information visit http://www.rsc.org/issues or write to the address below.

For further information please contact:
Sales and Customer Care, Royal Society of Chemistry, Thomas Graham House, Science Park, Milton Road, Cambridge, CB4 0WF, UK
Telephone: +44 (0)1223 432360, Fax: +44 (0)1223 426017, Email: booksales@rsc.org
Visit our website at www.rsc.org/books

ISSUES IN ENVIRONMENTAL SCIENCE AND TECHNOLOGY

EDITORS: R.E. HESTER AND R.M. HARRISON

44
Environmental Impacts of Road Vehicles
Past, Present and Future

THE QUEEN'S AWARDS
FOR ENTERPRISE:
INTERNATIONAL TRADE
2013

Issues in Environmental Science and Technology No. 44

Print ISBN: 978-1-78262-892-7
PDF eISBN: 978-1-78801-022-1
EPUB eISBN: 978-1-78801-176-1
ISSN: 1350-7583

A catalogue record for this book is available from the British Library

© The Royal Society of Chemistry 2017

All rights reserved

Apart from fair dealing for the purposes of research for non-commercial purposes or for private study, criticism or review, as permitted under the Copyright, Designs and Patents Act 1988 and the Copyright and Related Rights Regulations 2003, this publication may not be reproduced, stored or transmitted, in any form or by any means, without the prior permission in writing of The Royal Society of Chemistry or the copyright owner, or in the case of reproduction in accordance with the terms of licences issued by the Copyright Licensing Agency in the UK, or in accordance with the terms of the licences issued by the appropriate Reproduction Rights Organization outside the UK. Enquiries concerning reproduction outside the terms stated here should be sent to The Royal Society of Chemistry at the address printed on this page.

Whilst this material has been produced with all due care, The Royal Society of Chemistry cannot be held responsible or liable for its accuracy and completeness, nor for any consequences arising from any errors or the use of the information contained in this publication. The publication of advertisements does not constitute any endorsement by The Royal Society of Chemistry or Authors of any products advertised. The views and opinions advanced by contributors do not necessarily reflect those of The Royal Society of Chemistry which shall not be liable for any resulting loss or damage arising as a result of reliance upon this material.

The Royal Society of Chemistry is a charity, registered in England and Wales, Number 207890, and a company incorporated in England by Royal Charter (Registered No. RC000524), registered office: Burlington House, Piccadilly, London W1J 0BA, UK, Telephone: +44 (0) 207 4378 6556.

For further information see our web site at www.rsc.org

Printed and bound in the United States of America

Preface

The subject of air pollution has recently come back onto the public agenda. This has been highlighted particularly by the Volkswagen emissions scandal, which has shown how vehicle manufacturers can defy the spirit of the law and, in the case of Volkswagen, even allegedly break the letter of the law, with a consequence of much higher emissions of air pollutants than the regulatory limits intended. It is therefore perhaps unsurprising that road traffic immediately comes into the public mind when air pollutants are mentioned, although there are many circumstances where road vehicles are not the main sources of pollutants affecting the local atmosphere. However, what is often ignored is the fact that road traffic causes pollution of water and soil, as well as creating noise, which can have adverse effects on human health.

The majority of vehicles currently on the road burn fossil fuels in an internal combustion engine, but this will not always be so. Already there are significant numbers of hybrid vehicles on the road that combine an electric motor with an internal combustion engine. Battery electric vehicles, which use solely electric power, are also now available, although sales in most parts of the world remain modest. There is also the option of using fuel cells with fuels such as hydrogen to generate electricity on board in order to power the vehicle. Although in some jurisdictions electric vehicles are referred to as 'zero-emission vehicles', this ignores the fact that pollution is created by the generation of electricity, and there are also environmental costs to the production of batteries and the final disposal of the vehicle. Such external implications of vehicles can be compared through the use of life-cycle analysis, which takes account of the implications of the vehicle right through from the extraction of raw materials for vehicle construction and the production of fuel to the pollutant emissions caused during operation.

In order to set the context for the subsequent chapters, the first chapter by Athanasios Tsolakis and co-authors describes road vehicle technologies and fuels, starting with the present, but also looking forwards to the future. The next two chapters deal with emissions to the atmosphere, with gaseous and particulate greenhouse emissions considered by Magin Lapuerta and co-authors and locally acting (*i.e.* toxic) air pollutant emissions considered by Qingyang Liu and Jamie Schauer. One important point that is made clearly in the latter chapter is that non-exhaust emissions of particles from the wear of tyres, brakes and the road surface are unregulated and are typically now larger sources of particle emissions than the exhaust of the vehicle. Such particles also arise from hybrid and electric vehicles.

The next two chapters deal with specific environmental and human health effects of road vehicles. Ashantha Goonetilleke and co-authors describe the water and soil pollution implications of road traffic, while the cardiovascular health effects of road traffic noise are the subject of the following chapter by Anna Hansell and co-authors. Both are shown to cause significant impacts, which are frequently given inadequate consideration in relation to road traffic.

Two further chapters look to the future. Billy Wu and Gregory Offer describe the environmental impacts of hybrid and electric vehicles. While superficially attractive, such vehicles can cause substantial pollutant emissions, although not necessarily at the point of operation. Hydrogen has frequently been mooted as a possibly cleaner fuel for road transport, which would most likely be used in fuel cells rather than directly combusted. A major benefit of this is likely to be in reducing greenhouse gas emissions, provided that the hydrogen can be generated using electricity from renewable sources. Angelina Ambrose and co-authors take an economic approach to evaluating the developmental implications for Malaysia of hydrogen use as a road transport fuel.

The final two chapters give further valuable perspectives, firstly on end-of-life vehicle recycling and secondly on life-cycle analysis of road vehicles. Jeongsoo Yu and co-authors provide case study information on the fate of end-of-life vehicle recycling in the Far East and show how well-intentioned legislation designed to protect the environment can have unintended consequences that are detrimental to the environment. As mentioned above, a complete evaluation of the impact of road vehicles can be conducted through life-cycle analysis; Michel Vedrenne and co-authors specify the principles of life-cycle analysis and give examples of real-world applications in comparing vehicles of different types.

We are pleased to have compiled a volume giving a very broad overview and perspective on the impacts of road vehicles on the environment. We believe that this will provide a valuable resource for students, practitioners and policymakers alike in seeking information on the key considerations associated with the use of vehicles upon our roads.

Ronald E. Hester
Roy M. Harrison

Contents

Editors	xii
List of Contributors	xiv
Road Vehicle Technologies and Fuels	1
A. Tsolakis, M. Bogarra and J. Herreros	

1	Background	2
	1.1 Fuels and Pollutants Emitted	3
2	Compression Ignition Engines	4
3	Spark Ignition Engines	5
4	Fuels for Transportation	8
	4.1 Fuel Properties	8
	4.2 Alternative Fuels	9
5	Market Share	14
6	Future Trends	16
	6.1 Advanced Combustion Strategies	16
	6.2 Cylinder Deactivation	18
	6.3 Variable Compression Ratio	19
	6.4 Variable Valve Actuation and Atkinson–Miller Cycles	19
	6.5 Stop–Start	20
References		20

Gaseous and Particle Greenhouse Emissions from Road Transport	25
M. Lapuerta, J. Rodríguez-Fernández and J. M. Herreros	

1	Introduction	25
2	Carbon Dioxide Emissions	27
3	Methane Emissions	31
4	Nitrous Oxide Emissions	32

5	Equivalent Carbon Dioxide Emissions	33
6	Particle Emissions	37
7	Future Trends	39
References		42

Local-acting Air Pollutant Emissions from Road Vehicles   46
Qingyang Liu and James J. Schauer

1	Introduction	47
2	Fuel Type, Fuel Quality, and Vehicle Technology	48
	2.1 Fuel Sulfur Reduction	48
	2.2 Fuel Additives, Including Tetraethyl-lead, Methylcyclopentadienyl Manganese Tricarbonyl, and Lube Oil Additives	49
	2.3 Tailpipe NO_x, CO, VOCs, and PM Emission Related to the Combination of Technology and Fuel	49
	2.4 After-treatment Controls for Modern Vehicles	51
	2.5 Fugitive VOC Emissions from Vehicles	52
	2.6 Non-tailpipe PM Emissions from Vehicles	53
	2.7 Electric and Fuel Cell Vehicles	53
3	Evolution of Roadway Emissions	54
	3.1 Primary and Secondary Pollutants	54
	3.2 Changes in Pollutant Concentrations Downwind of Roadways	55
	3.3 Key Air Pollutants Associated with Roadway Emissions	57
4	Impacts on Human Health	60
	4.1 Health Impacts of Near-roadway Exposures and Urban Air Pollution from Traffic Emissions	60
	4.2 The Contributions of Mobile Sources to PM and O_3 in Cities around the World	63
5	Impacts on the Natural and Built Environments	65
6	Impacts on Remote Sites	68
7	Global Trends in Emissions	71
8	Future Technologies and Projected Trends	74
Acknowledgements		77
References		77

Water and Soil Pollution Implications of Road Traffic   86
Ashantha Goonetilleke, Buddhi Wijesiri and Erick R. Bandala

1	Introduction	87
2	Primary Pollutants from Road Traffic	88
	2.1 Pollutant Sources	88
	2.2 Influential Factors in Pollutant Generation	89
	2.3 Primary Pollutants	92

3	Pollutant Processes		94
	3.1	Pollutant Build-up	95
	3.2	Pollutant Wash-off	96
	3.3	Impact of Climate Change on Pollutant Processes	97
	3.4	Pollutant–Particulate Relationships and Mobility of Particle-bound Pollutants	99
4	Impacts of Traffic Pollutants		100
5	Conclusions		101
References			103

Cardiovascular Health Effects of Road Traffic Noise   107
Anna Hansell, Yutong Samuel Cai and John Gulliver

1	Introduction		107
	1.1	Biological Mechanisms	108
2	Assessment of Traffic Noise Exposure in Epidemiological Studies		110
3	Health Studies of Cardiovascular Disease in Adults		115
	3.1	Hypertension	115
	3.2	Cardiovascular Disease Incidence, Morbidity and Mortality	118
	3.3	Cardiovascular Risk Factors	122
	3.4	Further Factors to Consider in the Interpretation of Epidemiological Studies: Confounding and Effect-modifying Factors	125
4	Conclusions		126
Acknowledgements			126
References			127

Environmental Impact of Hybrid and Electric Vehicles   133
Billy Wu and Gregory J. Offer

1	Introduction		134
2	Energy Storage and Conversion Technologies		138
3	Hybrid Vehicles		141
4	Impact of Different Usage Cases		142
5	Life Cycle Assessment		145
	5.1	Battery Utilisation	147
	5.2	Vehicle-to-grid	148
	5.3	Battery Lifetime and Degradation	149
	5.4	Recycling and Second Life	151
6	Conclusion		152
References			153

Development Implications for Malaysia: Hydrogen as a Road Transport Fuel   157
Angelina F. Ambrose, Rajah Rasiah and Abul Quasem Al-Amin

1	Introduction	158
2	Energy Demand, Economic Growth and CO_2 Emissions	158
3	Hydrogen Fuel Cell Vehicles and Hydrogen Pathways	161
4	Concepts in Fostering Hydrogen in Transportation	162
5	Simulation Experiments	164
	5.1 Dynamic Computable General Equilibrium Model	164
	5.2 Malaysian Social Accounting Matrix	165
	5.3 Model Specifications	165
6	Scenarios and Results	167
7	The Way Forward	171
	References	173

Latest Trends and New Challenges in End-of-life Vehicle Recycling   174
Jeongsoo Yu, Shuoyao Wang, Kosuke Toshiki, Kevin Roy B. Serrona, Gengyao Fan and Baatar Erdenedalai

1	Introduction	175
2	Legislation on End-of-life Vehicle Recycling and Its Implications	175
	2.1 Background on the Evolution of Legal Systems	175
	2.2 Comparison of EPR-based ELV Recycling Laws	176
3	Popularization of Next-generation Vehicles and Their Impact on Vehicle Recycling	180
	3.1 Significant Developments in the Popularization of Next-generation Vehicles	180
	3.2 Trends in NGV Popularization	181
	3.3 Effective Utilization of Waste Batteries from Next-generation Vehicles	184
	3.4 Limitations on the Reuse and Recycling of Batteries	185
4	Effects of Second-hand Vehicle Exportation on International Resource Circulation and Emerging Cross-border Environmental Problems	186
	4.1 The Two Sides of Second-hand Vehicle Exportation	186
	4.2 Conditions and Characteristics of Second-hand Vehicle Exportation in Japan	187
	4.3 Analysis of the Condition of Second-hand Vehicle Imports in Mongolia	190
	4.4 Effect of Second-hand Vehicle Imports on Resource Recycling and the Environment	192

5	Environmental Pollution Caused by Improper End-of-life Vehicle Processing in Developing Countries: A Case Study on Lead Battery Recycling in Mongolia	197
	5.1 Potential of Serious Environmental Damage	197
	5.2 Overview of Field Investigations and Their Results	198
	5.3 Challenges from a Case Study	204
6	Environmental Problems Associated with the Proliferation of Used Vehicles in Metro Manila, Philippines	205
	6.1 Current State of Used Vehicles in the Philippines	205
	6.2 Existing Legislation	206
	6.3 Current Proposals to Undertake ELV Recycling	207
	6.4 Future of ELV Recycling in Metro Manila	207
	6.5 Challenges in Undertaking ELV Recycling in the Philippines	208
7	Recommendations and Challenges for the Future	208
Notes and References		209

Life Cycle Assessment of Road Vehicles   214

Michel Vedrenne, Javier Pérez, María Encarnación Rodríguez, Julio Lumbreras and Rafael Borge

1	Life Cycle Assessment: A General Concept	214
	1.1 Definition of the Goal and Scope of the Assessment	215
	1.2 Life Cycle Inventory	216
	1.3 Life Cycle Impact Assessment	217
	1.4 Interpretation of Results and Conclusions	218
2	Life Cycle Analysis: Review of the State-of-the-art	219
3	Life Cycle Analysis of Road Vehicles	220
	3.1 Material Life Cycle of Vehicles	222
	3.2 Fuel Life Cycle of Vehicles	224
	3.3 Vehicle Use Phase	230
4	Uncertainties and Limitations	233
5	Practical Example of Life Cycle Assessment: Comparison of Fuel Types for Cars	234
6	Life Cycle Assessment and the Role of the Road Transport Sector in Urban Air Quality	236
7	Concluding Remarks	238
References		238

Subject Index   243

Editors

Ronald E. Hester, BSc, DSc (London), PhD (Cornell), FRSC, CChem

Ronald E. Hester is now Emeritus Professor of Chemistry in the University of York. He was for short periods a research fellow in Cambridge and an assistant professor at Cornell before being appointed to a lectureship in chemistry in York in 1965. He was a full professor in York from 1983 to 2001. His more than 300 publications are mainly in the area of vibrational spectroscopy, latterly focusing on time-resolved studies of photoreaction intermediates and on biomolecular systems in solution. He is active in environmental chemistry and is a founder member and former chairman of the Environment Group of the Royal Society of Chemistry and editor of 'Industry and the Environment in Perspective' (RSC, 1983) and 'Understanding Our Environment' (RSC, 1986). As a member of the Council of the UK Science and Engineering Research Council and several of its sub-committees, panels and boards, he has been heavily involved in national science policy and administration. He was, from 1991 to 1993, a member of the UK Department of the Environment Advisory Committee on Hazardous Substances and from 1995 to 2000 was a member of the Publications and Information Board of the Royal Society of Chemistry.

Roy M. Harrison, BSc, PhD, DSc (Birmingham), FRSC, CChem, FRMetS, Hon MFPH, Hon FFOM, Hon MCIEH

Roy M. Harrison is Queen Elizabeth II Birmingham Centenary Professor of Environmental Health in the University of Birmingham. He was previously Lecturer in Environmental Sciences at the University of Lancaster and Reader and Director of the Institute of Aerosol Science at the University of Essex. His more than 500 publications are mainly in the field of environmental chemistry, although his current work includes studies of human health impacts of atmospheric pollutants as well as research into the chemistry of pollution phenomena. He is a past Chairman of the Environment Group of the Royal Society of Chemistry for whom he edited 'Pollution: Causes, Effects and Control' (RSC, 1983;

Editors

Fifth Edition 2014). He has also edited "An Introduction to Pollution Science", RSC, 2006 and "Principles of Environmental Chemistry", RSC, 2007. He has a close interest in scientific and policy aspects of air pollution, having been Chairman of the Department of Environment Quality of Urban Air Review Group and the DETR Atmospheric Particles Expert Group. He is currently a member of the DEFRA Air Quality Expert Group, the Department of Health Committee on the Medical Effects of Air Pollutants, and Committee on Toxicity.

List of Contributors

Abul Quasem Al-Amin, Institute of Energy Policy and Research (IEPRe), Universiti Tenaga Nasional (UNITEN), Malaysia

Angelina F. Ambrose, Faculty of Economics and Administration, University of Malaya, 50603 Kuala Lumpur, Malaysia. Email: angelina_ambrose@siswa.um.edu.my

Erick R. Bandala, Division of Hydrologic Sciences, Desert Research Institute, Las Vegas, NV, USA

M. Bogarra, University of Birmingham, Department of Mechanical Engineering, Edgbaston, Birmingham, B15 2TT, UK

Rafael Borge, Department of Chemical & Environmental Engineering, Technical University of Madrid, (UPM), c/ José Gutiérrez Abascal 2, 28006 Madrid, Spain

Yutong Samuel Cai, MRC-PHE Centre for Environment and Health, Department of Epidemiology and Biostatistics, St Mary's Campus, Imperial College London, Norfolk Place, Paddington, London, W2 1PG, UK

Baatar Erdenedalai, Graduate School of International Cultural Studies, Tohoku University, Department of International Environment and Resources Policy, 41 Kawauchi, Aoba, Sendai City, Miyagi, 9808576, Japan. Email: baatar.erdenedalai.t5@dc.tohoku.ac.jp

Gengyao Fan, Graduate School of International Cultural Studies, Tohoku University, Department of International Environment and Resources Policy, 41 Kawauchi, Aoba, Sendai City, Miyagi, 9808576, Japan. Email: fan.gengyao.r5@dc.tohoku.ac.jp

Ashantha Goonetilleke, School of Civil Engineering and Built Environment, Queensland University of Technology (QUT), Australia. Email: a.goonetilleke@qut.edu.au

John Gulliver, MRC-PHE Centre for Environment and Health, Department of Epidemiology and Biostatistics, St Mary's Campus, Imperial College London, Norfolk Place, Paddington, London, W2 1PG, UK

M. Lapuerta, University of Castilla–La Mancha, Edificio Politécnico, Avda. Camilo José Cela, s/n, 13071 Ciudad Real, Spain. Email: magin.lapuerta@uclm.es

Qingyang Liu, Civil and Environmental Engineering, College of Engineering, University of Wisconsin-Madison, Madison, WI, USA and Nanjing Forestry University, Nanjing, China

List of Contributors

Julio Lumbreras, Department of Chemical & Environmental Engineering, Technical University of Madrid, (UPM), c/José Gutiérrez Abascal 2, 28006 Madrid, Spain

Anna Hansell, MRC-PHE Centre for Environment and Health, Department of Epidemiology and Biostatistics, St Mary's Campus, Imperial College London, Norfolk Place, Paddington, London, W2 1PG, UK and Directorate of Public Health and Primary Care, Imperial College Healthcare NHS Trust, London, UK. Email: a.hansell@imperial.ac.uk

J. Herreros, Coventry University, Faculty of Engineering, Environment & Computing, Gulson Road, Coventry, CV1 2TL, UK

Gregory J. Offer, Department of Mechanical Engineering, Imperial College London, London, UK

Javier Pérez, Department of Chemical & Environmental Engineering, Technical University of Madrid, (UPM), c/José Gutiérrez Abascal 2, 28006 Madrid, Spain

Rajah Rasiah, Faculty of Economics and Administration, University of Malaya, 50603 Kuala Lumpur, Malaysia

María Encarnación Rodríguez, Department of Chemical & Environmental Engineering, Technical University of Madrid, (UPM), c/José Gutiérrez Abascal 2, 28006 Madrid, Spain

J. Rodríguez-Fernández, University of Castilla–La Mancha, Edificio Politécnico, Avda. Camilo José Cela, s/n, 13071 Ciudad Real, Spain

James J. Schauer, Civil and Environmental Engineering, College of Engineering, University of Wisconsin-Madison, Madison, WI, USA and Wisconsin State Laboratory of Hygiene, University of Wisconsin-Madison, WI, USA. Email: jjschauer@wisc.edu

Kevin Roy B. Serrona, World Bank – Manila, 26th Floor, One Global Place 5th Ave. Corner 25th St. Bonifacio Global City, Taguig City, 1634, Philippines. Email: wastesoc@gmail.com

Kosuke Toshiki, Faculty of Regional Innovation, University of Miyazaki, 1-1, Gakuenkibanadainishi, Miyazaki, 8892192, Japan. Email: toshiki.k@cc.miyazaki-u.ac.jp

A. Tsolakis, University of Birmingham, Department of Mechanical Engineering, Edgbaston, Birmingham, B15 2TT, UK. Email: a.tsolakis@bham.ac.uk

Michel Vedrenne, Department of Chemical & Environmental Engineering, Technical University of Madrid, (UPM), c/José Gutiérrez Abascal 2, 28006 Madrid, Spain and Air & Environment Quality, Ricardo Energy & Environment, 30 Eastbourne Terrace, London W2 6LA, UK. Email: m.vedrenne@upm.es

Shuoyao Wang, Graduate School of International Cultural Studies, Tohoku University, Department of International Environment and Resources Policy, 41 Kawauchi, Aoba, Sendai City, Miyagi, 9808576, Japan. Email: wang.shuoyao.q7@dc.tohoku.ac.jp

Buddhi Wijesiri, School of Civil Engineering and Built Environment, Queensland University of Technology (QUT), Australia

Billy Wu, Dyson School of Design Engineering, Imperial College London, London, UK. Email: billy.wu@imperial.ac.uk

Jeongsoo Yu, Graduate School of International Cultural Studies, Tohoku University, Department of International Environment and Resources Policy, 41 Kawauchi, Aoba, Sendai City, Miyagi, 9808576, Japan. Email: jeongsoo.yu.d7@tohoku.ac.jp

Road Vehicle Technologies and Fuels

A. TSOLAKIS,* M. BOGARRA AND J. HERREROS

ABSTRACT

Road vehicles are an indispensable part of human daily lives. Compression and spark ignition powertrains have been continuously evolving towards more efficient and cleaner technologies. The social awareness of the impact on the environment and human health of the toxic pollutants emitted during the combustion of fossil fuels has led to the introduction of legislation that restricts the emission limits of road vehicles. Vehicle manufacturers are researching and rapidly developing technologies that can offer both reduced fuel consumption and low emission of nitrogen oxides (NO_x), particulate matter (PM), carbon dioxide, carbon monoxide and unburnt hydrocarbons. This chapter provides an overview of the basic road vehicle transportation concepts from the past to the future trends, from the development of the precise fuel injection systems to recent research in new near-zero NO_x-PM emission combustion modes. Apart from the engine itself, alternative fuels can have benefits in pollution depletion. Bioalcohols, liquefied petroleum gas, compressed natural gas or hydrogen for spark ignition engines and fatty acid methyl esters or hydrotreated vegetable oil for diesel engines are under research. The benefits and barriers of these alternative fuels have been discussed. The inherent trade-off between pollutants and high-efficiency engines and the use of after-treatment systems to reduce engine-out emissions are also explored. The current state of the market share as well as a forecast for the near future are also parts of this chapter.

*Corresponding author.

Issues in Environmental Science and Technology No. 44
Environmental Impacts of Road Vehicles: Past, Present and Future
Edited by R.E. Hester and R.M. Harrison
© The Royal Society of Chemistry 2017
Published by the Royal Society of Chemistry, www.rsc.org

1 Background

Energy demand is forecast to increase by a third by 2040, as reported in the World Energy Outlook by the International Energy Agency in 2015.[1,2] Currently, energy demand is primarily fulfilled by fossil fuels (86% of the total energy required for the global demand[3] in 2014), despite the considerable efforts in promoting the use of renewable energy sources. The increment in the global energy demand is mainly driven by countries that are not members of the Organisation for Economic Co-operation and Development. Therefore, it is expected that fossil fuels are going to continue to play a significant role in the worldwide energy sector.

The transportation sector has a considerable impact on fuel security, as well as on quality of life.[4] Ischaemic heart disease, stroke, lung cancer, chronic obstructive pulmonary disease and acute lower respiratory tract infection caused by ambient air pollution represented 6.7% of all deaths in 2012.[4] Therefore, concerns regarding energy security and the adverse impact on climate change and air quality have motivated regulatory bodies to impose increasingly strict emission limits and the methodologies to quantify them. Currently, vehicle emission and performance evaluation procedures are required to be carried out in a laboratory-controlled environment using a chassis dynamometer. The vehicle emission limits and procedures are dependent on the type of vehicle and geographical region. In Europe, the new European driving cycle is currently being used for this purpose, while in the USA, driving cycles such as FTP7 or US06 are those that are used for emission standards. The above cycles will be replaced by the worldwide harmonised light vehicles test procedure (WLTP) in order to reduce the gap between official and real-word emissions.[5] In Europe, this will be implemented in 2017. There is some scepticism regarding whether the WLTP will actually represent real emissions, and therefore the real driving emissions (RDE) test is planned to be imposed between 2017 and 2021. Instead of laboratory testing, in the RDE, emissions will be analysed using portable emissions measurement systems.[5]

The impact of the transportation sector on the environment and fuel security depends on the vehicle, driver behaviour and transport and mobility patterns. From a vehicular point of view, the materials used to build the vehicle (light weighting), the strategies introduced for enhancing the handling and riding of the vehicle and the energy conversion systems (the main scope of this chapter) are the main factors that affect fuel economy and exhaust emissions. Original equipment manufacturers are offering an enormous selection of vehicle types, modular vehicle configurations and endless energy and emission management strategies to fulfil the needs and requirements of society. Road vehicles are classified based on their application, propulsion system and energy supply/fuel type.

(i) Application: different vehicle categories are adopted depending on the geographical region, weight of the vehicle, number of passengers, utilisation and specific regulation. In road vehicles, two main groups are defined: light and heavy duty.

(ii) Propulsion system: conventional: spark ignition (SI) and compression ignition (CI); hybrid not off-vehicle charging: micro, mild and full-hybrid; hybrid off-vehicle charging: plug-in hybrid and range extender; pure electric vehicle (e-vehicle).

(iii) Energy supply/fuel type: crude oil: gasoline, diesel, natural gas, liquefied petroleum gas (LPG); bio-ethanol; biodiesel; hydrogen; synthetic fuels; power station/power grid.

In this chapter, a review of the current engine technologies in addition to the future trends in the automotive sector is performed, as well as the main alternative fuels that have been researched.

1.1 Fuels and Pollutants Emitted

The current road vehicle fleet is powered predominantly by fossil fuels, with a small proportion being electric vehicles. The main fossil fuels in use are gasoline (petrol), comprising mostly aliphatic and aromatic hydrocarbons (HCs), and diesel, which contains a less volatile mixture of HCs. Alcohols may be blended into gasoline and biodiesel into diesel fuel. Also in use are LPG—mainly propane and butane, and liquefied natural gas (LNG) or compressed natural gas (CNG)—composed mostly of methane.

Atmospheric emissions include the following:

(i) Unburned fuel, derived from evaporation, leakage or inadequate combustion.

(ii) Major combustion gases, carbon dioxide and water vapour:

$$e.g.\ C_8H_{18} + 12.5O_2 \rightarrow 8CO_2 + 9H_2O.$$

(iii) Minor combustion gases; these include products of incomplete combustion (*e.g.* carbon monoxide) and partial oxidation of HCs (*e.g.* aldehydes and ketones). Benzene derives from both unburned fuel and thermal breakdown of other aromatic compounds.

(iv) Compounds synthesised at high temperatures such as nitric oxide (a major component of NO_x) whose main source is from combustion of atmospheric nitrogen and oxygen in the engine:

$$N_2 + O_2 \rightarrow 2NO.$$

(v) Particulate matter (PM), generally measured in the atmosphere as $PM_{2.5}$ and PM_{10}, which is formed from HC fuels in the combustion process.

The emissions of greenhouse gases (carbon dioxide, methane and nitrous oxide) are considered in Chapter 2, and toxic, locally acting air pollutants are discussed in Chapter 3.

2 Compression Ignition Engines

In a CI engine, the fuel–air mixture is auto-ignited due to high temperatures and pressures in the combustion chamber. They have inherently higher thermal efficiencies with respect to their counterpart SI engines due to their higher compression (increased thermodynamic efficiency) and expansion ratios (minimised waste thermal energy discharged to the exhaust[6,7]), less pumping losses (there is no need for a throttle to regulate the load) and closer operation to the ideal cycle.[8] Technological improvements such as the use of high-pressure common rail direct injection (DI) systems and advanced forced induction techniques (*e.g.* variable geometry turbochargers) have raised the demand for CI engines also in light duty vehicles, particularly in Europe. Common rail injection systems overcome some of the limitations of older injection technologies as they enable both high fuel injection pressures and multiple fuel injections at low engine speeds in order to facilitate improved fuel atomisation, vaporisation and fuel–air mixing. Forced air induction allows a larger mass of fuel to be burnt, producing more power for the same size engine or enabling engine downsizing (high power-to-weight ratio). Variable-geometry turbocharger systems offer the possibility to recover part of the waste energy present in the exhaust gas, as well as simultaneously producing low speed boost and low end torque,[6] reducing pumping and friction losses at part-load operation and improving fuel economy and CO_2 emissions overall.

However, conventional CI engines emit higher levels of particles and oxides of nitrogen (NO_x) emissions compared to conventional SI engines. The presence of local rich-in-fuel heterogeneous air–fuel regions (due to the short time available for air–fuel mixing[7]) and the locally high flame temperature are responsible for the formation of soot or PM. Those high flame temperatures and the presence of oxygen and nitrogen are also responsible for NO_x emission formation. Exhaust gas recirculation (EGR) has been applied as a strategy to control NO_x emissions, reducing the oxygen availability in the combustion chamber (dilution effect) as well as increasing the overall heat capacity of the cylinder by adding CO_2, water vapour (H_2O) and N_2, which reduces the local flame temperature (thermal effect) and thus NO_x formation.[9] Although, EGR is effective at reducing NO_x formation, it also increases soot formation (PM–NO_x trade-off) as a result of the reduced oxygen availability in the combustion chamber (dilution effect).[10] Therefore, simultaneous reductions of particles and NO_x emissions in the engine cylinder when applying conventional combustion strategies (*e.g.* multiple injections, forced induction and EGR) have been always a challenge due to the well-known NO_x–particulate emissions trade-off.[11] This trade-off has motivated the search for cleaner fuels as well as exhaust gas after-treatment technologies to meet the increasingly strict emission regulations. An overview of the transformation of the diesel engine technologies to control pollutants as a consequence of the European legislation is presented in Figure 1.[12]

A combination of different catalysts (after-treatment systems) is required for the simultaneous removal of the pollutants. Diesel oxidation catalysts (DOCs)

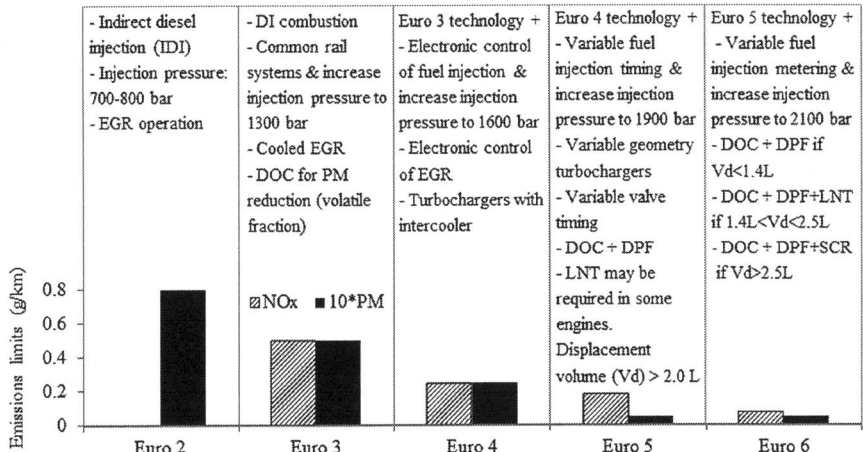

Figure 1 Diesel technology evolution as a consequence of European legislation. Adapted with permission from ref. 12.

are commonly used to oxidise CO, gas-phase HCs and the organic fraction of diesel particles to CO_2 and H_2O. Additionally, the oxidation of NO to NO_2 (which has been reported to facilitate passive diesel particle filter [DPF] regeneration and the performance of some selective catalytic reduction [SCR] catalysts) also occurs over the DOC. DPFs are used for the control of particle emissions together with a combined strategy of engine modification (*e.g.* high-pressure pumps, increased number of smaller injector nozzles and multiple injections).[13] Although this has been successfully introduced, energy-efficient and controlled filter regeneration methods are still being developed in order to maintain their long-term operation. NO_x emission abatement during lean exhaust conditions is much more difficult than in SI engines, especially under low exhaust gas temperatures. Currently, the legislative NO_x emissions of diesel vehicles have been met through engine control methods alone. However, with future legislation limits becoming more stringent, after-treatment NO_x control methods are recommended. There are two distinct approaches being developed. These include NO_x-SCR to N_2 with ammonia (NH_3-SCR) or HC (HC-SCR) and NO_x trapping with periodic reductive regeneration (lean NO_x trap/NO_x adsorber catalysts). In addition, advanced in-cylinder abatement techniques that rely on a combustion process at relatively low local temperatures and the absence of heterogeneous local rich-in-fuel air–fuel regions are also being researched and discussed (low-temperature combustion).

3 Spark Ignition Engines

Since the inception of the first SI engine more than a century ago, gasoline technology has witnessed several transformations towards better efficiency and lower emissions. Unburnt HCs, carbon monoxide and nitrogen oxide have traditionally been the problematic emissions of SI engines. Three-way catalysts (TWCs) have been used since the 1970s as the only after-treatment device for gasoline,[14] as they are capable of simultaneously controlling total hydrocarbons (THCs), CO and nitrogen oxides with efficiencies up to 99% by limiting the presence of oxygen in the exhaust stream under stoichiometric engine operation. Therefore, any technological advance in SI engines has always been influenced by the presence of the TWC. Probably one of the most notable examples of gasoline development is the evolution of the fuel delivery system.[15] Carburettors were the main fuel delivery systems in the late 1970s and early 1980s.[15] However, with carburettors, as the air/fuel ratio was not maintained at close to stoichiometric levels, the use of a TWC for emission control was not viable.[16] In the early 1980s, carburettors were replaced with the first fuel injection system,[15] throttle body injection, which became widespread between 1990 and 2005, particularly for trucks.[15] However, there were still issues associated with slower transport of fuel than air from the upstream of the throttle plate to the cylinder.[6] Between 2000 and 2015, the fuel delivery system started changing to port fuel injection (PFI). In PFI, the fuel is injected in the intake manifold and the fuel quantity is more accurately controlled per cylinder and per the amount of air needed.[6] The required air/fuel ratio for efficient TWC operation was maintained close to stoichiometric levels using a lambda sensor in the engine exhaust. The advantages of PFI over carburettors are the higher achievable engine volumetric efficiency that leads to increased power and torque output, more uniform fuel distribution, precise control of lambda during cold start and engine warm-up, fuel quantity being controlled per cylinder, more rapid engine response to throttle position changes and lower noise.[6,16] Despite these improvements, PFI engines cannot meet the current targets for CO_2 emissions (130 g km^{-1} in 2015 and 95 g km^{-1} by 2020) and the stricter pollutant emission legislation.[8,16] In the early 1990s, gasoline DI (GDI) engines were successfully developed and started replacing PFI systems.[15] The first engine produced using DI was developed by Mitsubishi in 1996.[17]

GDI engines usually operate at stoichiometric conditions near full load with early injection in the admission stroke to obtain a homogeneous air/fuel ratio in the whole combustion chamber. At part and low engine loads, the fuel injection timing could be delayed to the compression stroke; this enables the mixture to be stratified and for there to be a rich-in-fuel region close to the spark plug to start the combustion, but the global mixture is lean (*i.e.* the combustion occurs with air excess).[8,17] Lean air/fuel mixtures burn more slowly and at lower pressure and temperature peaks than stoichiometric combustion, leading to reduced knock tendency (a characteristic noise produced in SI engines when the end-gas auto-ignites, producing high-pressure waves that can damage the piston) and allowing increased compression ratios.[8] GDI advantages are reduced fuel consumption and thus lower tank-to-wheel CO_2 emissions. Injecting the fuel directly into the combustion chamber improves the accuracy of the fuel quantity and therefore the control of air/fuel ratios during dynamics. The charge cooling increases the volumetric efficiency and reduces knock probability, allowing for an increase in the compression ratio. Other advantages are the reduced throttle losses of the gas exchange, higher thermal efficiency due to the possibility of stratified operation, lower heat losses, rapid starting, better cold start performance and drive control.[16–18] Depending on the air–fuel mixture preparation, GDI engines can be classified as wall guided, air guided or spray guided. These three follow different approaches to achieve the same objective: create an ignitable mixture before the spark event. Wall- and air-guided systems are considered the first-generation GDI engines. In wall-guided engines, the fuel is directed towards a specially shaped piston bowl that moves the flow to the spark plug. This system has several inefficiencies derived from the fuel impingement in the piston resulting in fuel consumption, CO, unburnt HCs and soot penalties. In air-guided systems, the air flow is in charge of moving the fuel to the spark plug. The fuel impingement in the piston is avoided and thus soot formation does not occur. However, this system is highly dependent on air motion and suffers from high cycle-by-cycle variations. Spray-guided systems are known as the second-generation GDI engines. In these cases, the fuel is injected near the spark plug and the injector is located between the intake and exhaust valves contrary to wall- and air-guided systems in which the injector is side mounted. The advantages of spray-guided systems are the reduced wall wetting, the performance at part and full loads being optimised simultaneously, the increased stratified range, the reduced sensitivity to cylinder-by-cylinder variations, the reduction in HCs and the better fuel economy. However, the harsher in-cylinder conditions promote the fouling of the spark plug and injectors, which can deteriorate combustion stability.[16,17,19] GDI engines can be also coupled with other technologies, such as turbocharging, downsizing and variable valve timing (VVT), which can further exploit the potential of DI to achieve higher efficiencies and further fuel economy improvements. For instance, engine downsizing shifts the engine operation to wide open throttle, avoiding the least efficient part load operation of SI engines and so improving fuel economy.[17] Furthermore, the reduced surface area lowers friction losses and the smaller engine decreases vehicle weight.[17] It has been reported that fuel consumption reductions of between 12.1% and 24.6% can be achieved.[20] The challenge associated with downsizing is combustion limitation, as these engines are more prone to knock, reducing the efficiency and transient operation.[21] The term 'rightsizing' is currently used by automotive manufacturers and is defined as the optimum combination of displacement, power and torque delivery, fuel consumption and operating characteristics for a given application.[22]

Despite the potential of GDI engines, the main drawbacks are the increased level of PM compared to PFI and injector deposit formation.[17,23–25] While injector deposits are detrimental for the fuel spray pattern and for the engine's efficiency, PM is known to be a toxic pollutant and any potential increase raises concerns about health effects. PM is formed from a carbonaceous core known as soot onto which HCs, including polyaromatic HCs, are adsorbed. PM has been classified by the International Agency for Research on Cancer as carcinogenic to humans,[26] and there are several respiratory and cardiovascular exacerbations related to its inhalation.[27] In addition, soot is a contributor to global warming.[28] For all of the above reasons, GDI vehicles have been included in the upcoming legislation to limit the particle number emitted. This is the case of the Euro 6c, in which the particle number limit is established at 6×10^{11} particles kg^{-1}.[29] Although some authors claim that with the optimisation of engine parameters this limit can be met,[30] others suggest that the introduction of gasoline particulate filters will be necessary to meet more stringent legislation.[31] Furthermore, as in lean conditions the TWCs cannot be employed, other technologies, such as EGR or lean NO_x traps, might be needed to meet NO_x limit legislation.

4 Fuels for Transportation

4.1 Fuel Properties

Transport and mobility demands and the needs of our society are mainly fulfilled with internal combustion engine vehicles powered by fossil fuels. As in the case of emission standards, the quality and methods for measuring fuel properties are strictly regulated in the majority of countries, and the limiting values are dependent on the fuel and the region. For example, in Europe, the EN228 and EN590 standards specify the fuel properties for gasoline and diesel fuel, respectively. Similarly, the ASTM D4814 (gasoline) and ASTM D975 (diesel) are the equivalent standards for fuels in the USA. In this section, the main fuel properties that are regulated in the current legislation will be discussed.

Conventionally, the fuel properties that determine fuel-powertrain suitability are the octane and cetane rating. Fuels with high octane ratings are desirable for SI engines in order to avoid abnormal fuel combustion such as pre-ignition and detonation (knock), which deteriorates combustion efficiency and could cause severe damage to the engine components. High-octane number fuels also enable the use of higher compression ratios, resulting in better thermal efficiency. On the other hand, fuels with good auto-ignition properties (high cetane rating) are more suitable for conventional CI engines, ensuring the ignition of the heterogeneous air–fuel mixture. With the development of advanced combustion technologies that hybridise compression and SI characteristics, fuels and/or fuel blends with intermediate cetane and/or octane numbers could be desirable. Fuel volatility, which is the fuel's capability to be vaporised, is also regulated as it is related to the engine's cold start performance. Low T10 (the temperature at which 10% of the fuel is vaporised) improves cold start, while high T90 (heavy HCs evaporating at high temperatures) will increase the level of PM and deposit formation.[32] Fuel lubricity, oxidation stability and cold flow properties are also regulated, as even if a fuel provides excellent performance and emissions behaviour, if the fuel lubricity, oxidation stability and cold filter plugging point are not acceptable, the usage of the fuel is not recommended unless some appropriate additives are incorporated. As the fuel injection systems are based on volumetric flow rates, fuel density is also considered as it influences the volumetric energy content of the fuel. Fuel viscosity is also regulated as fuels with high viscosity will offer a high resistance to flow, resulting in a high demand of energy to transport the fuel from the tank to the injector. Depending on the origin of the crude oil, the sulphur content can vary. Sulphur is linked with acid rain and catalyst poisoning, and therefore its content is strictly limited in the countries following the European or American standards. For instance, fuel standards in Europe impose a limit of 10 mg kg^{-1} of fuel. There are fuels that cannot be directly used in the engine because they do not fulfil some of the properties previously mentioned (*e.g.* high viscosity of vegetable oils), simply because of the regulations. In these cases, some modifications to the engine and/or the fuels have to be made. A simple approach is to directly blend the alternative fuel with the fuel that is commonly used with this specific powertrain technology. If this approach is adopted, it is recommended to test fuel miscibility and blend stability in order to avoid the risk of blend separation.

As a result of current targets for CO_2 reductions, the European Directive 2009/28/EC and the American Renewable Fuels Standard (RFS) have promoted the use of biofuels in transportation, and the use of renewable fuels has been extended. Currently, 7% of renewable oil must be blended in diesel and gasoline fuels. In Europe, this percentage must be increased to 10% by 2020.[33] According to the RFS, renewable fuel production must be 36 billion gallons by 2022, a 22.9 billion gallon increase compared to 2009.[34] The following sections present a selection of alternative fuels.

4.2 Alternative Fuels

4.2.1 Alcohols. Alcohols can be obtained from fossil fuels including natural gas, crude oil and coal, as well as from renewable feedstock. Renewable alcohols are usually classified as first generation when the feedstocks are edible materials and as second generation when non-edible feedstocks are used. Taking into account the alcohol characteristics such as their high octane rating, low cetane number, poor lubricity, high flame speed, wide flammability limits and high volatility, they are more suitable for SI engines than for CI engines. When alcohols are used as fuels in SI engines, extensive vehicle modifications are not necessary. However, issues with material compatibility in the fuel system due to the corrosive characteristics of alcohols and at cold start, part load and transient operation due to their low vapour pressure need to be addressed. Also, the lower energy content of alcohols than gasoline is a concern.[35] They are also hygroscopic (tendency to absorb water), leading to mixture separation, so special care in handling and storage is required. To take advantage of the higher octane rating compared to petrol, higher compression ratios can be employed.[36] The use of alcohols in CI engines is attractive for reducing CO_2 and PM emissions, but such use requires changes in the fuel (fuel additives to improve the cetane rating or emulsification of the alcohol in diesel),[37,38] in the powertrain to adapt the CI engine (alcohol fumigation, duplication of the injection system or the use of spark or glow plugs) and/or in the nature of the combustion process (advanced combustion operation).

Conventional studies are focused on the use of blends in diesel fuel using relatively low alcohol percentages because no major modifications in the engine are required.

Methanol is the simplest alcohol not containing carbon-to-carbon bonds. Methanol has a lower carbon/hydrogen ratio than gasoline and diesel and provides better energy efficiency, resulting in lower CO_2 emissions. Bioethanol presents some challenges relative to the high cost of production feedstock and the production process. First-generation bioethanol can be obtained from sugar cane, sugar beet, corn and other grains, but due to the issues relative to first-generation biofuels, great efforts are being made to use non-edible materials such as woody and herbaceous crops (lignocellulosic biomass) to address those issues and to develop new production processes to reduce the cost of bioethanol production.[39] The energy density of ethanol is lower than gasoline and diesel but higher than methanol, so the reduction in the vehicle's range is lower than in the case of methanol. Butanol is another primary alcohol that is also used in internal combustion engines. Until recently, not much attention was paid to butanol, possibly due to its higher production costs compared to other primary alcohols, but currently there is an increased interest in the use of butanol (particularly *n*-butanol) as a new, alternative fuel due to the possibility of using renewable feedstocks (biobutanol). In the case of butanol, one of its traditional disadvantages compared to bioethanol is its higher production costs, but new production processes utilising fermentation of agricultural feedstocks by cellulosic enzymes are being developed in order to reduce and optimise the production costs. Comparing butanol and ethanol properties, the former provides some general advantages; for example, butanol is less corrosive than ethanol, has a higher energy density, is less prone to water contamination and has a higher flame speed.

4.2.2 Biodiesel. Biodiesel is the name given to fatty acid methyl or ethyl esters produced from virgin or used vegetable oils and animal fat *via* a transesterification process.[40,41] Edible crops with high oil content were first used to produce biodiesel, but there were some issues associated with the use of arable land to grow energy crops. Thus, the use of residual feedstocks such as waste cooking oil and grease tallow has also been considered. New feedstock and production technologies have been developed based on the transformation of lignocellulosic material to liquid fuel by means of thermochemical or biological processes, as well as the use of aquatic species such as algae. For given vehicle operating conditions, the use of biodiesel increases the volumetric fuel consumption when compared to diesel due to the lower heating value of the biodiesel.[42–45] In agreement with that, most of the published studies have reported equal thermal efficiency compared to diesel.[42,46,47] PM,[40,42,48] total HC[42,45,49] and CO[42,45,50] emissions with biodiesel are usually lower. The oxygen content[51–54] and the absence of aromatic compounds[53,55] in biodiesel are the most reported factors to justify the decrease in PM, THC and CO emissions. On the other hand, the majority of the studies have reported a slight increase in NO_x with the use of biodiesel,[56–58] even though some discrepancies can be found depending on the engine technology, engine operating conditions, injection system, timing and engine maintenance and biodiesel composition, such as chain length and unsaturation level.

4.2.3 Synthetic Fuels. The scarcity of fossil fuels, the economic, technical and social issues associated with vegetable oils and the need to find cleaner and efficient liquid fuels has motivated intensive research to develop processes that are able to synthesise liquid fuels from different carbon sources. The production of synthetic liquid fuels generally consists of four steps starting from: (i) the identification of the source(s) of carbon; (ii) the production of the synthetic gas (which is a mixture of carbon monoxide, hydrogen and HC components); (iii) the synthesis process from the synthetic gas to liquid fuel; and (iv) the purification process. Synthetic fuels are categorised depending on the state of the final product as power to liquid or power to gas; depending on the source of carbon as coal to liquid, gas to liquid or biomass to liquid; and depending on the synthesis process as Fischer–Tropsch or biological power to liquid. Fischer–Tropsch is the most common process of synthesising fuels catalytically, producing a range of saturated liquid HCs from synthetic gas. The resultant liquid product is compatible for use in CI engines, being able to simultaneously reduce NO_x and carbonaceous emissions with respect to the combustion of conventional diesel fuel. Furthermore, the non-aromatic content leads to a reduction in PM.[59,60]

4.2.4 Hydrotreated Vegetable Oil. Apart from transesterification, an alternative process for converting biomass to liquid biofuels is through hydrotreating. In the hydrotreated vegetable oil (HVO) process, the hydrogen removes the oxygen from the triglycerides (vegetable oils), and as a result a mixture of paraffinic alkanes free of impurities (sulphur, metal and aromatic HCs) is produced. HVO is known as 'renewable diesel' in order to differentiate it from the biodiesel obtained by transesterification.[61] The by-product is LPG, water and CO_2, but no glycerol, as in the case of the esterification process.[61] Different feedstocks such as food vegetable oils (rapeseed oil, sunflower, soybean and palm oil), non-food vegetable oils (jatropha and algae oils) and waste animal fats can be used in the process without affecting the quality of the final fuel.[61] HVO meets diesel standards (EN590) and has similar properties to commercial diesel, although it has a lower density than pump diesel and lubricity must be improved with additives.[61,62] HVO's cetane number is between 84 and 99, compared to 53 for diesel.[62] Furthermore, HVO has higher resistance to oxidation than biodiesels.[62] HVO can be used in current diesel vehicles without engine modification. Engine tests have shown reductions of 6% and 35% in NO_x and PM, respectively, using 100% HVO in a heavy-duty diesel engine.[61] From 2007 to 2010 in Helsinki, HVO was tested in urban buses, and a 30% reduction in PM and a 10% reduction in NO_x were achieved.[63] Also, in passenger cars, considerable reductions in PM of 39% on average have been reported.[64]

4.2.5 Hydrogen. Hydrogen has been seen as a promising energy carrier for both CI and SI internal combustion engines due to its high energy density by weight (143 MJ kg^{-1}), wide flammability limits, low ignition energy and high-speed flames.[65,66] However, in terms of volume, its energy density is 3000-times lower than gasoline.[65] Hydrogen can be produced using a wide variety of methods such as methane steam reforming and coal gasification (from fossil sources or nuclear-assisted), as well as electrolysis of water from solar or wind energy.[67,68] The low energy density together with the high risk of explosion makes hydrogen storage for road transport a challenge. Several alternatives to hydrogen storage such as mechanical techniques (compressed and liquefied), chemical hydrides and adsorption materials have been investigated, but these methods are still under research as they do not yet meet governmental targets.[65] An alternative for solving the hydrogen storage challenge is its on-board production by the catalytic conversion of hydrogen carriers such as HCs[69] or ammonia.[70]

Hydrogen shows benefits in both CI and SI engines. As hydrogen is a non-carbon fuel, its addition results in lower CO, CO_2, HC and PM emissions.[69] The smaller quenching distance of hydrogen allows the flame to reach the cylinder crevices, resulting in a more complete combustion and a lower level of unburnt HCs.[71] However, the higher in-cylinder flame temperature of hydrogen combustion during normal diesel operation leads to a NO_x penalty.[69] This can be palliated with the use of EGR, as hydrogen combustion is more stable on account of the lower minimum ignition energy, which can also ease engine cold start, allowing higher EGR percentages.[71,72]

4.2.6 Natural Gas. Natural gas is an abundant resource that can be directly extracted or can be produced from coal and from biomass (biogas).[73] Methane is the main component of natural gas, with its use in the transportation sector being limited by its low energy density. In Germany, billions of euros were invested to promote the utilisation of natural gas in automotive applications. However, despite this, the initiative was not well received in the markets. The lack of infrastructure is thought to be the reason for this.[74] Similarly to the case of the German market, CNG utilisation in the American and Japanese markets was not well adopted.[74] Natural gas can be stored as compressed gas (CNG) or cryogenically liquefied (LNG) at temperatures below −161.5 °C. Natural gas is most commonly used in SI engines due to its high octane rating. It can be used in bi-fuel systems, which combine the use of petrol and natural gas. CNG and LNG readily mix with the air when they are injected, creating a complete and uniform mixing of fuel and air, which facilitates the combustion process, particularly engine warm-up. On the other hand, there is a loss of volumetric efficiency because of the air being displaced in the cylinder,

Road Vehicle Technologies and Fuels   13

resulting in a 10–20% power loss, unless DI of CNG is applied. It can also be used in dedicated natural gas engines, which can be optimised using a high compression ratio thanks to the high octane rating of natural gas and lean in-fuel operation (lambda in the region of 1.3–1.5) due to the wide flammability limits of natural gas. These factors, together with the low C/H ratio of methane and the absence of aromatics, will result in improvements in fuel economy, CO_2 emissions and PM, CO and non-methane HC species with respect to conventional SI engines operated with gasoline. However, NO_x and methane emissions are much higher than those with conventional liquid fuels in dedicated natural gas engines. Due to the low cetane rating of natural gas, its use in CI engines is less common. A pilot diesel injection (dual-fuel engine) or the addition of an ignition system and spark plugs to initiate the injection of natural gas are required.[75]

4.2.7 Liquefied Petroleum Gas. LPG is composed of a limited number of short-chain HC species such as propane and butane, with some traces of propylene, butylene, ethane and pentane, whose final composition depends on the country. In some countries, LPG has received government tax incentives, making it a relatively cheap, and it is also incentivised by encouraging the recovery of LPG from oil wells. However, governments have started to reduce the tax differential between LPG and conventional fuels, and further reductions in these tax incentives are planned over the coming years. As in the case of natural gas, a large pressurised fuel tank (around 14 bar) is required due to its reduced energy density with respect to conventional liquid fuels. LPG is instantly vaporised, facilitating the air–fuel mixing, and its consistent/simple fuel composition increases combustion efficiency, but it also displaces part of the intake air, decreasing volumetric efficiency and then power output unless forced induction (*e.g.* supercharging and turbocharging) and/or direct LPG injection are used. Due to the characteristics of LPG, including its high octane rating, it is mostly used in SI engines. LPG can be used in dual-fuel gasoline–LPG engines with the added complication of two injection systems, or in SI engines converted to work exclusively with LPG, but they cannot be fully optimised. Ideally, the engine should be designed for LPG usage using a higher compression ratio than conventional SI engines, adequate forced induction and ignition systems and optimal injection strategy. In optimised engines, and taking into account the low C/H ratio of LPG, due to the absence of aromatic compounds and the possibility of operating the engine in overall lean conditions (wide flammability limits), LPG will produce higher thermal efficiencies and lower CO_2, CO, unburnt THC and particle emissions. However, the propane and butane emissions will be higher, and there is not a clear trend in NO_x emissions, which could be equal, higher or lower depending on the operation. In the case of CI engines, only the alternative of dual-fuel technologies is feasible. The drawbacks are similar with this type of technology, such as the difficulty in optimisation and the need to maintain two fuel and injection systems.

5 Market Share

The automotive industry is a major economic and industrial force[76] and vehicle sales is a synonym of economic growth.[77] Since the economic crisis in 2009, vehicle sales have again increased 35% in the last 5 years, 46% of which are concentrated in China and the USA. By 2030, in Europe, it is expected that 297 million vehicles, including passenger cars, light commercial vehicles and heavy-duty vehicles, will be on the road, producing a total of 1022 million tons year^{-1} of CO_2 emissions. The implementation of strategies resulting in improvements in fuel economy (*e.g.* light weighting and drag reduction) and the progressive adoption of low-carbon vehicle systems (*e.g.* electric and fuel cell vehicles), being partially motivated by greenhouse emissions policies, are reflected in the CO_2 emissions downward trend,[77] as well as in the market share of road vehicles. In 2013, for the first time, the average CO_2 vehicle emissions fell below 130 g km^{-1}, and in 2014, they reached 123.3 g km^{-1}.[76] The reduction of CO_2 emissions is more noticeable in SI vehicles since 2005 due to the introduction of the GDI technology, as well as the smaller gap between the SI and CI engines' efficiencies (126 g km^{-1} and 123 g km^{-1}, respectively). In the case of passenger cars, it seems to be challenging to reach the target of 95 g km^{-1} by 2020,[76] while light commercial vehicles have already accomplished the 2017 target in 2013 of 175 g km^{-1}; the future target for 2020 will be 147 g km^{-1}.[76] The European Union is the foremost research and development investor, with 41.5 billion euros spent in 2013, followed by Japan and the USA, which invested 24 and 12.5 billion euros, respectively.[77]

In 2014, CI propulsion systems were the preferred propulsion systems in Europe, accounting for 53% of the sales in passenger vehicles and reaching 97% for light commercial vehicles.[76] In India, an emerging market, 50% of the passenger vehicles were also CI.[76] The trend was different in the USA, China and Japan, with petrol-powered vehicles dominating the markets. Gasoline dominates the US market, with a 94.6% share in 2015,[15] although the number of passenger vehicles decreased by approximately 12% from 2010 to 2014. On the other hand, the E85 (85% ethanol in gasoline by volume)-fuelled vehicle market share increased substantially in those years, and the presence of such vehicles on American roads is greater than diesel passenger cars. The E85 fleet in 2014 was around 200 000 vehicles.[20] Due to the lower CO_2 emissions of GDI engines, this technology has been widely accepted worldwide. The overall market share in 2014 in Europe was 35%.[76] GDI was first introduced in 2007 and, by 2015, 46% of vehicles had incorporated this technology in the US market,[15,20] with it being thought that this technology has not reached its maximum penetration rate yet. It has also to be noted that, in 2015, 85% of GDI cars were also turbocharged.[15]

The number of the new registrations of passenger electric vehicles, including plug-in hybrid electric vehicles (PHEVs) and battery electric vehicles (BEVs), increased by 70% between 2014 and 2015, with over 550 000 sales worldwide.[78] Hybrids were first introduced in 2000 in the USA.[15] In 2015, the market share of gasoline hybrid vehicles was 3% of all new vehicles,[15,76] although sales have dropped since 2013.[20] In Japan, 20% of all new sales are hybrid.[76] China overtook the USA in hybrid vehicle registrations in 2015.[78] In Europe, 1.4% of total vehicle sales were hybrid-electric, although there are substantial differences between Member States, as sales are dependent on government incentives.[76] In Norway, the electric vehicle market, including PHEVs and BEVs, accounted for 13.8% of sales in 2014, and in the first quarter of 2015, sales rose to 22.9%, making Norway the world's leading market in terms of market share. In The Netherlands, the taxation scheme based on high reductions for vehicles emitting less than 50 g km^{-1} of CO_2 has made the electric market grow significantly; 3.1% of new vehicles were PHEVs and 0.9% were BEVs in 2013. However, in 2014, electric vehicle sales decreased due to a change in the rebate scheme. Despite BEVs and PHEVs being commercially available, there are still issues regarding the battery technology, and it is thought that a 'battery breakthrough' will be needed to achieve practicality.[79] A summary of the market share by fuel is reported in Figure 2.[76]

Registrations of heavy-duty vehicles, including trucks, vans and buses, were recorded at 18.4 million in 2014: 26.9% in America, 13.3% in Europe, 54.2% in Asia and 5.6% in the Middle East.[77] Truck and bus sales in the EU in 2014 were 0.3 million, which was 30% lower than sales before the economic crisis, but showing an increasing trend over the preceding 5 years. In USA, sales of trucks also increased notably from 2011 to 2015.[20] Class 3 (light-duty trucks weighing between 4536 and 6350 kg) and Class 8 (heavy-duty trucks whose weight exceeds 14 969 kg) sales increased by 45%, as well as sales of Class 4–7 trucks increasing by 47%. Most of the vehicles corresponding to Classes 3–8 were powered by CI engines (74% in 2014).[20] Almost 100% of Class 8 trucks were diesel, while the use of diesel for Class 4 trucks has fallen to 66%. Significant efforts are being made to reduce fuel consumption in heavy-duty vehicles in the USA. Electrification during truck idle stops can reduce fuel consumption significantly.[20] For Class 8 trucks, aerodynamic drag is a large source of energy loss. Thus, devices such day cab roof deflectors, sleeper roof fairing, chassis skirts or cab extenders have been adopted. Fuel consumption improvements of between 2% and 10% have been achieved.[20]

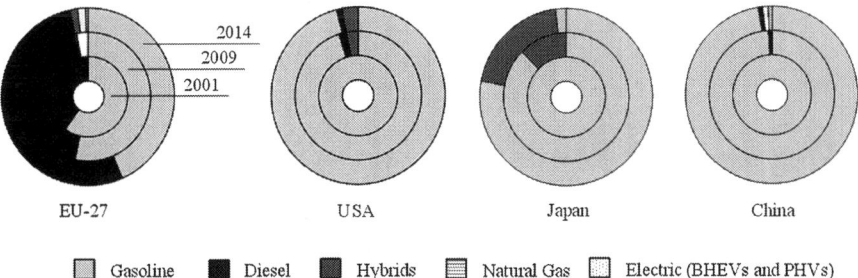

Figure 2 Market share by fuel in passenger cars.
Reproduced with permission from ref. 76.

Amongst the technologies used to achieve better fuel economy, cylinder deactivation[15,20] accounted for 13% of production, while stop–start was 7% for non-hybrid vehicles and 9% when hybrids are considered.[15] Turbocharged engines achieved 18% of the market share in 2015.[15] Cooled exhaust-gas recirculation accounted for 5% and turbocharging and downsizing for between 12.1% and 24.7%, depending on the vehicle attributes.[20] In 2015, almost 20% and 12% of cars and trucks, respectively, were turbocharged.[15] For the near future, two models of fuel cells are expected for 2016–2018 with similar ranges to conventional vehicles.[20]

6 Future Trends

6.1 Advanced Combustion Strategies

The classical concept of combustion is currently changing and moving towards hybrid combustion between conventional CI (diesel) and SI (gasoline) engines. Maintaining the high efficiency and high compression ratio of the CI engine combined with the premixed/ideally homogeneous charge of the SI engine to avoid soot formation without the need for spark and throttle is the aim of the new, advanced combustion strategies. In order to achieve this aim, the combustion must occur in low-temperature combustion regions where the formation of soot and NO_x is limited.

The first attempt to apply this combustion concept to four stroke engines is known as homogeneous charge CI (HCCI) or controlled auto-ignition, which began in 1970,[80,81] although it was not until the 1980s that this strategy became popular.[82,83] In HCCI, the air and fuel are premixed to form a homogeneous mixture, as in SI engines, with overall lean local equivalence ratios, as in CI engines, typically lower than 0.5, which is auto-ignited during the compression stroke. To produce lean charge, high percentages of EGR are used. The combustion temperature must be maintained at between 1500 and 1800 K; temperatures higher than 1500 K ensure the full conversion of CO to CO_2, while temperatures lower than 1800 K limit NO_x formation.[84,85] The efficiency of HCCI is similar to that of CI engines, and while NO_x and PM are drastically reduced, HC and CO emissions are increased equivalently due to the low in-cylinder temperatures.[86] Moreover, as the combustion is dominated by chemical kinetics, controlling the start of combustion and the rate of heat release is a major challenge, resulting in a narrow operation range. At low loads, the engine suffers from ignition difficulties and low combustion efficiency due to the high-dilution and low-temperature conditions. On the other hand, at high load, the fast heat release leads to intense cylinder pressure rates, which could produce ringing and engine damage.[19,80,84–86] HCCI can be performed with both diesel and gasoline fuels.[84] Gasoline is more volatile, promoting mixture homogeneity; however, it is highly resistant to auto-ignition, in contrast to diesel.[87] Therefore, an optimal fuel that facilitates mixture formation and optimal auto-ignition properties is needed for successful HCCI operation.

To overcome HCCI obstacles while maintaining the high thermal efficiency and reduced NO_x and PM, several approaches have been initiated. Partially premixed combustion or premixed charge CI (PCCI) assists in the control of the combustion event. Multiple injection events could be also applied, in which part of the fuel is injected early during the compression stroke to precondition the combustion chamber with a homogeneous fuel–air mixture, and a late second injection creates a stratified region in order to start the ignition. This heterogeneity is needed to ensure combustion.[88] However, the high reactivity (high auto-ignition tendency quantified by cetane number) of diesel obstructs the homogeneous premixing of the charge, and high rates of EGR—more than 70%—are required.[80] Fuels with high octane numbers, or gasoline-like fuels, have been found to be more suitable for PCCI.[87,88] The performance of gasoline and ethanol in PCCI combustion mode has been investigated,[89] with 3.3% and 10.3% reductions in fuel consumption achieved with gasoline and ethanol fuelling, respectively, as well as significant reductions in NO_x and soot. Also, moderate blends of butanol (50–70% butanol in diesel) show potential for PCCI.[90] The high volatility of butanol favours the premixing between the air and the fuel, considerably reducing soot emissions below Euro 6 levels; however, NO_x emissions are increased above the standards. CO and HC levels were similar to diesel over the whole operation range and indicated that efficiency up to 50% can be achieved with butanol–diesel blends.[90] PCCI can improve the engine's efficiency; nevertheless, it is not exempt from challenges, and research is ongoing.

A different combustion strategy is the so-called reactivity-controlled CI. This combines the separate injection of two fuels with different reactivity levels (*i.e.* diesel-like fuels with high auto-ignition propensity and gasoline-like fuels with high volatility for good mixture formation but low auto-ignition propensity). First, the low-reactivity fuel is port injected to create a homogeneous mixture between fuel, air and the recirculated exhaust. Then, the high-reactivity fuel is directly injected in the combustion chamber in one or more injection events to control the start of the ignition.[91] This strategy provides better combustion control than HCCI or PCCI and the engine can achieve a thermal efficiency of up to 60%.[91] Current research is focused on the optimisation of the fuels used to achieve fuel stratification for controlled combustion and low emissions in light- and heavy-duty vehicles.[92,93]

If available commercial gasoline is used in CI engines, the combustion strategy is known as gasoline CI (GCI).[94,95] GCI faces the same load limitation as HCCI: at low loads it suffers from misfiring, while at high loads knocking is produced.[95] The fuel injection strategy varies between studies, ranging from multiple injections[96] to single injection.[94] Also, different injectors have been used: high-pressure injectors (typically from CI engines)[94] and GDI injectors.[96] At low injection pressures, less wall wetting is produced, which is an important source of soot formation. Furthermore, lower injection pressures reduce parasitic losses and fuel injection system costs.[96]

6.2 Cylinder Deactivation

Deactivation of one or more cylinders at low engine load conditions has also been researched in order to increase the load in the 'active/running' cylinders and shift the overall operation towards minimal fuel consumption areas,[97–99] as shown in Figure 3.[97] Deactivating some of the cylinders, usually half,[98] forces the active cylinders to operate at higher mean effective pressures in order to reach the torque demand, thus leading to reduced pumping losses.[98,100] The fuel injection and ignition systems are cut off in the deactivated cylinders, which only compress and expand the intake air, reducing friction and heat losses.[98] The most effective cylinder deactivation strategy is to also disable moving parts such as valves and pistons in order to further reduce friction losses.[97,101] This is normally applied to engines with more than four cylinders, as noise vibration and harshness (NVH) remains an issue for these engines, according to vehicle manufacturers.[98] Cylinder deactivation is a technology that is currently used by several vehicle manufacturers. Cylinder deactivation is reported to reduce fuel consumption by between 4.7% and 6.5%.[20] It is normally combined with valve phasing technologies such as single overhead camshafts, dual overhead camshafts or overhead valves to achieve further fuel consumption reductions.[98]

Cylinder deactivation can be used in both gasoline and CI engines, although the benefits for each engine type are different. For SI engines, cylinder deactivation reduces pumping losses at low loads, improving the brake-specific fuel consumption. On the other hand, in CI engines, apart from the benefits in fuel consumption, the exhaust temperature increases when only some of the cylinders are fired, improving the after-treatment performance.[100] Cylinder deactivation technology will be improved in terms of the dynamic deactivation of individual cylinders.[99] Instead of deactivating the cylinders symmetrically in order to avoid NVH and torque fluctuations, new systems currently under development change the active

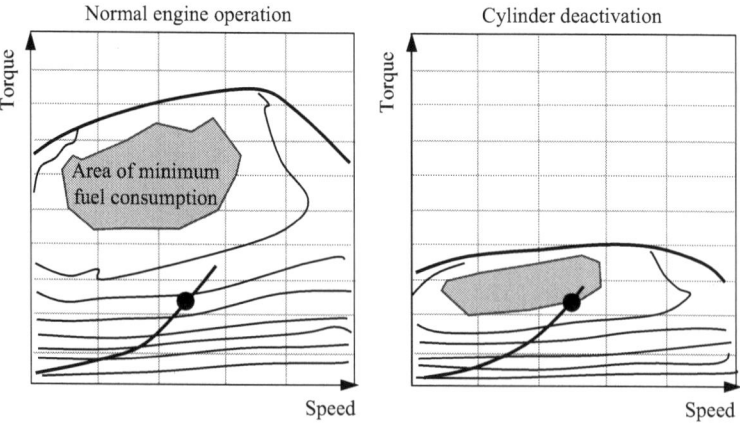

Figure 3 Effect of cylinder deactivation on engine map and fuel consumption. Adapted with permission from ref. 97.

cylinders continuously. The advantages of these systems over conventional cylinder deactivation are the uniformity of engine operation temperatures, operation with a fully open throttle, reduced NHV and extended use of cylinder deactivation to small engines with an odd number of cylinders.[99]

6.3 Variable Compression Ratio

The compression ratio is defined as the ratio of maximum volume to minimum volume (volume of the combustion chamber).[6] Typical compression ratio values are between 8 to 12 in SI engines and between 12 and 24 in CI engines.[6] Higher compression ratios are desirable as more power for the same air–fuel mixture can be extracted due to the higher thermal efficiency.[6] However, there is a maximum compression ratio achievable for both gasoline and CI engines. For SI engines, knock limits the compression ratio. In diesel engines, the constraints come from the material resistance to harsher temperature and pressure conditions. Today, the compression ratio is fixed depending on the cylinder geometry.[20] However, new developments have been able to alter the engine geometry, increasing the compression ratio under partial and light-load conditions or reducing it under heavy-load conditions, thus increasing efficiency.[20]

6.4 Variable Valve Actuation and Atkinson–Miller Cycles

Variable valve actuation is designed to control the lift, duration and timing of the intake and exhaust valves. There are two approaches: VVT and variable valve lift (VVL). These technologies can allow the unthrottled operation of SI engines and can aid in depleting pollutants coupled with other technologies, such as Atkinson cycles.[99,102]

VVT alters the timing or phase of the intake and/or exhaust valves.[81,103] The advantage of this is the reduction in pumping losses, which leads to higher volumetric efficiency, increasing the torque output while reducing fuel consumption.[103] Control of the internal residual exhaust gases is also achieved.[81] According to the US Department of Transportation, in 2010, 86% of passenger vehicles and light trucks had VVT.[81] Currently, almost all vehicle manufacturers have adopted this technology.[15] VVL regulates the height lift of the valves. Further advantages in reduced pumping losses at low loads can be achieved with VVL when compared to VVT.[103]

With the development of VVT and VVL, it has been possible to recover 'old ideas' that seemed inefficient years ago. James Atkinson (1846–1914) developed a long expansion engine that could provide high efficiency, but also suffered from several mechanical issues.[6] The Atkinson or over-expanded cycle attempts to recover the energy contained in the exhaust gas at the time of exhaust valve opening, thus increasing the indicated work per cycle.[6] Currently, the over-expansion is obtained through the use of VVT or VVL. The intake valve remains opened longer during the intake stroke and is closed during the compression, increasing the work extracted per cycle. However, the piston pushes part of the charge back to the intake manifold, increasing the geometric compression ratio as the trapped mass of charge is reduced, lowering torque and power output.[6,99,104] To maintain vehicle performance, it is necessary to increase the engine displacement or to use full-hybrid systems in which the electric propulsion motor re-establishes the torque.[99] Miller cycles, developed in 1957 by Ralph Miller, use the concept of the Atkinson cycle in supercharged engines.[105]

6.5 Stop–Start

In real driving conditions, idling contributes to up to 25% of the total emissions. Vehicle idling is associated with populated places such as roadways, traffic lights, bus stops, rest areas, drive-through restaurants and schools.[106] Vehicles during idle operation emit CO, HC, PM and NO_x to the atmosphere, as well as consuming a significant quantity of fuel.[106] Stop-starts reduce the idling time by shutting the engine down when the vehicle is stopped in traffic, reducing fuel consumption and emissions. The engine automatically restarts when the driver releases the brake pedal or presses the clutch.[99,107] The system is inherent in hybrid vehicles, but it is also used in internal combustion engine-powered vehicles.[15,108] The objectives of the stop–start system are to reduce fuel consumption and pollutant emissions, to achieve fast and smooth engine restart, to minimise the impact on driver's experience and to ensure the functionality of auxiliary systems.[108] Fuel economy can be improved by 3.5%.[108] The main challenges that stop–start systems face are the need for a starter motor and battery that are suitable for higher-duty cycles, safety issues relating to unwanted engine restarting, guaranteeing engine restart when required (driver's change of mind) and low NVH.[107,108] A 12 V electrical system is required for auxiliary systems, such as A/C systems, while the engine is stopped.[107] Furthermore, the cost associated with stop–start must be justified by the CO_2 reductions.[108] In 2015, the stop–start technology achieved a market share of 7.4% for passenger cars and 5.5% for light trucks in the USA.[20]

References

1. IEA, *World Energy Outlook 2015 Factsheet - Global energy trends to 2040*, 2015.
2. IEA, *World Energy Outlook 2015 - Executive Summary*, 2015.
3. IEA, *Energy and Air Pollution*, 2016.
4. W. H. Organisation, *Global Health Observatory (GHO) data*, 2015.
5. *The future of vehicle emissions testing and compliance. How to align regulatory requirements, customer expectations, and environmental performance in the European Union*, International Council of Clean Transportation, 2016.
6. J. Heywood, *Internal Combustion Engine Fundamentals*, McGraw-Hill Education, 1988.
7. G. L. Borman and K. W. Ragland, *Combustion Engineering*, McGraw Hill, 1998.

8. R. Stone, *Introduction to Internal Combustion Engines*, 1999.
9. M. Zheng, G. T. Reader and J. G. Hawley, *Energy Convers. Manage.*, 2004, **45**, 883–900.
10. N. Ladommatos, S. Abdelhalim and H. Zhao, *Int. J. Engine Res.*, 2000, **1**, 107–126.
11. K. Akihama, Y. Takatori, K. Inagaki, S. Sasaki and A. M. Dean, *SAE Technical Paper*, 2001, **2001-01-0655**.
12. ICCT, *Estimated Cost of Emission Reduction Technologies for Light-Duty Vehicles*, 2012.
13. M. V. Twigg, *Appl. Catal., B*, 2007, **70**, 2–15.
14. J. Kašpar, P. Fornasiero and N. Hickey, *Catal. Today*, 2003, **77**, 419–449.
15. EPA, *Light-Duty Automotive Technology, Carbon Dioxide Emissions, and Fuel Economy Trends: 1975 Through 2015*, 2015.
16. M. B. Çelik and B. Ozdalyan, Gasoline Direct Injection, in *Fuel Injection*, ed. D. Siano, InTech, 2010, ch. 1.
17. H. Zhao, *Advanced Direct Injection Combustion Engine Technologies and Development: Diesel Engines*, 2010, pp. 1–19.
18. F. Zhao, M. C. Lai and D. L. Harrington, *Prog. Energy Combust. Sci.*, 1999, **25**, 437–562.
19. A. C. Alkidas, *Energy Convers. Manage.*, 2007, **48**, 2751–2761.
20. ORNL, *Vehicle Technologies Market Report*, 2015.
21. N. Fraser, H. Blaxill, G. Lumsden and M. Bassett, *SAE Int. J. Engines*, 2009, **2**, 991–1008.
22. Audi Technology Portal, Mobility for the Future Rightsizing, 2016, http://www.audi-technology-portal.de/en/mobility-for-the-future/audi-future-lab-mobility_en/audi-future-engines_en/rightsizing_en, accessed 02/04/2017.
23. R. Bahreini, J. Xue, K. Johnson, T. Durbin, D. Quiros, S. Hu, T. Huai, A. Ayala and H. Jung, *J. Aerosol Sci.*, 2015, **90**, 144–153.
24. D. Arters, E. Bardasz, E. Schiferl and D. Fisher, *SAE Technical Paper*, 1999, **1999-01-1498**.
25. H. Song, J. Xiao, Y. Chen and Z. Huang, *Fuel*, 2016, **180**, 506–513.
26. *Press Release N°221*, 2013.
27. K. H. Kim, E. Kabir and S. Kabir, *Environ. Int.*, 2015, **74**, 136–143.
28. E. E. A. (EEA), *Air quality in Europe - 2015 report*, 2015.
29. *Commission Regulation (EU) No 459/2012 of 29 May 2012*.
30. W. Piock, G. Hoffmann, A. Berndorfer, P. Salemi and B. Fusshoeller, *SAE Int. J. Engines*, 2011, **4**, 1455–1468.
31. T. W. Chan, E. Meloche, J. Kubsh, D. Rosenblatt, R. Brezny and G. Rideout, *SAE Int. J. Fuels Lubr.*, 2012, **5**, 1277–1290.
32. M. Lapuerta, J. Rodríguez-Fernandez, R. García-Contreras and M. Bogarra, *Fuel*, 2015, **139**, 171–179.
33. *European Directive 2009/28/EC*, 2009.
34. *Energy Independence and Security Act of 2007, Public Law 110-140, Title II, Subtitle A, Sec. 202. Renewable Fuel Standard, December 19, 2007*.
35. R. A. Stein, J. E. Anderson and T. J. Wallington, *SAE Int. J. Engines*, 2013, **6**, 470–487.

36. H. S. Yücesu, A. Sozen, T. Topgül and E. Arcaklioğlu, *Appl. Therm. Eng.*, 2007, **27**, 358–368.
37. M. Lapuerta, O. Armas and R. García-Contreras, *Fuel*, 2007, **86**, 1351–1357.
38. M. Lapuerta, R. García-Contreras, J. Campos-Fernández and M. P. Dorado, *Energy Fuels*, 2010, **24**(8), 4497–4502.
39. M. Balat and H. Balat, *Appl. Energy*, 2009, **86**, 2273–2282.
40. M. S. Graboski and R. L. McCormick, *Prog. Energy Combust. Sci.*, 1998, **24**, 125–164.
41. A. K. Agarwal, *Prog. Energy Combust. Sci.*, 2007, **33**, 233–271.
42. M. Lapuerta, O. Armas and J. Rodriguezfernandez, *Prog. Energy Combust. Sci.*, 2008, **34**, 198–223.
43. M. Alam, J. Song, R. Acharya, A. Boehman and K. Miller, *SAE paper*, 2004, **2004-01-3024**.
44. M. Lapuerta, J. Rodríguez-Fernández, F. Oliva and L. Canoira, *Energy Fuels*, 2009, **23**, 121–129.
45. *Assessment and Standards Division. A comprehensive analysis of biodiesel impacts on exhaust emissions*, Office of Transportation and Air Quality of the U.S. Environmental Protection Agency, 2002.
46. A. Tsolakis, *Energy Fuels*, 2006, **20**, 1418–1424.
47. M. Dorado, *Fuel*, 2003, **82**, 1311–1315.
48. M. Cardone, M. Mazzoncini, S. Menini, V. Rocco, A. Senatore, M. Seggiani and S. Vitolo, *Biomass Bioenergy*, 2003, **25**, 623–636.
49. F. G. Staat and P. Gateau, *SAE paper*, 1995.
50. C. D. Rakopoulos, D. C. Rakopoulos, D. T. Hountalas, E. G. Giakoumis and E. C. Andritsakis, *Fuel*, 2008, **87**, 147–157.
51. M. A. Lapuerta, O. Armas and R. Ballesteros, *SAE paper*, 2002, **2002-01-1657**.
52. M. Lapuerta, J. M. Herreros, L. L. Lyons, R. García-Contreras and Y. Briceño, *Fuel*, 2008, **87**, 3161–3169.
53. M. J. Haas, K. M. Scott, T. L. Alleman and R. L. McCormick, *Energy Fuels*, 2001, **15**, 1207–1212.
54. O. Armas, J. Hernandez and M. Cardenas, *Fuel*, 2006, **85**, 2427–2438.
55. S. H. Yoon, H. K. Suh and C. S. Lee, *Energy Fuels*, 2009, **23**, 1486–1493.
56. M. S. Graboski, R. L. McCormick, T. L. Alleman and A. M. Herring, *The Effect of Biodiesel Composition on Engine Emissions from a DDC Series 60 Diesel Engine*, 2003.
57. Y. Di, C. S. Cheung and Z. Huang, *Sci. Total Environ.*, 2009, **407**, 835–846.
58. A. Schönborn, N. Ladommatos, J. Williams, R. Allan and J. Rogerson, *Combust. Flame*, 2009, **156**, 1396–1412.
59. M. Lapuerta, O. Armas, J. J. Hernández and A. Tsolakis, *Fuel*, 2010, **89**, 3106–3113.
60. S. S. Gill, A. Tsolakis, K. D. Dearn and J. Rodríguez-Fernández, *Prog. Energy Combust. Sci.*, 2011, **37**, 503–523.
61. H. Aatola, M. Larmi, T. Sarjovaara and S. Mikkonen, *SAE Int. J. Engines*, 2009, **1**, 1251–1262.
62. M. Lapuerta, M. Villajos, J. R. Agudelo and A. L. Boehman, *Fuel Process. Technol.*, 2011, **92**, 2406–2411.

63. *Clean Buses - Experiences with Fuel and Technology Options*, Clean Fleets project, 2014.
64. T. Hartikka, M. Kuronen and U. Kiiski, *SAE Technical Paper*, 2012, **2012-01-1585**.
65. D. J. Durbin and C. Malardier-Jugroot, *Int. J. Hydrogen Energy*, 2013, **38**, 14595–14617.
66. S. Verhelst and T. Wallner, *Prog. Energy Combust. Sci.*, 2009, **35**, 490–527.
67. F. Suleman, I. Dincer and M. Agelin-Chaab, *Int. J. Hydrogen Energy*, 2016, **41**, 8364–8375.
68. *Hydrogen Production Using Nuclear Energy*, International Atomic Energy Agency, 2013.
69. A. Megaritis, A. Tsolakis, S. E. Golunski and M. L. Wyszynski, in *Advanced Direct Injection Combustion Engine Technologies and Development: Diesel engines*, ed. H. Zhao, 2010, pp. 543–561.
70. W. Wang, J. M. Herreros, A. Tsolakis and A. P. E. York, *Int. J. Hydrogen Energy*, 2013, **38**, 9907–9917.
71. R. Stone, H. Zhao and L. Zhou, *SAE Technical Paper*, 2010, **2010-01-0580**.
72. D. Fennell, J. Herreros, A. Tsolakis and H. Xu, *SAE Technical Paper*, 2013, **2013-01-0537**.
73. S. J. Curran, R. M. Wagner, R. L. Graves, M. Keller and J. B. Green, *Energy*, 2014, **75**, 194–203.
74. D. P. von Rosenstiel, D. F. Heuermann and S. Hüsig, *Energy Policy*, 2015, **78**, 91–101.
75. M. L. Poulton, *Alternative Fuels for Road Vehicles*, Computational Mechanics Publications, Southampton, United Kingdom, 1994.
76. ICCT, *European Vehicle Market Statistics. Pocketbook 2015/2016*.
77. ACEA, *The Automobile Industry Pocket Guide*, 2015/2016.
78. OECD/IEA, *Global EV Outlook 2016. Beyond one mission electric cars*, 2016.
79. CAR, *The U.S. Automotive Market and Industry in 2025*, 2011.
80. A. B. Dempsey, N. R. Walker, E. Gingrich and R. D. Reitz, *Combust. Sci. Technol.*, 2014, **186**, 210–241.
81. NHTSA, *Corporate Average Fuel Economy for MY 2017-MY 2025 Passenger Cars and Light Trucks*, 2011.
82. P. Najt and D. Foster, *SAE Technical Paper*, 1983, **830264**.
83. R. Thring, *SAE Technical Paper*, 1989, **892068**.
84. M. C. Drake and D. C. Haworth, *Proc. Combust. Inst.*, 2007, **31**, 99–124.
85. J. E. Dec, *Proc. Combust. Inst.*, 2009, **32**, 2727–2742.
86. S. Saxena and I. D. Bedoya, *Prog. Energy Combust. Sci.*, 2013, **39**, 457–488.
87. G. Kalghatgi, *SAE Technical Paper*, 2005, **2005-01-0239**.
88. J. Chang, G. Kalghatgi, A. Amer and Y. Viollet, *SAE Technical Paper*, 2012, **2012-01-0677**.
89. V. Manente, B. Johansson and P. Tunestal, *SAE Technical Paper*, 2009, **2009-01-0944**.

90. C. A. J. Leermakers, P. C. Bakker, L. M. T. Somers, L. P. H. de Goey and B. H. Johansson, *SAE Int. J. Fuels Lubr.*, 2013, **6**, 217–229.
91. R. D. Reitz and G. Duraisamy, *Prog. Energy Combust. Sci.*, 2015, **46**, 12–71.
92. A. B. Dempsey, S. Curran and R. D. Reitz, *SAE Int. J. Engines*, 2015, **8**, 859–877.
93. D. A. Splitter, R. M. Hanson, S. L. Kokjohn and R. D. Reitz, *SAE Technical Paper*, 2011, **2011-01-0363**.
94. C. Kolodziej, J. Kodavasal, S. Ciatti, S. Som, *et al.*, *SAE Technical Paper*, 2015, **2016-01-0832**.
95. K. Hiraya, K. Hasegawa, T. Urushihara, A. Iiyama and T. Itoh, *SAE Technical Paper*, 2002, **2002-01-0416**.
96. M. Sellnau, M. Foster, K. Hoyer, W. Moore, J. Sinnamon and H. Husted, *SAE Int. J. Engines*, 2014, 7, 835–851.
97. A. Ihlemann and N. Nitz, *Schaeffler Technologies AG & Co. KG*, 2014.
98. *Corporate Average Fuel Economy (CAFE) for MY 2012-MY 2016 Passenger Cars and Light Trucks*, U.S. Department Of Transportation National Highway Traffic Safety Administration, 2009.
99. M. Wilcutts, J. Switkes, M. Shost and A. Tripathi, *SAE Int. J. Engines*, 2013, **6**, 278–288.
100. S. Pillai, J. LoRusso and M. Van Benschoten, *SAE Technical Paper*, 2015, **2015-01-2809**.
101. A. Boretti and J. Scalco, *SAE Technical Paper*, 2011, **2011-01-0368**.
102. IEA/OECD, *Transport, Energy and CO2: Moving Toward Sustainability*, 2009, https://www.iea.org/publications/freepublications/publication/transport2009.pdf.
103. *Energy Technology Systems Analysis Programme*, International Energy Agency, 2010.
104. D. Luria, Y. Taitel and A. Stotter, *SAE Technical Paper*, 1982, **820352**.
105. M. Hitomi, J. Sasaki, K. Hatamura and Y. Yano, *SAE Technical Paper*, 1995, **950974**.
106. I. Shancita, H. H. Masjuki, M. A. Kalam, I. M. Rizwanul Fattah, M. M. Rashed and H. K. Rashedul, *Energy Convers. Manage.*, 2014, **88**, 794–807.
107. T. Wellmann, K. Govindswamy and D. Tomazic, *SAE Int. J. Engines*, 2013, **6**, 1368–1378.
108. X. Wang, R. McGee and M. Kuang, *SAE Technical Paper*, 2013, **2013-01-0347**.

Gaseous and Particle Greenhouse Emissions from Road Transport

M. LAPUERTA,* J. RODRÍGUEZ-FERNÁNDEZ AND J. M. HERREROS

ABSTRACT

This chapter discusses the sources and impacts of greenhouse and particle emissions from road transport, as well as future pathways to mitigate their environmental impacts. The main greenhouse gas species considered in greenhouse inventories by the United Nations Framework Convention on Climate Change, such as carbon dioxide, methane and nitrous oxide, are first independently and then conjointly (equivalent carbon dioxide emissions) presented in the different sections of this chapter. Tailpipe greenhouse emissions produced by the propulsion and/or after-treatment systems (direct tank-to-wheel) and those emitted during the life cycle of fuels and vehicles (indirect well-to-tank) are both accounted for. Particle emissions are also addressed from a broad environmental standpoint. Improvements in vehicle fuel economy, behavioural changes in society in order to avoid unnecessary journeys and a higher market penetration rate of low-carbon vehicle technologies and energy carriers are the pathways required to achieve an overall reduction in greenhouse gas emissions and the weather-related disasters associated with them.

1 Introduction

The transport sector remains a major source of air pollutants and greenhouse gases, despite the many policy and technology advances achieved.

---

*Corresponding author.

Transport emissions can be categorised as exhaust emissions (produced during combustion), abrasion emissions (produced by the wear of tyres, brakes, clutches and road surfaces and by vehicle corrosion) and evaporative emissions (from fuel evaporation). Transport is responsible for 23% of the anthropogenic carbon dioxide (CO_2) emissions worldwide and for more than half of all energy-related nitrogen oxide emissions (56 Mt in 2015), and is an important source of primary particulate matter (PM; around 10% of the total energy-related primary PM2.5 emissions). Among the transport subsectors, road transport is by far the largest source of NO_x and primary PM2.5 emissions (58% and 73% of the totals worldwide, respectively). Road vehicles are driven intensively in populated regions, which tend to concentrate emissions in urban areas, leading to inhabitant exposure to pollutants. However, some of these emissions also have global effects.

The main global effect of transport is the emission of CO_2, either directly from the engine exhaust or during the life cycle of fuels and vehicles. Similarly to water vapour, CO_2 is emitted in major concentrations by internal combustion (IC) engines (up to 12% molar under stoichiometric conditions), whereas other gases or particles are emitted in minor concentrations. With respect to water vapour, although emitted in major concentrations by vehicles, its emission is not believed to affect the average global concentration of water vapour, and by contrast, the emission of other greenhouse gases (GHGs) may affect the hydrologic cycle. In the case of CO_2, its relative contribution to radiative forcing has grown from 55% in 1990 to 76% in 2014, and the increasing importance of transport is partly responsible for this growth. In fact, the contribution of CO_2 with high traffic intensity is even higher. Other gases with significant contributions to radiative forcing are methane (CH_4), nitrous oxide (N_2O), tropospheric ozone (O_3) and halocarbons. Among them, methane (produced from anaerobic decomposition of organic matter in biological systems, but also released during the production and distribution of fossil fuels), nitrous oxide (mainly emitted from agricultural soils) and ozone (formed photochemically from hydrocarbons and NO_x emissions) contribute to life cycle GHG assessments of fuels and biofuels, while halocarbons also contribute in refrigerant assessments, with both fuels and refrigerants being consumed by vehicles. Other gases emitted by vehicles are not accounted for in greenhouse inventories by the United Nations Framework Convention on Climate Change, but they are known to affect radiation forcing, such as carbon monoxide (CO), hydrogen (H_2), nitrogen dioxide (NO_2) and sulphur dioxide (SO_2), as far as sulphur is not removed from fuel and lube oil composition. Among the hydrocarbons emitted from incomplete combustion, lubricant degradation or as evaporative losses, some of them have considerable warming potential, such as ethane, propane, butane, isobutene, pentane, iso-pentane and ethylene. Finally, PM emissions from vehicles, composed of solid carbonaceous particles or liquid droplets, may remain suspended in the atmosphere or may be deposited on the earth's surfaces. In the first case, they can affect the absorptive and scattering characteristics of the atmosphere and can play a role in radiative forcing, either directly or through the formation and lifespan of clouds. In the second case, they can affect the reflectivity of surfaces, with consequences for radiative forcing, and speed up snow melt (especially in spring), with consequences for the hydrological responses of snow reservoirs and glaciers.

In the following sections, the importance and the global effects of these emissions are described in more detail.

2 Carbon Dioxide Emissions

CO_2 is a colourless, odourless gas existing in the atmosphere at a concentration of around 0.04% by volume. It is produced by both natural and anthropogenic means. Natural sources include its release from carbonate rocks and from volcanoes and other eruptions. Deforestation and the combustion of fossil fuels are the main contributors to anthropogenic CO_2. CO_2 is the most significant GHG, accounting for 80% of the global GHG emissions. Although its global warming potential (GWP; see definition in Section 5 of this chapter) is low compared to other gases, the large quantity of CO_2 emitted every year makes CO_2 the most relevant cause of global warming. The radiative forcing of CO_2 (defined in ref. 1 as the balance of incoming and outgoing energy in the Earth atmosphere) was 1.909 $W\,m^{-2}$ in 2014, the largest recorded level for a GHG and four times that of the next, methane. These numbers agree well with other published calculations.[2]

The Keeling curve (in recognition of the work of Prof. Dr C. Keeling) represents the atmospheric CO_2 concentration since 1958. Measurements are taken at the Mauna Loa Observatory, Hawaii. The curve represents a uniform background CO_2 level in the Earth atmosphere since the equipment is at a remote location, 3397 m altitude, and far from the influence of vegetation and human activity.†[4] Figure 1 shows the historical trend of atmospheric CO_2, along with an accepted pre-industrial CO_2 level. Despite the reservations about the procedure followed for obtaining all of these data (smoothing, selection/discarding of some values, ignoring direct readings and giving preference to other techniques),[5] it is evident that the CO_2 increase is accelerating (in the 1960s, atmospheric CO_2 increased by less than 1 ppm annually, while currently it increases more than 2 ppm per year). Moreover, in the period 2010–2015, the increase rate of atmospheric CO_2 concentration was 0.55% $year^{-1}$, faster than the increase of methane (0.4% $year^{-1}$) or nitrous oxide (0.3% $year^{-1}$).

By sector, road transport is the second largest CO_2 emitter, just behind the energy sector (which includes the burning of fossil fuels for electricity and heat generation). Road transport is responsible for more than 20% of the world's CO_2 emission, most of it (15%) from light-duty vehicles including

---

†Nevertheless, atmospheric CO_2 is not evenly distributed all over the Earth: CO_2 concentrations are higher in the northern hemisphere, originating around South East Asia, North America and Europe (CO_2 sources), but spreading out due to atmospheric phenomena as revealed, for example, by space-based measurements performed under the Orbiting Carbon Observatory-2 project.[3]

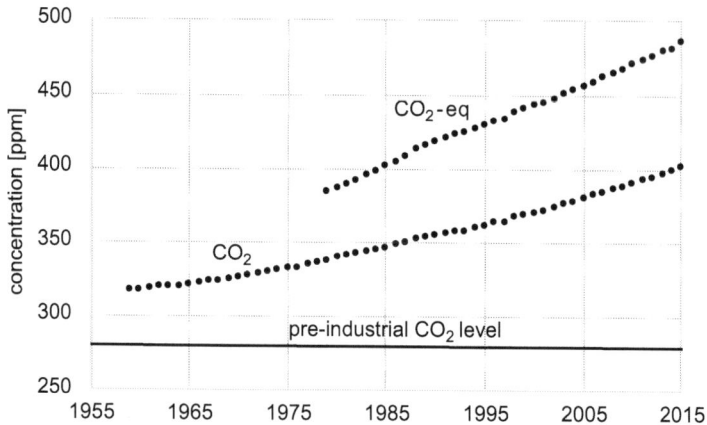

Figure 1 CO_2 and equivalent CO_2 (see Section 5 of this chapter) atmospheric concentrations.

both cars (12%) and vans (3%).[6] This share will probably increase in the coming years, as the main fuels for road transport come from fossil crude oil.

CO_2 emissions from road transport, and also other emissions discussed in this chapter, can be classified as direct (or tailpipe) and indirect. Direct emissions are produced by the combustion of a carbon-containing fuel in the vehicle, as soon as the IC engine is started. Obviously, vehicles that run without an IC engine, such as battery electric vehicles, and/or that use hydrogen as the energy carrier emit zero direct CO_2. Indirect emissions are generated during the fuel cycle (feedstock production, transport, processing, distribution, *etc.*) and the vehicle cycle (production, assembly, maintenance, disposal, *etc.*). In the case of a particular fuel, the life cycle analysis (or well-to-wheel analysis) of CO_2/GHG considers both direct and indirect emissions (for more detail, see Section 5). Tailpipe CO_2 depends on the type of fuel and the vehicle fuel economy. Simple stoichiometry indicates that the direct CO_2 emission (in kg CO_2 kg fuel^{-1}) of a fuel with the molecular formula $C_nH_mO_p$ depends on the m/n and p/n ratios only (the higher these ratios, the lower the emissions). Therefore, methane, the major component of natural gas, emits less CO_2 per fuel mass than either diesel or gasoline because of the higher m/n ratio of methane (Table 1); the emissions of ethanol, biodiesel and diesel rank according to the p/n ratio (although they show different m/n values, the p/n ratio is dominant in the calculation). Because fuel consumption is widely indicated on a volume basis (per litre or per gallon), the CO_2 emissions per fuel volume (instead of mass) is a useful measure as well. As observed in Table 1, diesel fuel emits similar CO_2 (per fuel mass) as gasoline, but much more on a volume basis because of the higher density of diesel. But diesel vehicles are about 20–30% more efficient than gasoline, which makes the CO_2 emissions per distance travelled lower for diesel vehicles.

Table 1 Tailpipe CO_2 emissions of some fuels for road transport.

Fuel	m/n	p/n	Density (kg m^{-3})	kg CO_2 kg^{-1}	g CO_2 MJ^{-1}	kg CO_2 L^{-1}
Methane	4.00	0	—	2.75	55.00	—
Diesel[a]	2.00	0	840	3.18	73.49	2.67
Gasoline	1.80	0	750	3.14	72.72	2.35
Ethanol	3.00	0.50	789	1.91	70.87	1.51
Butanol	2.50	0.25	810	2.37	68.96	1.92
Biodiesel	1.87	0.11	885	2.82	76.20	2.50

[a]Data for diesel and gasoline fuels agree well with those proposed in the Code of Federal Regulations (USA), which suggests 8.887 and 10.180 kg CO_2 per gallon of gasoline and diesel, respectively.

Different measures and timetables of application have been introduced worldwide to tackle CO_2 from road transport, not only by promoting sustainable biofuels (this is explained in more detail in Section 5), but also by encouraging the use of more environmentally friendly vehicles. In the European Union (EU), the European Commission adopted in 1995 a strategy to cut CO_2 emissions from cars proposing that vehicle manufacturers should assume voluntary CO_2 targets. A goal of 120 g CO_2 km^{-1} was set by 2005, a number that had been first proposed by Germany in 1994. The target was postponed to 2010 first and to 2012 later. In response, European and Asian automobile associations subsequently adopted a non-binding commitment to reduce the average CO_2 emissions of the cars they sold in the EU to 140 g km^{-1} by 2008–2009. In 2007, the European authorities evaluated the partial results and concluded that the progress made would not be enough to meet the 120 g km^{-1} limit by 2012. The legislative framework was then decisively modified,[7] setting: (i) a mandatory reduction of CO_2 to 130 g km^{-1} that the car industry must achieve by improvements in the vehicle motor technology; and (ii) an additional reduction of 10 g km^{-1} by other means (including biofuels). The new limit was phased in between 2012 (when the average emissions of only 65% of the cars registered in the EU needed to comply with the target) and 2015 (100%). Each manufacturer was assigned an individual target based on the average mass of its registered cars in 1 year (heavier vehicles were allowed higher emissions), ensuring that the global objective of 130 g km^{-1} would be achieved if all of the manufacturers complied with their individual targets. This mechanism for CO_2 assignment depending on the average vehicle mass was termed 'the limit value curve' (although it is not a curve, but a straight line). The curve was defined in order to request from the heavier cars a larger effort than from the lighter ones. To give some flexibility to the car industry and satisfy the expectations of all final users, the European authorities allowed the manufacturers to average emissions over all of their cars, rather than making every new car meet the CO_2 target.

As shown in Figure 2, the targets for 2015 were already achieved by 2013. The next target in Europe[7] is 95 g CO_2 km^{-1} by 2021 (originally, this was proposed for 2020). Cars with CO_2 emissions below 50 g CO_2 km^{-1} will be counted more than once during 2020–2022. Eco-innovations (*i.e.* CO_2 saving

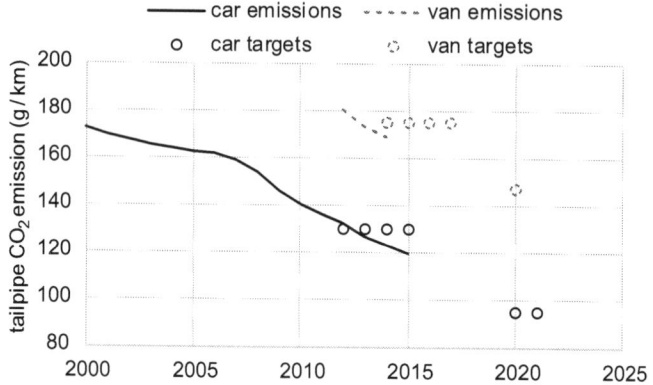

Figure 2 CO_2 emissions and targets for light-duty vehicles in the EU.

technologies whose contributions must be accountable, verified and not covered by the test cycle used for CO_2 measurement) are permitted at up to 7 g km^{-1} in order to achieve the objectives. If the car manufacturers do not meet their targets for a particular year, they must pay a penalty: €5 per vehicle for the first gram exceeding the limit, €15 per vehicle for the second gram, €25 per vehicle for the third and €95 per vehicle for every gram thereinafter. From 2019, the penalty will be €95 per vehicle for every gram exceeding the target.

In the case of vans, there is a similar scheme of targets in the EU: a first one of 175 g km^{-1} by 2017 (but phased in from 2014) that was already achieved in 2013, and a second one of 147 g km^{-1} by 2020. Eco-innovations are also allowed, and penalty payments are due if van manufacturers fail. For heavy-duty vehicles, EU actions are much more recent and there are no explicit limits on CO_2 emissions yet. Due to the urgency of curbing CO_2 resulting from transport and the impact that heavy-duty vehicles have on it (25% of total CO_2 emissions from road transport in the EU), several measures are under study. In the short term, measures consist of certifying and monitoring emissions. In the medium term, the authorities are considering setting limits on CO_2 emissions and promoting the use of alternative fuels.[6]

Finally, CO_2 road transport goals in the USA are less ambitious than in the EU, as diesel technology has lower penetration in the USA and vehicles are heavier. There are two related standards: a fuel economy standard (Corporate Average Fuel Economy Standard) developed by the Department of Transport and a GHG Standard adopted by the Environmental Protection Agency. In 2010, both sets of standards were harmonised for the period 2012–2016, with the aim of achieving 250 g CO_2 mile^{-1} by 2016. The agreement was later extended to cover the period 2017–2025, with a final objective of 163 g CO_2 mile^{-1} and 54.5 mpg (US gallons) of fuel economy. The mechanism consisted of limit curves of CO_2 emission depending on the vehicle footprint (instead of the vehicle mass used in Europe) that allow vehicles with a larger footprint to have higher CO_2 emissions. The fuel economy objective was fixed assuming a future market composed mainly of cars, but the composition still remains approximately 50% cars/50% other road vehicles such as sport utility vehicles or small trucks, making the attainment of the objective more difficult.[8]

3 Methane Emissions

Methane has 25 times more greenhouse effect per unit mass than CO_2 (over a 100 year period), but its concentration in the atmosphere is over 200 times less than that of CO_2 (*e.g.* around 1750 ppb). This concentration is more than double what it was 150 years ago. The anthropogenic increase in methane concentration started around 5000 years ago, when humans moved from hunters to agricultural farmers, but the rate of increase became higher from the industrial revolution.[9] Methane is estimated to contribute around 16% to anthropogenic global warming. This contribution could be even higher because, in addition to the direct warming effect, methane oxidises to CO_2 and water in the stratosphere, leading to water and ice clouds that are very efficient at enhancing the warming effect.

As mentioned in the Section 1, methane is mainly produced from anaerobic decomposition of organic matter in biological systems, specifically from bacteria that decompose organic materials in the absence of oxygen. Among the biological sources, livestock is the main one, and smaller sources include oceans, wetlands, sediments, volcanoes and wildfires. However, human-related sources create the majority of methane emissions, accounting for 64% of global methane emissions.

Related to transport, methane is released during the production and distribution and use of fossil fuels. It is released during the production, processing, storage, transmission and distribution of natural gas, and because gas is often found alongside oil, the production, refinement, transportation and storage of crude oil is also a source of methane emissions. The contribution of fossil fuels to anthropogenic methane emissions is estimated to be around 33%, whereas that of biofuels is only around 4%, and this also includes solid biofuels. European directive 28/2009/CE[10] promotes methane sequestration from biodiesel production processes such as those used in biodiesel derivation from palm oil. Finally, a minor contribution, although not negligible,[11] is associated with incomplete combustion processes in vehicle engines, especially in natural gas engines.[12] In addition, methane, as a short-chain saturated hydrocarbon species, presents high resistance to being depleted in oxidation catalysts.[13,14]

Despite the greenhouse effect of methane, US emissions regulations have restricted the limits on non-methane organic gases for light-duty vehicles, in recognition of the fact that the atmospheric chemistry of methane is slow enough to assume that methane has a negligible impact on local ground-level ozone levels, but ignoring its enhanced greenhouse contribution.

4 Nitrous Oxide Emissions

The main natural sources of N_2O emissions are oceans, tropical forests, other forests and grassland, while anthropogenic N_2O is derived from agriculture, industry and energy supply. Besides its effect on global warming, this gas is an ozone-depleting molecule.[15] The tailpipe N_2O emissions from road transport originate as by-products of fuel combustion as well as from the malfunctioning of after-treatment catalysts. Nevertheless, a much larger amount of N_2O is emitted during the cultivation and growing of crops for biofuel production, and must be taken into account when a biofuel life cycle analysis of GHG is performed (see next section). Some authors[16] state that N_2O can contribute significantly (up to 80%, but highly variable) to the total life cycle GHG emissions of biofuels. The atmospheric concentration of N_2O was around 270 ppb in pre-industrial times, but currently the concentration has increased to 330 ppb, with an annual increase of approximately 0.7 ppb occurring for the last 50 years. According to ref. 17, the main cause of this trend is the expansion of agricultural lands and the use of fertilisers.

Nitrous oxide from gasoline engines is mainly produced in the three-way catalyst light-off within a temperature range starting from the catalyst activation temperature[18] up to the temperature at which the catalyst reaches around 95% conversion efficiency.[18,19] At higher temperatures, NO, NO_2 and N_2O are successfully decomposed.[20] The actual temperature range depends on catalyst composition (type of catalyst and loading), exhaust composition (*e.g.* presence of hydrocarbon species and oxygen[20]), residence time of exhaust gas in the catalyst and the state of the catalyst (*e.g.* fresh *vs.* aged[20]). It has been reported that new spark ignition vehicles have lower N_2O emissions than older vehicles due to the improved catalyst light-off strategies minimising the time during which the catalyst produces N_2O emissions, as well as the use of new catalyst-coating materials and decreases in fuel sulphur.[19]

Nitrous oxide in compression ignition engines is mainly formed in the after-treatment system. Heavy-duty compression ignition engines commonly require the operation of a selective catalytic reduction (SCR) catalyst for NO_x abatement and in combination with engine operation strategies, it is possible to simultaneously reduce fuel consumption, GHG and NO_x emissions. SCR catalysts need the upstream addition of NH_3 to convert NO and NO_2 to nitrogen, producing a trade-off between de-NO_x performance and NH_3 slippage, demanding an additional ammonia slipping catalyst (ASC). N_2O can be formed *via* the direct oxidation of NH_3 in the ASC as well as *via* ammonium nitrate decomposition in copper-based SCR catalysts.[21] Depending on the after-treatment catalyst configuration, N_2O could also be formed at temperatures between 250 and 350 °C in a downstream-catalysed diesel particulate filter[22] and/or diesel oxidation catalyst.[23]

It has been reported that advanced combustion modes such as homogeneous charge compression ignition could produce higher concentrations of N_2O than conventional combustion modes. It is thought that N_2O is formed in the operating conditions, resulting in the simultaneous presence of hydrocarbons and NO due to the presence of temperature inhomogeneity, poor mixing and slow combustion.[24]

With regards to N_2O emission during the cultivation step in the biofuel chain, this gas is emitted in traces as a consequence of: (i) microbial activity in soils that convert fixed nitrogen into N_2O (direct emission); and (ii) losses, leaches and volatilisation (indirect emission). Although N_2O is emitted naturally, using nitrogen-containing fertilisers or extending nitrogen-fixing plants contribute to increasing overall emissions. The estimation of the level of N_2O emitted is inherently difficult as it depends strongly on the agricultural practices used, but its inclusion in the analysis is imperative for a correct methodology. Usually, authors use the so-called N_2O emission factor (EF), a parameter that quantifies how much of the nitrogen added to soils is emitted as N_2O. Values as high as 3–5% have been suggested,[25] which would lead to an increase of GHGs with the use of some biofuels. Previous estimations by the International Panel on Climate Change[26] suggest a lower range of 1–3%. Another analysis[27] of corn and soybean production in the USA concludes EF values to be in an intermediate range of 2.0–2.5%. This variability relates to the number of factors affecting the agricultural N_2O release: cropping system, allocation methodology and climate, among others.[16] In ref. 28, the cultivation of 15 potential biofuel feedstocks in different world regions was examined in terms of N_2O emissions. In conclusion, sugarcane cultivated in Brazil emitted 1.6–2.0 g equivalent CO_2 (CO_2-eq.) MJ^{-1} (much higher in the USA and China); oil palm cultivated in Indonesia emitted 6.1–8.0 g CO_2-eq. MJ^{-1} and in Malaysia emitted 4.1–6.0 g CO_2-eq. MJ^{-1}; and soybean cultivated in Argentina emitted lower N_2O (2.2–5.0 g CO_2-eq. MJ^{-1}) than in the USA or Brazil (5.1–7.0 g CO_2-eq. MJ^{-1}).

Despite the discrepancy of values, increasing the competitiveness of biofuels in terms GHG emissions will require improvements in their nitrogen efficiency: enhanced use of fertilisers, improved water practices or the addition of nitrification inhibitors have been suggested.[27] The transition from food crop-derived biofuels to other alternatives may reduce N_2O emissions as well. In this way, cultivating switchgrass and hybrid poplar for biofuel feedstocks instead of corn and soybean increased the GHG savings compared to fossil fuels by threefold.[29]

5 Equivalent Carbon Dioxide Emissions

CO_2-eq. is used to group the emissions from different GHGs based on their GWP values. CO_2 is used as the reference substance ($GWP_{CO_2} = 1$), and the values of the rest of the gases is assigned to express the mass of CO_2 that would lead to the same warming effect in a period of 100 years ($GWP_{CH_4} = 25$, $GWP_{N_2O} = 298$). To calculate the CO_2-eq. of a set of gases, the emissions (in mass basis) are multiplied by their GWP values, one by one, and then all are added up. The CO_2-eq. trend in the atmosphere is graphed in Figure 1. In road transport, CO_2 is the largest contributor to global warming, even larger than in other sectors, but it is not the only such gas. That is why the CO_2-eq. units are preferable.

When different fuels for road transport are to be evaluated in terms of global warming, a full life cycle approach must be applied (*i.e.* accounting for emissions during extraction/cultivation, collection, distribution, processing, transport, final use, *etc.*). However, this entails some technical issues, such as deciding on what steps in the fuel chain must be included, allocation methodology, technology improvements over time, references and databases used, *etc.*, which leads to final results that are highly dispersed depending on the assumptions made. The following paragraphs present the attempts that have been made worldwide to harmonise the methodology and to propose some standardised values.

In the European Union, the Parliament approved on 17 December 2008 the Renewable Energy Directive (RED) 2009/28/CE.[10] It proposed a series of binding targets for each Member State to increase the share of renewable energy in its gross energy consumption to meet a global objective of 20% by 2020. This is one of the three targets marked in the 2020 climate and energy package, together with a 20% improvement in energy efficiency and a 20% reduction of GHGs from 1990 levels. The RED set a mandatory target of 10% for the share of renewable energy in transport, equal for all Member States. However, the main novelty (related to transport) of the RED was not the setting of renewable energy objectives (a previous biofuel directive, 2003/30/CE, compelled Member States to introduce biofuels for up to 2% of the energy consumed by 2005 and 5.75% by 2010), but the inclusion of sustainability requirements and a precise methodology for the calculation of CO_2-eq. savings.

About the sustainability requirements of biofuels, the RED took the following actions:

- The GHG[‡] emission reduction from the use of biofuels must be 35% at least. From 2017, this requirement is elevated to 50%, and to 60% from 2018 if the production of the installation started in 2017 or later.
- If biofuels are to be counted in the achievement of the renewable energy targets and/or as benefitting from financial support, raw materials should not come from lands with rich biodiversity, such as primary forests and other wooded lands, areas under nature protection programmes or that contain protected ecosystems or species, grasslands, *etc.*
- Similarly, raw materials for biofuels should not come from lands with high carbon stock, such as wetlands or forested areas; neither should they come from peatlands, unless the drainage of soil is impeded.

[‡]Only CO_2, CH_4 and N_2O are accounted for in the calculations.

For the calculation of the CO_2-eq. emissions of biofuels and fossil fuels, the RED proposes a method consisting of aggregating the emissions (g CO_2-eq. MJ^{-1} §) from all of the steps in the fuel chain:

- Extraction/cultivation of raw material (e_{ec}), including the emissions derived from harvesting, wastes, leakages and the production of pesticides/fertilisers used (machinery is not taken into account).
- Carbon stock modification by land-use change (e_l), which is calculated by dividing the total emissions equally over a period of 20 years. A bonus of 29 g CO_2-eq. MJ^{-1} is deducted if the biomass is obtained from restored degraded land.
- Fuel processing (e_p), including the emissions from wastes, leakages, chemicals or products used for processing and the emissions associated with the electricity used (obtained from average values of the region or particular values if the production plant is not connected to the electricity grid).
- Transport and distribution (e_{td}).
- Fuel combustion (e_u) in the engine, which will be zero for biofuels.
- If there is a proved reduction of GHG emissions due to an enhanced soil carbon accumulation, carbon capture and storage/replacement or excess electricity from cogeneration, they are accounted for as well.

Applying this methodology, the RED proposes default values of CO_2-eq. savings for some typical raw materials and processes (see Figure 3), and a

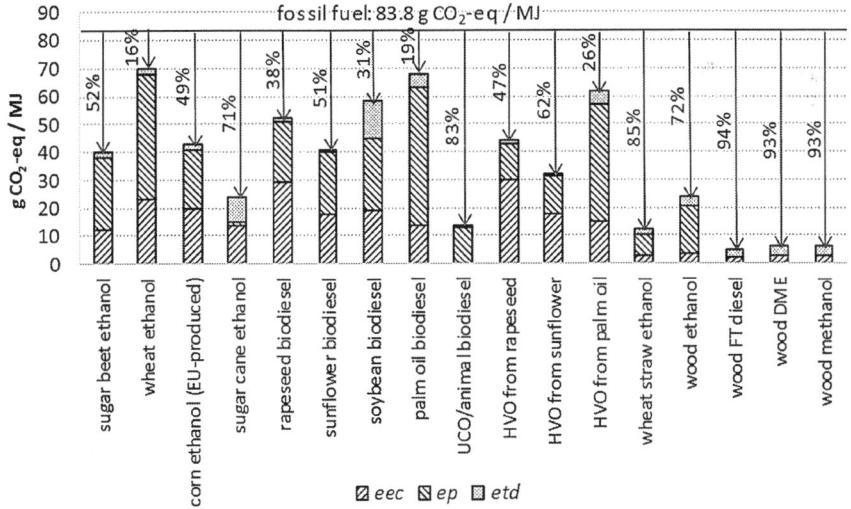

Figure 3 CO_2-eq. savings of biofuels compared to fossil fuels.

---

§If there is evidence of an improvement in vehicle efficiency when operating with a particular biofuel, emissions can be expressed in g CO_2 km^{-1} in order to represent the useful work done in the vehicle.

biofuel manufacturer is able to use them to prove the fulfilment of sustainable requirements. Nevertheless, the default values are very conservative estimations of the CO_2-eq savings with biofuels. In fact, for the same raw materials and processes shown in Figure 3, the RED proposes typical CO_2-eq savings which are higher than the default values. This creates an incentive for manufacturers to calculate the GHG balance on their own. Besides, the emission of fossil fuel is taken as 83.8 g CO_2-eq. MJ^{-1}, and this has given rise to some criticism because it may be an underestimation. Indeed, it seems contradictory that the recent Directive 2015/652/CE on the calculation methods and reporting requirements relating to the quality of petrol and diesel fuels set a value of 94.1 g CO_2-eq. MJ^{-1} (even higher numbers have been calculated by others[30]).

As present, the RED explicitly prohibits of the use of lands with high biodiversity or carbon stocks for biofuel production (this would be a 'direct' land-use change). However, the 'indirect' land-use change (ILUC) was not adequately addressed here. ILUC refers to the fact that lands that are currently used for agricultural crops may be changed in order to cultivate the raw materials for biofuels; as the crops are still needed for food, it is possible that new agricultural land needs to be created on land with high biodiversity. This is a serious concern for conventional biofuels (those derived from food crops). Thus, the recent Directive 2015/1513 (the 'ILUC Directive')[31] amends the RED and introduces the following changes, among others:

- The contribution of conventional biofuels towards the 2020 target (10%) is capped at 7%, with the aim that other alternatives with no or minor ILUC fill the remaining 3%. These alternatives benefit from a multiple-counting scheme: advanced biofuels (those derived from lignocellulosic feedstocks, non-food crops or wastes) and biofuels from cooking oil and animal fats are double-counted, whereas renewable electricity is counted from 2.5-times (in rail) to 5-times (in road vehicles) greater rates. In addition, the ILUC Directive sets an indicative target of 0.5% for the share of advanced biofuels by 2020.
- The calendar for the introduction of sustainability requirements is modified: a 60% reduction of GHGs is demanded from new installations from October 2015.

In the case of the USA, the legislative framework covering biofuel sustainability comprises a number of regional standards in force in particular states, and the Renewable Fuel Standard (RFS) is applicable in the entire USA. The metric used is volumetric (annual biofuel volume targets), but an estimation can be made of expected GHG (in CO_2-eq.) savings. In 2005, the first version of the RFS (RFS1) set a minimum target of 7.5 billion gallons of biofuels by 2012 to substitute a part of gasoline consumption. In 2007, a second version (RFS2)[32] expanded the temporal framework to 2022, and mandated an ambitious target of 36 billion gallons of biofuels by that date. With this penetration into the marketplace, biofuels are expected to reduce the amount of CO_2-eq. produced each year by about 138 million metric tons.

The RFS2 also expanded the list of fossil fuels affected in the substitution scheme to both diesel and gasoline. The RFS2 addresses the sustainability of biofuels by creating four nested categories of biofuels (*i.e.* a biofuel may qualify for more than one category), with each category having specific GHG saving targets compared to a 2005 baseline and biomass feedstock requirements:

- Total renewable fuels, with a target of 36 billion gallons by 2022. They must reduce GHGs by 20% at least compared to the 2005 baseline of fossil fuels. From 2015, the volume of corn-starch ethanol has been capped at 15 billion gallons in order to promote other more sustainable alternatives.
- Advanced biofuels, with a target of 21 billion gallons by 2022, must reduce GHGs by 50% at least and must not be produced from corn-starch.¶
- Cellulosic biofuels derived from cellulose, hemicellulose or lignin, including agricultural waste. The target for this category is 16 billion gallons in 2022. In order to qualify here, a biofuel must reduce GHGs by a minimum of 60%.
- Biomass-based biodiesel, defined as any diesel fuel produced from renewable biomass. The mandate for 2016 is 1.9 billion gallons and is expected to grow slightly over the coming years. The GHG reduction is 50%.

Similarly to European regulation, the RSF2 evaluates the GHG savings of some typical biofuel production routes, including all stages in the fuel life cycle and explicitly considering direct and indirect emissions. As a result of this evaluation, palm oil-derived diesel does not meet the mandated 20% GHG saving, for example.

6 Particle Emissions

Particles emitted by road vehicles are different to particles sampled from ambient air. Both diesel engines and gasoline direct injection engines emit small particles (with sizes mostly below 500 nm and a significant number of ultrafine particles below 100 nm). By contrast, secondary particles obtained from ambient sampling are in general larger as a consequence of coagulation processes and have a variety of sizes. Environmental scientists classify particles as PM10, PM2.5 and PM1. Since all particles from transport are below 1 μm in size, this classification is not adequate for transport particles.

As described in Chapter 1, particles emitted from vehicles are mainly composed of soot agglomerates comprising primary particles that are

---

¶Therefore, the definition of 'advanced biofuel' is different in US and EU regulations. Sugarcane ethanol and food crop-derived biodiesel are not advanced biofuels in the EU, but they are considered so in the USA if they meet the 50% GHG reduction mandate.

approximately spherical and with an inherent density of around 1.850 g cm^{-3}, and they have variable contents of organic compounds with lower densities, usually adsorbed on the soot surface. Since soot agglomerates have high light absorption efficiencies in the visible wavelength range, transport particles are generally identified as black carbon by environmentalists. In fact, they usually estimate the black carbon mass concentration from the light extinction at 880 nm by means of aethalometers using the Beer–Lambert law. Although at this infrared wavelength light absorption from soot agglomerates is lower than in the visible range, they absorb more than 90% of the light extinguished (the rest is scattered), in contrast to other materials. Therefore, it can be assumed that all light-absorbing material is black carbon and that its absorption coefficient is known and independent of the particle size.[33] With regards to the organic material, it has a higher light scattering efficiency but a lower absorption efficiency, and it is therefore identified as brown carbon. However, in case it is adsorbed on the soot surface, the absorption efficiency of soot can even be enhanced by the presence of a transparent coating, which acts as a lens focusing radiation into the soot core. Since this occurs in the visible range rather than in the infrared range, the presence of these organic materials leads to sharper rates of decrease of the absorption efficiency with increasing wavelengths, and can be quantified from the Ångström exponent. This is defined as the logarithmic ratio between the difference of absorption efficiencies at two different wavelengths and the difference of wavelengths. Brown carbon is known to have much larger Ångström exponents than black carbon.[34]

Once particles are emitted from vehicles, they remain suspended in the ambient air for a period of time during which aerosols may undergo regional and intercontinental transport, until they are finally deposited to the ground. The rate of deposition is, in general, faster as the size of the particle increases because the sedimentation mechanisms are faster and impaction processes are more probable. However, very small particles are subjected to Brownian diffusion, which also increases their rate of deposition. The average atmospheric lifetime is estimated at about 1 week. Deposition can be either dry through impaction, sedimentation, thermophoresis, electrophoresis and different types of diffusion processes, or wet when particles are scavenged by raindrops, snow or ice crystals. Some studies have shown that among the particle components, the water-soluble organic compounds have the highest scavenging efficiencies by wet deposition.[35]

Recently, climate models have shown that transport particles affect climate change if they either remain suspended as black carbon aerosols or if they are deposited to the earth's surface.[36] Black carbon aerosols absorb direct solar radiation, nucleate water from vapour and interact with both liquid and ice clouds, modifying the droplet number and size. The warming effect of these phenomena is very variable depending on the geographical coordinates and the type of Earth surface. When black carbon aerosols are located above a reflective surface, such as clouds or snow, they also absorb solar radiation reflected from that surface. Because black carbon absorbs much more light than it reflects, the consequence is the heating of the atmosphere, together with a reduction in sunlight reaching the surface, which can alter the hydrological cycle through changes in convection and large-scale circulation patterns. The particle size also has significant effects. As the sizes of particles increase, the fraction of scattered light (single scattering albedo) increases and the fraction of absorbed light decreases.[37] This means that small particles would contribute more to warming the atmosphere, while larger ones contribute more to preventing radiation reaching the ground and thus cooling it. Globally, it has been estimated that the absorption from black carbon aerosols leads to important positive radiative forcing, thereby warming the climate. From the start of the industrial era (1750), only CO_2 is estimated to have experienced greater forcing. However, due to the short lifetime of aerosols, and unlike CO_2 radiative forcing, sustained reductions in soot emissions could quickly decrease positive climate forcing and hence climate warming.

Among the different ecosystems affected by particle emissions and deposition, snow surfaces are especially sensitive. Particles deposited on snow contribute to reductions of the snow albedo (the ratio between reflected radiation and incident radiation) as a consequence of their high absorption efficiencies (usually higher than 1, meaning that they absorb more radiation than that focusing on their cross-section). Therefore, they heat the snow or ice surfaces and speed up melting. In addition, heating increases the snow/ice grain size, which further lowers the albedo, especially in the near infrared range (1000–2000 nm), and speeds up the melting of the seasonal snowpack. The consequence of this that the lower albedo of the surface below the snowpack is exposed earlier, leading to additional heating.[38] This phenomenon also contributes to modifying the radiative forcing,[39] the regional climate, the water resources and the recession of glaciers.[40]

Finally, it must be acknowledged that the uncertainties in net climate forcing from black carbon aerosols are substantial, due to a lack of knowledge about cloud interactions with both black carbon and co-emitted organic carbon, about the interactions between black carbon and dust and about the effect of soot on the microphysical properties of snow and ice, among other uncertainties.[41]

7 Future Trends

The scarcity of fossil fuels, the weather-related disasters associated with GHG emissions, the human health and environment effects of other pollutant emissions and the stringent fuel and vehicle regulations influence the current and future trends in road vehicles. CO_2 is the most abundant GHG, with its contribution being particularly large within the transportation sector. Total direct CO_2 emissions (please refer to Section 2) depend on the number of vehicles, the kilometres travelled by each vehicle and tank-to-wheel CO_2 emitted per vehicle and kilometre travelled. Reductions of any of those will contribute to controlling the anthropogenic CO_2 level emitted by road transport. Improvements in fuel economy (energy intensity), the implementation of low-carbon technologies and the introduction of low-carbon fuels (carbon intensity) have resulted in the reduction of the mass of CO_2 emitted per kilometre and vehicle. However, the continuous increase in the total number of registered vehicles in emerging countries[42] and the increased scale of their use[43] have diluted such improvements, resulting in larger total CO_2 emissions in the world.[44] The increase is especially noticeable in developing and emerging countries due to the expected growth in demand for faster transport modes and new infrastructures. On the other hand, the non-existence of an established infrastructure provides an opportunity to meet transport needs using low-carbon technologies.

The promotion of alternatives to car usage,[45,46] behavioural changes in society to avoid unnecessary journeys,[47] further improvements in vehicle fuel economy, new low-carbon transport systems and a larger market share of those technologies are required to decrease worldwide CO_2 emissions and then mitigate the weather modification effects of road vehicles (see Figure 4, which shows a simplified diagram of the pathways to reduce direct CO_2 emissions). From a wider perspective, not only should direct CO_2 emissions be considered, but also the CO_2 emitted in all of the processes involved in the whole life cycle of the vehicle systems (*e.g.* construction of infrastructure, manufacture of vehicles and provision of fuels) should be evaluated. Roadmaps and pathways have been proposed in order to meet GHG emission regulations.[47–49] The regulatory limits differ between the type of vehicle

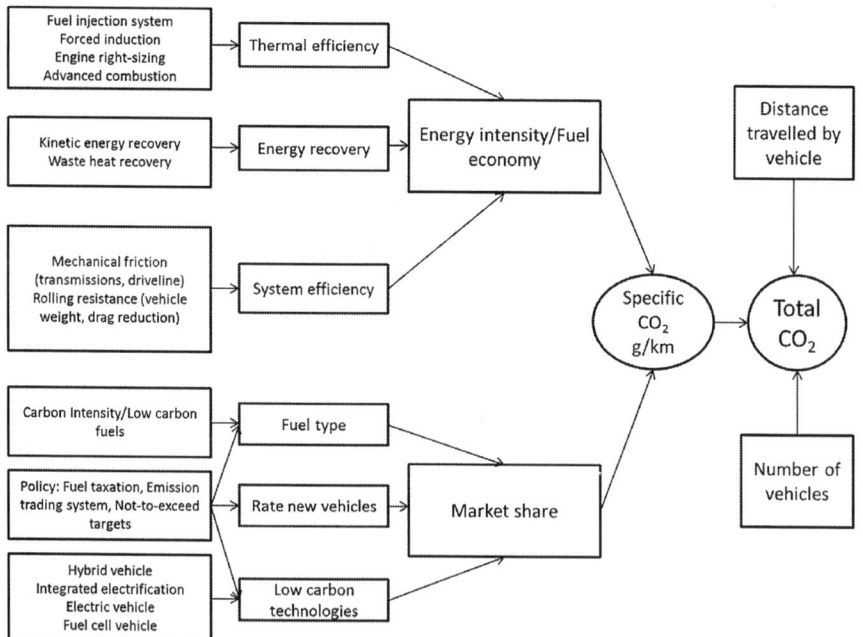

Figure 4 Simplified pathways to reduce direct CO_2 emissions.

(e.g. heavy duty and light duty) and the geographical region, and thus those roadmaps could be different in terms of specific technologies and timeframes depending on vehicle application and country. However, as the collective trend is the required reduction in the CO_2 emitted per vehicle and unit of travel distance, technologies that are able to offer improvements in fuel economy and the sustainable penetration of low-carbon technologies into the market are the main pillars such roadmaps have in common.

Improvements in thermal efficiency, enhancements in overall vehicle efficiency and the application of energy recovery technologies seem to be the main approaches to improving vehicle fuel economy and thus reducing CO_2 emissions. The adoption of optimised injection (e.g. multiple and controlled injections in common rail systems and gasoline direct injection) and forced induction systems (e.g. variable-geometry turbochargers), as well as extending engine operation in high-efficiency engine speed-load operations (e.g. down-speeding, right-sizing and cylinder deactivation), are the main approaches to increasing thermal efficiency.[47] It should be noted that road vehicles need to not only fulfil CO_2 emission regulations, but also the standards related to other pollutant emissions, such as NO_x, PM, CO and total hydrocarbons (THC). Therefore, advanced combustion technologies combining the advantages of spark and compression ignition engines (e.g. homogeneous charge compression ignition engines, partially premixed compression ignition engines, reactivity-controlled compression ignition and gasoline compression ignition) that are able to break the traditional trade-offs of PM–NO_x–thermal efficiency are being researched[50–52] in order to simultaneously meet both standards. Vehicle efficiency could be improved through the reduction of mechanical losses in the vehicle driveline, rolling resistance and aerodynamics losses. Reductions in the mass of the vehicle and rolling and drag coefficients could be utilised in order to improve overall vehicle efficiency. The utilisation of low-weight materials (e.g. aluminium, magnesium, titanium and composites) in the vehicle systems, as well as the optimisation of vehicle architecture, are the main approaches to reducing the mass of the vehicle. The usage of low-friction materials in the driveline and optimised tyre rubber materials and inflation pressures, as well as vehicle aerodynamics, are strategies for minimising vehicle friction losses. In addition to reduced vehicle thermal and friction losses, the overall fuel economy could be also enhanced through the recovery of parts of the thermal and friction losses. The implementation of energy recovery techniques such as kinetic energy recovery systems that enable the recovery of braking energy, exhaust waste heat recovery through organic fluids in a Rankine cycle, thermoelectric generators[53] and thermochemical processes,[54] as well as developments in thermal energy storage,[55] are the main alternatives for improving fuel economy through energy recovery.

Alternative fuels with a low- or zero-carbon content are envisaged as approaches to reducing the carbon footprint of road vehicles.[47–49] Fuels with zero carbon (e.g. hydrogen) or lower carbon contents (e.g. methane) than conventional fossil fuels will result in lower direct (tank-to-wheel) CO_2

emissions. However, indirect CO_2 emissions (well-to-tank) over the whole life cycle of the fuel (*e.g.* production and transportation) should also be taken into account. For instance, some of the routes in the production process of some synthetic fuels, such as dimethyl ether and synthetic power-to-gas and power-to-liquid fuels, suggest the production of hydrogen through electrolysis, as well as the recovery/sequestration of CO_2, as sources of carbon for catalytic direct hydrogenation and/or synthetic gas production. Therefore, the total effect (direct and indirect) on CO_2 emissions highly depends on the sources of energy utilised in the production process.

The progressive market penetration of low-carbon propulsion technologies is also considered to be a main driver for achieving a reduction in overall CO_2 emissions.[47–49] The usage of electric vehicles powered by batteries or fuel cells provides evident tank-to-wheel CO_2 as well as other pollutant emission reductions. However, electricity and hydrogen are energy carriers that are first produced from primary energy sources, and thus a whole life cycle analysis is required. Therefore, the real impact of electrification in road transport depends on the primary energy source and the efficiency of the production process utilised. Additionally, electric vehicles currently face the high price of energy storage devices (*i.e.* batteries), driving range anxieties and recharging times.[47] Further improvements in these areas are needed in order to incentivise the penetration of electric vehicles into the road transport market. Hybrid drivetrains combine two or more propulsion systems with the intention that each of the propulsion systems works in regions of maximum efficiency, and this approach was able to reduce tank-to-wheel CO_2 emissions overall by up to 91 $g\,km^{-1}$ in 2014.[56] Apart from the necessary technological improvements in these low-carbon fuel and vehicle technologies, there are other measures that could affect their market penetration and impact on CO_2 emissions. For instance, fuel policies influencing fuel taxation, the inclusion of road vehicles in the emission trading system and the establishment of worldwide not-to-exceed targets (which are already established in the USA) are believed to also be able to influence CO_2 emissions.[49]

References

1. *Intergovernmental Panel on Climate Change. Climate Change 2007: Synthesis Report. November 2007*, 2007.
2. *International Energy Agency. Part III: Greenhouse Gas Emissions. CO2 emissions from fuel combustion, Edition 2012*, 2012.
3. *Jet Propulsion Laboratory of the California Institute of Technology. Orbiting Carbon Observatory –2*.
4. *National Oceanic & Atmospheric Administration (US Department of Commerce – Global Monitoring Division of the Earth System Research Laboratory. Mauna Loa Hawaii Observatory*.
5. T. Ball, *Measurement of Pre-Industrial CO2 Level*, 2008.

6. *Road Transport: Reduction CO2 emissions from vehicles.* http://ec.europa.eu/clima/policies/transport/vehicles/index_en.htm, European Commission, Climate Action.
7. *Regulation (EC) No 443/2009 of the European Parliament and of the Council of 23 April 2009 setting emission performance standards for new passenger cars as part of the Community's integrated approach to reduce CO2 emissions from light-duty vehicles.* 2009, European Parliament, 2009.
8. Air Resources Board, National Highway Traffic Safety Administration. *Midterm evaluation of Light-Duty Vehicle Greenhouse Gas Emission Standards and Corporate Average Fuel Economy Standards for Model Years 2022-2025*, Environmental Protection Agency, 2016.
9. W. F. Ruddiman, *Clim. Change*, 2003, **61**, 261–293.
10. European Parliament. *Directive 2009/28/EC of the European Parliament and of the Council of 23 April 2009 on the promotion of the use of energy from renewable sources and amending and subsequently repealing Directives 2001/77/EC and 2003/30/EC,* 2009.
11. L. A. Graham, G. Rideout, D. Rosenblatt and J. Hendren, *Atmos. Environ.*, 2008, **42**, 4665–4681.
12. T. Grigoratos, G. Fontaras, G. Martini and C. Peletto, *Energy*, 2016, **103**, 340–355.
13. F. Diehl, J. Barbier Jr., D. Duprez, I. Guibard and G. Mabilon, *Appl. Catal., B*, 2010, **95**, 217–227.
14. J. M. Herreros, S. S. Gill, I. Lefort, A. Tsolakis, P. Millington and E. Moss, *Appl. Catal., B*, 2014, **147**, 835–841.
15. A. R. Ravishankara, J. S. Daniel and R. W. Portman, *Science*, 2009, **326**, 123–125.
16. L. Reijnders and M. A. J. Huijbregts, *Curr. Opin. Environ. Sustain.*, 2011, **3**, 432–437.
17. *Climate Change 2001: the scientific basis (in contribution of Working Group I to the 3rd Assessment Report.* Cambridge University Press, Cambridge, UK, International Panel on Climate Change, 2001.
18. M. Odaka, N. Koike, H. Ishii and H. Suzuki, *SAE Technical Paper*, 2002, **2002-01-1717**.
19. T. Chan, E. Meloche, J. Kubsh, R. Brezny, D. Rosenblatt and G. Rideout, *SAE Int. J. Fuels Lubr.*, 2013, **6**, 350–371.
20. N. Koike, M. Odaka and H. Suzuki, *SAE Technical Paper*, 1999, **1999-01-1081**.
21. D. Clark and T. Pauly, *SAE Int. J. Engines*, 2016, **9**(3), 1623–1629.
22. J. Girard, G. Cavataio and C. Lambert, *SAE Technical Paper*, 2007, **2007-01-1572**.
23. A. Vressner, P. Gabrielsson, I. Gekas and E. Senar-Serra, *SAE Technical Paper*, 2010, **2010-01-1216**.
24. P. Amnéus, F. Mauss, M. Kraft, A. Vressner et al., *SAE Technical Paper*, 2005, **2005-01-0126**.
25. P. J. Crutzen, A. R. Mosier, K. A. Smith and W. Winiwarter, *Atmos. Chem. Phys.*, 2008, **8**, 389–395.

26. *Guidelines for National Greenhouse Gas Inventories (Chapter 11: N2O Emissions for Managed Soils and CO2 Emissions From Lime and Urea Application)*, International Panel for Climate Change, 2006.
27. S. M. Ogle, S. J. Del Grosso, P. R. Adler and W. J. Parton, *Biofuels, Food Feed Tradeoffs*, 2009, 11–18.
28. R. Koeble, *Quantification of N2O emissions from biofuel feedstock cultivation*, European Commission Joint Research Center and Institute for Environment and Sustainability, 2011.
29. P. R. Adler, S. J. Del Grosso and W. J. Parton, *Ecol. Appl.*, 2007, **17**, 675–691.
30. *Greenhouse gas impact of marginal fossil fuel use. Project Number BIENL14773*, ECOFYS, 2014.
31. European Parliament. *Directive (EU) 2015/1513 of the European Parliament and of the Council of 9 September 2015 amending Directive 98/70/EC relating to the quality of petrol and diesel fuels and amending Directive 2009/28/EC on the promotion of the use of energy from renewable sources*, 2015.
32. *Regulation of Fuels and Fuel Additives: Changes to Renewable Fuel Standard Program. Federal Register 58*, Environmental Protection Agency, 2010.
33. J. C. Chow, *J. Air Waste Manage. Assoc.*, 2012, **45**, 320–382.
34. D. A. Lack and J. M. Langridge, *Atmos. Chem. Phys.*, 2013, **13**, 10535–10543.
35. M. Cerqueira, C. Pio, M. Legrand, H. Puxbaum, A. Kasper-Giebl, J. Alfonso, S. Preunkert, A. Gelencser and P. Fialho, *J. Aerosol Sci.*, 2010, **41**, 51–61.
36. T. C. Bond, S. J. Doherty, D. W. Fahey, P. M. Forster, T. Berntsen and B. J. De Angelo, *J. Geophys. Res., D: Atmos.*, 2013, **118**, 5380–5552.
37. H. Moosmüller, R. K. Chakrabarty and W. P. Arnott, *J. Quant. Spectrosc. Radiat. Transfer*, 2009, **110**, 844–878.
38. O. L. Hadley and T. W. Kirchstetter, *Nat. Clim. Change*, 2012, **2**, 437–440.
39. H. V. Jacobi, S. Lim, M. Ménégoz, P. Ginot, P. Laj, P. Bonasoni, P. Stocchi, A. Marinoni and Y. Arnaud, *Cryosphere*, 2015, **9**, 1685–1699.
40. W. S. Lee, R. L. Bhawar, M. K. Kim and J. Sang, *Atmos. Environ.*, 2013, **75**, 113–122.
41. C. G. Gertler, S. P. Puppala, A. Panday, D. Stumm and J. Shea, *Atmos. Environ.*, 2016, **125**, 404–417.
42. *Global Health Observatory (GHO) data provided by World Health Organization* http://www.who.int/gho/road_safety/registered_vehicles/number_text/en/.
43. *Road traffic, vehicles and networks, Environment at a Glance 2013 OECD Indicators*.
44. *BP Energy Outlook 2035*, 2014.
45. *European vehicle market statistics, The International Council of Clean Transportation*, 2014.
46. *Recent trends in car usage in advanced economies – slower growth ahead?*, International Transport Forum 2013.
47. O. Edenhofer, R. Pichs-Madruga, Y. Sokona, E. Farahani, S. Kadner, K. Seyboth, A. Adler, I. Baum, S. Brunner, P. Eickemeier, B. Kriemann,

J. Savolainen, S. Schlömer, C. von Stechow, T. Zwickel and J. C. Minx, *IPCC, 2014: Climate Change 2014: Mitigation of Climate Change. Contribution of Working Group III to the Fifth Assessment. Report of the Intergovernmental Panel on Climate Change.* Cambridge University Press, Cambridge, United Kingdom and New York, NY, USA. International Panel for Climate Change., 2014.
48. *UK Automotive council roadmaps.*
49. *Reducing CO2 emissions from road transport in the European Union: An evaluation of policy options*, The International Council on Clean Transportation, 2016.
50. R. D. Reitz and G. Duraisamy, *Prog. Energy Combust. Sci.*, 2015, **46**, 12–71.
51. P. B. Mark, P. C. M. Musculus and L. M. Pickett, *Prog. Energy Combust. Sci.*, 2013, **39**, 246–283.
52. *Advanced combustion engines. 2015 Annual Report. Energy Efficiency and Renewable Energy*, U.S. Department of Energy, 2015.
53. R. Saidur, M. Rezaei, W. K. Muzammil, M. H. Hassan, S. Paria and M. Hasanuzzaman, *Renewable Sustainable Energy Rev.*, 2012, **6**, 5649–5659.
54. D. Fennell, J. M. Herreros and A. Tsolakis, *Int. J. Hydrogen Energy*, 2014, **39**, 5153–5162.
55. H. Zhang, J. Baeyens, G. Cáceres, J. Degrève and Y. Lv, *Prog. Energy Combust. Sci.*, 2016, **53**, 1–40.
56. *European Vehicle Market Statistics. The International Council on Clean Transportation. Pocketbook.* 2016.

Local-acting Air Pollutant Emissions from Road Vehicles

QINGYANG LIU AND JAMES J. SCHAUER*

ABSTRACT

This chapter reviews the local impacts of air pollution emissions from roadways on air quality, human health, and the natural and built environments. These impacts are a global issue affecting urban areas around the world, including developed and developing nations, emphasizing the need to reduce emissions from roadway transport sectors around the world. Air pollutants emitted from vehicles, including particulate matter, nitrogen oxides, carbon monoxide, and hydrocarbons, are reactive (*i.e.* they have larger impacts on human health per unit of emissions compared to many other air pollution sources due to the proximity of the emissions and the chemical and physical natures of the emissions). These impacts affect public health, visibility, material damage, the surrounding ecosystems, and the quality of life for populations living near roadways. Although roadway emissions vary from region to region depending on the state of local fuel quality, vehicle technology, and emissions standards, general trends of roadway emissions and roadway impacts are summarized in this chapter. This summary includes the chemical composition of emissions from roadways, the contribution of mobile sources to local and regional air quality, the impacts on cultural heritage, and the health impacts of mobile sources. Existing regulations and projected trends in national standards for vehicle emissions are discussed and the implications of roadways for local environments are considered.

*Corresponding author.

Issues in Environmental Science and Technology No. 44
Environmental Impacts of Road Vehicles: Past, Present and Future
Edited by R.E. Hester and R.M. Harrison
© The Royal Society of Chemistry 2017
Published by the Royal Society of Chemistry, www.rsc.org

1 Introduction

Air pollution is the leading environmental health risk factor for disease in developing and developed countries.[1] A 2010 Health Effects Institute report suggests that 3.2 million premature deaths are caused by outdoor air pollution, which includes the impacts of roadways.[2] Vehicle emissions can be divided into either tailpipe emissions or non-tailpipe emissions. Non-tailpipe emissions include road wear, brake wear, tire wear, fugitive emissions of fuel leakage, evaporative emissions of fuel, and the resuspension of road dust. Tailpipe emissions include particulate matter (PM) and gaseous emissions, such as nitrogen oxides (NO_x), sulfur dioxide (SO_2), carbon monoxide (NO), and volatile organic compounds (VOCs).[3] PM consists of a fine particle fraction ($PM_{2.5}$), which is defined as particles having an aerodynamic diameter of less than 2.5 μm, and a coarse fraction (PM_{10}), which is particles with diameters ranging between 2.5 and 10 μm. Both particle fractions ($PM_{2.5}$ and PM_{10}) contain toxic components such as polycyclic aromatic hydrocarbons (PAHs) and transition metals.[4-7]

Laboratory experiments have shown that gaseous organic compounds in both gasoline engine (spark ignition [SI]) and diesel engine exhaust (compression ignition [CI]) can form secondary organic aerosols (SOAs) and photochemical smog, including ozone (O_3).[8] Older diesel engines that do not have after-treatment technologies produce significant amounts of NO_x and soot pollutants compared to modern diesel and gasoline engines with effective emission controls.[7,9] Thus, growing concerns about the impacts of transport emissions on air quality have resulted in the implementation of strict emission regulations in many economically developed countries, such as the USA, Japan, and some Western European countries, staring in the mid-1970s for gasoline engines and the 2000s for diesel engines.[10,11] To address vehicle emissions in countries with extreme air pollution and countries that are rapidly developing, these emission control technologies are becoming more widely used around the world.

Despite great improvements in the technology of exhaust reductions for modern vehicles, global road emissions are projected to increase over the next 30 years due to the expected growth in vehicle ownership worldwide (*i.e.* between 2 and 3 billion vehicles by 2050).[12,13] An improved understanding of various aspects of road emissions, including chemical characteristics, source contributions, and other associated impacts, is therefore vital for policy makers working to implement updated traffic emission standards.[14,15]

This review presents a global perspective on the local impacts of roadway emissions. Section 2 of this chapter focuses on the changes in road emissions due to developing technology and addresses both tailpipe and non-tailpipe emissions. The review focuses on pollutants from mobile sources that have significant local impacts. Section 3 of the report focuses on key pollutants emitted from roadways that undergo chemical and physical transformations as they transport across the urban scale, with an emphasis on air toxics. Section 4 provides the latest epidemiological evidence for mobile source impacts and discusses the differences and linkages between health impacts of proximal road exposures and exposures to mobile sources as components of the urban air pollution mixture. This section also covers recent studies of the contributions of mobile sources to PM and O_3 in cities around the world. Section 5 of the report summarizes several case studies of the effects of mobile source emissions on the deterioration of historical heritage sites and their influence on the air pollution-derived discoloration of the built environment. Section 6 of the report summarizes the impacts of mobile sources on regional air quality and presents data from around the world. Section 7 presents a summary of recent mobile source emission trends, including a comparison between emissions in developed and developing countries and their underlying differences resulting from fossil fuel quality and vehicle technology. Finally, Section 8 summarizes the current regulatory requirements for emission reductions around the world and the projected trends for the coming decades.

2 Fuel Type, Fuel Quality, and Vehicle Technology

Vehicle emissions are composed of a range of chemical pollutants including PM, carbon monoxide (CO), carbon dioxide (CO_2), NO_2, VOCs, and SO_2. These pollutants, which are directly emitted from air pollution sources, are called primary pollutants. Gasoline-powered vehicles are a major source of VOCs, and heavy-duty diesel engines are a major source of NO_x in many urban areas. VOCs and NO_x react together, in the presence of sunlight, to form O_3 and other components of photochemical smog, which are called secondary pollutants.[16,17] It is important to recognize that although O_3 is not directly emitted from roadways, O_3 is formed in the atmosphere because of emissions from roadways. The primary mitigation tools used to reduce emissions of primary pollutants and the precursors for secondary pollutants from roadways are changes in fuel quality and vehicle technology.

2.1 Fuel Sulfur Reduction

Sulfur is a naturally occurring component of crude oil and is found in both gasoline and diesel. When these fuels are burned, sulfur is predominantly emitted as SO_2 or sulfate PM.[19] Any reduction in fuel sulfur immediately reduces these sulfur compounds as well as the total pollutant emissions.[20] Reducing sulfur levels in fuels can reduce vehicle emissions by directly reducing SO_2 and sulfate PM and achieving better performance from emission control systems in existing vehicles.[21]

Globally, 90% of new gasoline vehicles are equipped with a three-way catalyst (TWC), which simultaneously controls emissions of CO, hydrocarbon (HC), and NO_x. The sulfur in gasoline is a poison for the TWC and reduces the catalyst effectiveness for the removal of gaseous emissions (*e.g.* CO, HC, and NO_x). For this reason, gasoline desulfurization is a necessary precursor to the adoption of TWC technology for gasoline-powered vehicles. The parallel is true for diesel fuel and modern after-treatment technologies discussed below for diesel engines. The sulfur content of crude oil varies by origin but is typically within the range of 100–33 000 ppm. Hydrotreating during fuel processing is a widely used technology by oil refineries when removing sulfur from gasoline and diesel. It can reduce sulfur levels to less than 10–15 ppm for gasoline and diesel compared to high-sulfur diesel fuels that can have more than 5000 ppm sulfur.[21] Because of this widely used hydrotreating technology in oil refining, low-sulfur fuels are already common and are required in many parts of the world, including the USA, Europe, and China,[21] which leads to a reduction in tailpipe SO_2, PM, and VOC emissions. In the USA, a reduction in fuel sulfur levels from 368 to 54 ppm cut out 4% of PM mass emissions.[21] In Denmark, tests on diesel vehicles demonstrated that a reduction in fuel sulfur levels from 440 to 0.7 ppm could lead to a 56% reduction in the number of particles.[21]

2.2 Fuel Additives, Including Tetraethyl-lead, Methylcyclopentadienyl Manganese Tricarbonyl, and Lube Oil Additives

Unleaded gasoline allows effective utilization of the vehicle emission control technologies in gasoline vehicles. The presence of lead in gasoline poisons catalytic converters and impairs their ability to remove CO, VOCs, and NO_x.[24] Methylcyclopentadienyl manganese tricarbonyl (MMT) was initially used as a replacement additive to increase the octane rating after leaded gasoline was phased out.[24] Results indicated that there are no significant differences in emissions (CO, HC, and NO_x) and fuel consumption for both catalyst and non-catalyst cars when MMT is used.[25]

Lubricating oil (lube oil) is known to be minimally consumed during normal operation of an internal combustion engine from an engine perspective, but is not insignificant from an emissions perspective. The sulfur content of a petroleum-based engine lube oil is typically three-times higher than a desulfurized fuel and can lead to an elevation in the level of SO_2 in vehicle emissions, but is typically low enough not to greatly impact after-treatment control technologies.[26] Recent studies suggest that trace metals and metallic nanoparticles emitted by properly operating gasoline internal combustion engines are derived mainly from the combustion of lubrication oil.[26]

2.3 Tailpipe NO_x, CO, VOCs, and PM Emission Related to the Combination of Technology and Fuel

Several studies have been conducted extensively to characterize the tailpipe emission rates and factors of air pollutants from gasoline and diesel vehicles.[6,7] The rates and amounts of gaseous and particulate emissions from gasoline and diesel vehicles are influenced by various factors, including vehicle age and mileage, engine load, fuel type and quality, and vehicle operating mode (*e.g.* cold start, hot stabilized, and driving cycle).[27] Diesel engines offer some emission benefits over gasoline engines.[28] Conventional diesels typically use only 70% of the amount of fuel required by a gasoline engine, significantly reducing per-mile CO_2 emissions.[9,29,30] Due to the nature of CI combustion, CO emissions from properly working diesel engines are much lower than those from gasoline engines.[6] After-treatment technologies that are widely used for SI engines are highly effective and allow CO emissions to be greatly reduced.[9]

Diesel fuel has the added benefit of low volatility, which virtually eliminates evaporative HC emissions.[5,7] Data from chassis dynamometer tailpipe emission tests of conventional light-duty gasoline vehicles without proper after-treatment show that regulated, low-mileage vehicles have intermediate NO_x and VOC emission rates and high CO and PM emission rates during both hot stabilized operation and relatively nonaggressive driving conditions.[31,32] In contrast, emission rates of NO_x for conventional diesel vehicles without proper after-treatment are higher than those of gasoline (Table 1).[33]

Studies of ethanol-based fuels and biodiesel as alternative fuels promote independence from imported petroleum sources and offer the advantage that these fuels can be derived from renewable resources.[27,34] These fuels have some properties similar to those of traditional fuels, such that they can be directly used in conventional vehicles with little or no engine modification. Zhai *et al.*[34] revealed that replacing gasoline with E85 (85% ethanol, 15% gasoline) in SI vehicles reduces CO emissions, slightly

Table 1 Relative emissions of key air pollutants from road vehicles as a function of vehicle technology, based on a selection of emission data. The relative magnitudes of the various tailpipe emissions vary with driving cycle and fuel.[5-7,9]

Vehicle technology	Tailpipe NO_x	Tailpipe VOCs	Tailpipe CO	Tailpipe PM	Non-tailpipe PM	Fugitive VOCs
Conventional gasoline without proper after-treatment	Medium	Medium	High	High	Medium low	High
Conventional gasoline with modern after-treatment	Low	Low	Low	Low	Medium low	Low
Conventional diesel without proper after-treatment	High	Low	Low	High	Medium low	High
Conventional diesel with modern after-treatment	Low	Low	Low	Ultra-low	Medium low	Low
Electric vehicles	Zero	Zero	Zero	Zero	Medium low	Zero
Fuel cell vehicles	Zero	Zero	Zero	Zero	Medium low	Zero

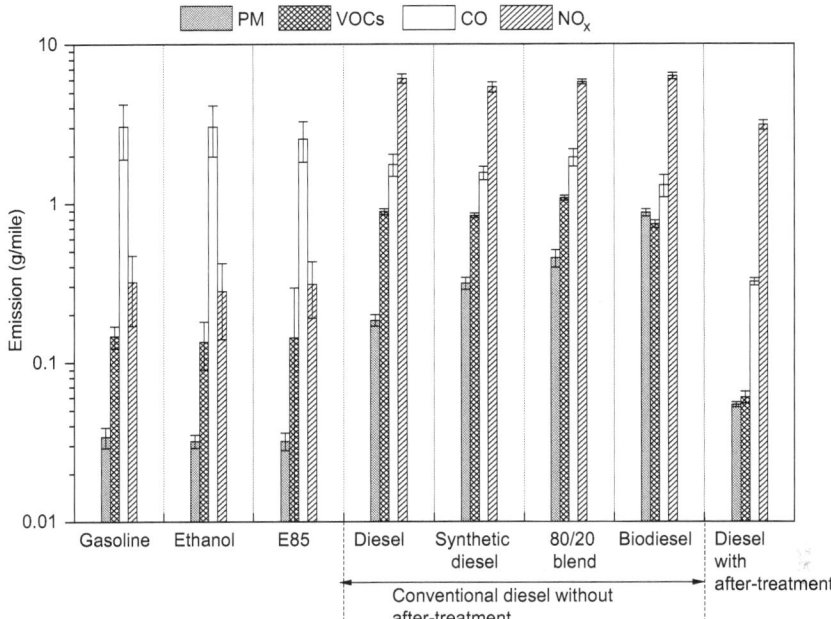

Figure 1 Comparative emissions of key pollutants for SI and CI engines as a function of fuel type and fuel quality. Emission data for SI and CI engines are reproduced from Zhai et al.[34] and with permission from Durbin et al.,[27] Copyright (2000) American Chemical Society, respectively. E85 is a blend of 85% ethanol and 15% gasoline. Synthetic diesel is produced using the Mossgas conversion of olefins to distillate process. The olefin byproducts include propene, butene, pentene, and hexane. The 80/20 blend is a blend of 80% diesel and 20% biodiesel. The biodiesel is obtained from Taurus Lubricants Corporation in Lowell, MA.

increases NO_x emissions, and increases VOC emissions. Durbin et al.[27] showed that biodiesel, as an alternative fuel for diesel engine generators, could increase combustion efficiency and reduce VOCs and CO, accordingly (Figure 1). Despite the recognized advantages of alternative fuels, widespread commercialization of biofuels has been limited, but the use of biofuels has been greatly increasing in several regions that have government policies either mandating or incentivizing their use.

2.4 After-treatment Controls for Modern Vehicles

The TWC has been mandatory in all new gasoline vehicles in the USA and Europe since the introductions of the Tier 1 and Euro 1 emission standards, respectively.[22,23] The TWC has proven to be effective at reducing emissions of NO_x, CO, and VOCs. Gasoline vehicles with TWC technology emit less total amounts of NO_x, CO, VOCs, and PM on a mass per mile basis in comparison to those without after-treatment technologies, so the emissions were low in gasoline fuels with modern after-treatment technologies (Table 1). The wide use of TWC for gasoline-powered mobile sources in the 1970s in the USA, Japan, and Europe arose much earlier than the introduction of parallel controls for diesel engines, which did not occur until the 2000s.[14] Modern diesel engines benefit from a combination of three technologies to reduce the most concerning emissions of diesel engines: PM and NO_x.[9] A diesel oxidation catalyst (DOC) is designed to oxidize HC and volatile PM to CO_2 and H_2O.[31] Typically, the conversion efficiency of HC is more than 95% for commercial diesel vehicles (Table 1). The extent of the conversion depends on the catalyst specification and the exhaust gas temperature.[8] DOCs do not directly target NO_x reduction, but can lead to the conversion of NO to NO_2 in the tailpipe, which thereby increases primary NO_2 emissions. NO_2/NO_x ratios range from ~10% for diesel vehicles without oxidation after-treatment to more than 50% for DOC-equipped vehicles.[23] A diesel particulate filter (DPF) is a device designed to remove diesel PM and soot particles from the vehicle, but offers no NO_x removal capability. A DPF is similar to a DOC in terms of its conversion of NO to NO_2, which results in more NO_2 emissions and a higher NO_2/NO_x ratio in the diesel tailpipe.[23]

Selective catalytic reduction (SCR) is currently the primary technology for enabling diesel vehicles to comply with NO_x emission standards. The method chemically converts NO_x into nitrogen with the aid of ammonia (NH_3) and a catalyst.[20,21,23,36] The NH_3 is generated from an aqueous solution of urea, which is degraded to NH_3.[20] The NH_3 reacts with NO to form N_2 and water by heterogeneous catalysis. SCR systems can reduce NO_x emissions with conversion efficiencies of 60–70%, meeting the US Tier 3 and Euro 6 emission standards, respectively.[18,20]

2.5 Fugitive VOC Emissions from Vehicles

In addition to tailpipe VOC emissions from vehicles, fugitive VOC emissions resulting from evaporative emissions and whole-fuel leakage and spillage are important sources that contribute to tropospheric photochemical O_3 formation and the release of toxics and carcinogens, which often cause adverse human health effects.[37] The contributions of fugitive VOC emissions to ambient air quality are usually calculated using field experiments, emission inventories, and receptor modeling, respectively.[38] Fujita et al.[39] compared emissions inventories from the entire state of Massachusetts with those from California's South Coast Air Basin and found that 45% (California) and 46% (Massachusetts) of ambient VOCs were from stationary emitters, including fuel combustion, waste burning, solvent use, petroleum processing storage and transfer, industrial processing, and miscellaneous processes. Meanwhile, 46% (California) and 34% (Massachusetts) of ambient VOCs were from on-road vehicles, including SI and diesel engine vehicles. Watson et al.[38] summarized that VOC fractions contributed to emissions by different sources in more than 20 urban areas, most of which were in the USA, and found that 50% or more of the ambient VOCs were from vehicle exhaust, liquid gasoline, and fugitive vehicle emissions.

2.6 Non-tailpipe PM Emissions from Vehicles

Recent source apportionment (SA) studies present results only for total contributions from road traffic to ambient PM, but sometimes the road dust component gets mixed in with other mineral/soil sources.[4,40–44] Nevertheless, air quality management is essential to understanding the individual source contributions.[4] PM contributions from vehicular traffic should be differentiated between tailpipe (exhaust) and non-tailpipe (non-exhaust) contributions.[4] Usually, non-tailpipe contributions should be further separated between road dust, brake dust, tire wear, and road wear contributions.[3] As discussed above, important technological innovations and policies have been made to reduce tailpipe emissions, but no actions and policies were made to lessen the burden of non-tailpipe emissions on air quality in cities.[3] The relative importance of these non-tailpipe emissions changes widely in space.[33–37] Spatially, Schauer et al.[40] demonstrated with a large set of mobile source profiles that brake wear and re-suspended road dust accounted for about 11–20% and 29–72% of road traffic PM_{10} emissions, respectively, in Milwaukee, WI. Ketzel et al.[41] estimated that 50–85% of road traffic PM_{10} emissions in Denmark, Germany, Austria, and Finland were from non-tailpipe emissions. Amato et al.[42] analyzed road traffic PM samples in Barcelona, Spain, and concluded that 37% of PM_{10}, 15% of $PM_{2.5}$, and 3% of PM_1 were related to non-exhaust emissions. Bukowiecki et al.[43] reported that non-exhaust emissions contributed to 50% and 59% of road traffic PM_{10} emissions in street canyons and motorways, respectively, in Zurich, Switzerland. In addition, Harrison et al.[4] reported that re-suspended road dust, brake dust, and tire dust contributed to $29.3 \pm 7.5\%$, $42.5 \pm 5.3\%$, and $8.2 \pm 1.7\%$ of the road traffic increment above background, respectively, in London, UK. Lawrence et al.[44] concluded that non-tailpipe emissions including brake wear, surface wear, and re-suspended road dust contributed to about 49% of the total traffic PM_{10} emissions in Hatfield, UK (Table 2). Table 2 provides an overview of the most important literature studies on calculated contributions of non-tailpipe emissions to local transport emissions. Discriminating source contributions within the non-exhaust is also of importance for health outcomes.[1,33] Brake and tire wear particles have higher oxidative potential than other traffic-related sources, and their effect is very local (50–100 m from the roadway).[33,45–47] Given this evidence, investigating the role of non-exhaust emissions in air quality impairment is essential for local authorities, mostly because such particles are emitted locally and are therefore easier to mitigate, thereby improving public health.

2.7 Electric and Fuel Cell Vehicles

Fleet electrification is one strategy for improving urban air quality because it removes combustion from the roadway. This process allows for reduced on-road air pollutant emissions compared to conventional gasoline or diesel vehicles.[31] It comprises a wide spectrum of technological options that range

Table 2 Relative contributions of non-tailpipe emissions to local transport emissions from the literature.[40–44]

References	Area	Contributions
Schauer et al. 2006[40]	Milwaukee, WI, USA	11–20% for brake wear and 29–72% for resuspended road dust of road traffic PM_{10} emission
Ketzel et al. 2007[41]	Denmark, Germany, Austria, Finland	50–85% (varying among sampling locations) of road traffic PM_{10} emission
Amato et al. 2009[42]	Barcelona, Spain	37%, 15%, and 3% for resuspended road dust of road traffic PM_{10}, $PM_{2.5}$, and PM_1, respectively
Bukowiecki et al. 2010[43]	Zurich, Switzerland	Street canyon: 21% for brake wear, 38% resuspended dust of road traffic PM_{10} emission Motorway: 3% brake wear, 56% suspended/abraded road dust of road traffic PM_{10} emission
Harrison et al. 2012[4]	London, UK	$29.3 \pm 7.5\%$ for resuspended dust, $42.5 \pm 5.3\%$ for brake dust, and $8.2 \pm 1.7\%$ for tire dust of the road traffic increment above background
Lawrence et al. 2013[44]	Hatfield, UK	11% for brake wear, 11% for surface wear, and 27% for resuspended dust of road traffic PM_{10} emission

from the early stages of electric vehicles to pure fuel cell vehicles, which are entirely propelled by stored electricity and have no direct on-road exhaust emissions (Table 1). However, due to the additional load on the electric power system, greater emissions from electricity generation would cause an increase in coal-fired capacity and result in modest increases of NO_x, SO_2, and PM emissions in certain areas of the world. It is important to note that these emissions are not on-road and have a very different local impact than on-road emissions.[31,32] With the exception of on-road emissions, some studies concerning non-tailpipe emissions show that they are relatively higher than those of conventional gasoline or diesel vehicles because electric vehicles are 24% heavier than their conventional counterparts, resulting in more brake wear and fugitive dust emissions.[32] Therefore, further improvement could be made to specifically target non-tailpipe PM emissions by encouraging innovations in vehicle weight reduction.

3 Evolution of Roadway Emissions

3.1 Primary and Secondary Pollutants

Road traffic emissions consist of PM and gaseous emissions, with active carbonaceous products present in both phases.[48] Pollutants from road traffic can be divided into two different groups: (1) primary pollutants such as CO, black carbon (BC), and ultrafine particles (UFPs), which are relatively unreactive in the atmosphere, and (2) NO_x, VOCs, and SO_2, which undergo chemical processing within roadways to generate secondary pollutants such as O_3 and secondary aerosol (*i.e.* SO_4^{2-} and NO_3^-).[49–51] Since both primary and secondary pollutants exhibit inhomogeneous distributions near roadways and vary substantially in abundance with time, it is essential to assess individual concentration gradients and chemical transformations in such air pollutants.[3,4]

3.2 Changes in Pollutant Concentrations Downwind of Roadways

It is estimated that 30–45% of people in large North American cities living near roadways are highly affected by traffic emissions.[52,53] In addition, roadways are where the majority of outdoor activities of the urban population occur and hence where substantial exposure to pollutants results for pedestrians and road users. Occupants of adjacent buildings may also be subject to emissions from the roadway environment.[54] The pedestrian level (breathing height) of streets is expected to experience particularly high levels of pollutants due to the proximity of vehicle emissions.[55] Pollutant concentrations within roadways frequently exceed those in wider urban backgrounds.[56–61] Air quality mitigation targeting for long-term averages (hours, days, or annual) may be imperfect when accounting for the exposure associated with actual nonlinear changes in pollutant concentrations near roadways, which have repeated peaks within a short period of time.[57,59,60] Therefore, acquiring both dynamic and chemical processes governing the spatial air pollutant gradients near roadways is essential to accurately assessing personal health risk and helping governments develop associated policies (*e.g.* roadway design and utility of green landscapes) to mitigate such health impacts.[62]

Recently, the numbers of air pollution measurements near roadways have been increasing in North America.[56–61] Overall, spatial pollutant gradients moving downwind from roadways have been recorded, but the trends over different periods have varied widely. Zhu *et al.*[60,61] found that the concentrations of UFPs, PM, BC, and CO were two- to five-times greater than background concentrations near the freeways in southern California and that pollution levels exponentially decreased to near-normal (background) levels by around 300 m away from the freeways. Both seasonal and diurnal variations were identified near Interstates 405 and 710 in California, USA. Clements *et al.*[57] investigated the spatial gradients of vehicular-emitted air pollutants of three highways (*i.e.* State Highway 71, Interstate 35 and Farm to Market Road 973) in Austin, TX, and found that concentrations of CO, NO, and NO_x returned to background levels within a few hundred meters of the roadway (Figure 2). Concentrations of most carbonyl species decreased with distance downwind of State Highway 71 (SH-71), while concentrations of acetaldehyde and acrolein increased further downwind of SH-71, suggesting

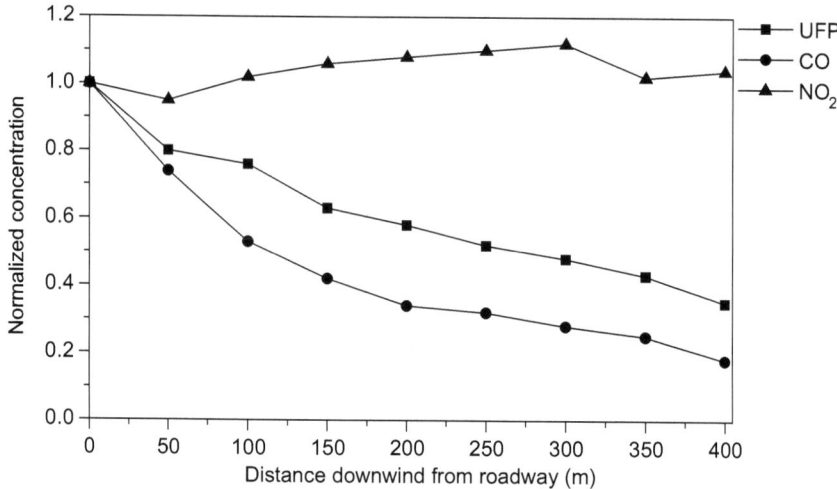

Figure 2 Schematic of pollutant changes during downwind transport from roadways. The data are obtained from and Zhu et al.[60] Reproduced from A. L. Clements, Y. Jia, A. Denbleyker, E. McDonald-Buller, M. P. Fraser, D. T. Allen, D. R. Collins, E. Michel, J. Pudota, D. Sullivan and Y. Zhu, Air pollutant concentrations near three Texas roadways, part II: Chemical characterization and transformation of pollutants, *Atmos. Environ.*, **43**(30), 4523–4534, Copyright (2009) with permission from Elsevier.[57] *Note*: downwind air pollutants extended about 2600 m from the freeway until reaching background levels during pre-sunrise hours.[75]

chemical generation from the oxidation of primary vehicular emissions (Table 3). Fine PM mass concentrations, UFPs, PAHs, hopanes, and elemental carbon (EC) generally exhibited concentrations that decreased with distance downwind of SH-71, regardless of wind speed (Table 3). Wang et al.[63] modeled the chemical evolution of NO_x with distance downwind of SH-71 and found that the results were comparable to field measurement data. Olson et al.[58] reported that the concentration of 55 VOCs (C_2–C_{12}) decreased exponentially with increasing distance (10–100 m) from a highway in Raleigh, NC (Table 3). These studies presented much sharper air pollutant gradients downwind of freeways, with levels above background concentrations extending only 300 m downwind of roadways during the day and up to 500 m at night. Hu et al.[64] observed a wide area (∼2600 m) downwind of Interstate 10 in West Los Angeles, CA, that was being impacted by air pollution during the pre-sunrise hours in both winter and summer seasons. This phenomenon could be attributed to strong atmospheric stability, low wind speeds (0–1 m s^{-1}), low temperatures (9–13 °C), or high humilities (61–79%), which together result in slower transport of air pollutants before dilution and dispersion to background levels. Durant et al.[56] concluded that the rapid changes in pollutant gradients were due to variations in highway traffic flow rate, wind speed, and surface boundary layer height. Before

Table 3 Key air toxics downwind of roadways and trends in transformations during urban-scale transport. The data are reproduced from a selection of urban field studies.[56–58]

Component	Transformation during transport	Concentration distribution with the horizontal distance from roadway
BTEX		
Benzene	No	
Toluene	No	Decreased exponentially to urban
Ethylbenzene	No	background levels
m,p-Xylene	No	
o-Xylene	No	
Carbonyl compounds		
Formaldehyde	Yes (photolysis)	Decreased exponentially to urban background levels
Propionaldehyde	Yes (photolysis)	
Acetaldehyde	Yes (oxidation)	Increased downwind from the roadway
Acrolein	Yes (oxidation)	
PAHs		
Fluoranthene	Yes (oxidation)	
Pyrene	Yes (oxidation)	
Chrysene	Yes (oxidation)	
Benzo(*b,k*)fluoranthene	Yes (oxidation)	Decreased to urban background levels
Benzo(*e*)pyrene	Yes (oxidation)	
Benzo(*a*)pyrene	Yes (oxidation)	
Perylene	Yes (oxidation)	
2-Butanone	Yes (photolysis)	Decreased exponentially to urban background levels
1,3-Butadiene	Yes (oxidation)	

sunrise and peak traffic flow rates, downwind concentrations of PM, CO_2, NO, and NO_2 were highest within 100–250 m of the highway. After sunrise, pollutant levels sharply declined and the gradients became less pronounced as wind speeds increased and the surface boundary layer rose, allowing mixing with cleaner air aloft. These results suggest that air quality policy makers need to have sufficient information about traffic patterns and local meteorology in order to make optimal regulatory decisions and protect public health from roadway emissions.

3.3 Key Air Pollutants Associated with Roadway Emissions

Pollutants emitted from road transport can be characterized by their origins as either tailpipe or non-tailpipe emissions.[3] Tailpipe emissions mainly contribute to UFPs (diameter <100 nm), VOCs and toxics, NO_x, CO, CO_2, O_3, and fine PM (<2.5 μm in aerodynamic diameter; $PM_{2.5}$), while non-tailpipe emissions predominately contribute to PM_{10} (<10 μm in aerodynamic diameter) and coarse PM (an aerodynamic diameter ranging from 2.5 to 10 μm; $PM_{2.5}$–PM_{10}), but UFPs, $PM_{2.5}$, and VOCs are also emitted as non-tailpipe emissions.[4]

Ultrafine aerosol refers to airborne particles having aerodynamic diameters less than 50 or 100 nm. By number, UFPs constitute the largest portion of ambient aerosol loading and only account for a small fraction of the total PM mass.[61] Elevated exposure to ultrafine aerosol from roadways has been shown to be highly associated with adverse cardiopulmonary effects.[65,66] In addition, UFPs could rapidly grow to larger sizes, and particles formed by these processes can account for up to half of the global budget for cloud condensation nuclei and thereby substantially influence the natural and built environment, ecosystem, and regional haze formation[67] (Table 4).

VOCs are abundant in urban atmospheres because they come from a wide range of sources, including point sources, mobile sources, and area sources.[58] In the urban atmosphere, VOCs affect air quality at different scales.[68] They are important O_3 precursors that contribute to high tropospheric concentrations of O_3, peroxyacyl nitrates, and other oxidants. Together, they lead to the formation of photochemical smog and SOAs all around the world, which are the main factors resulting in haze formation in some developing countries.[35,68,69] Numerous VOCs associated with traffic emissions have been identified and quantified, including BTEX (composed of benzene, toluene, ethylbenzene, and xylenes), carbonyl compounds (*e.g.* formaldehyde, propionaldehyde, acetaldehyde, and acrolein), and gaseous alkenes (*e.g.* acetylene, ethane, propylene, *etc.*).[30] Some of the compounds from traffic emissions are also classified as hazardous air pollutants by the US Environmental Protection Agency, including benzene, toluene, ethylbenzene, *m,p*-xylene, *o*-xylene, formaldehyde, propionaldehyde, acetaldehyde, and acrolein[58] (Table 3). Several toxicological studies have concluded that these gaseous compounds can irritate human eyes and lungs, stimulate the

Table 4 Key pollutants emitted from roadways, location of highest concentrations, and local impacts of the pollutant, based on a selection of field observations.[56–58,60,61]

Key pollutants	Location of highest concentrations	Human health	Regional haze	Material damage	Ecosystem damage
PM_{10}	Near roadway	X			X
$PM_{2.5}$	Urban center	X	X	X	X
UFP	Near roadway	X		X	X
CO	Near roadway	X			
CO_2	Urban center				X
NO	Near roadway	X	X	X	
NO_2	Urban center	X	X	X	
NO_x	Urban center	X	X	X	
O_3	Urban center	X	X	X	X
VOCs	Urban center	X	X		X
BC	Near roadway	X	X	X	X
OC	Urban center	X	X	X	
Metals	Near roadway	X	X	X	X
NO_3^-	Urban center	X	X	X	X
SO_4^{2-}	Urban center	X	X	X	X

human respiratory system, cause central nerve injury, and increase the risk of cancer.[11,45,70] In addition, VOCs can damage crops and reduce agriculture yield[71,72] (Table 4).

NO_x emissions to the atmosphere have multiple environmental impacts, including the production of tropospheric O_3 and the formation of particulate and gaseous nitrogen deposition products (*i.e.* particulate NO_3^- and gaseous HNO_3).[52,73–75] CO is a spatially variable trace gas and is toxic to hemoglobic animals when the ambient concentration is above about 35 ppm, while CO_2 is an important greenhouse gas.[76,77] A substantial number of studies have indicated that NO_x and CO from traffic emissions are highly associated with adverse health effects, including premature mortality and morbidity from cardiovascular and respiratory causes.[78] The US Environmental Protection Agency and European Commission have established limit values for NO_2 and CO in the urban atmosphere to protect public health[79] (Table 4).

Surface O_3 concentrations tend to be prevalent in both urban and rural environments, where O_3 is produced photochemically in the presence of NO_x, VOCs, and sunlight, particularly during the summer months when UV sunlight intensity is greatest.[80,81] Elevated O_3 concentrations have also been observed in more remote areas that exhibit long-range transport of O_3 from urban source regions or downward mixing of O_3-rich air from above.[82] Surface O_3 plays a central role in atmospheric chemistry and is the third most important greenhouse gas contributing to the warming of Earth.[82] At ground level, high levels of O_3 have adverse effects on human health and ecosystem productivity (Table 4). Therefore, in order to protect human health and ecosystems, the US Environmental Protection Agency has set limit values for ground-level O_3 in National Ambient Air Quality Standards, which are reviewed every 5 years to assess and incorporate the best available scientific evidence. Following recent reviews, the O_3 level has been lowered over the past decades from 0.080 ppm in 1997 to 0.070 ppm in 2015.[80] The European Union Framework Directive on Air Quality Assessment and Management (Directive 2008/50/EC) defines a target value for daily 8 hour mean concentrations of O_3 to be 120 $\mu g \, m^{-3}$.[81]

Tailpipe $PM_{2.5}$ from road traffic comprises a variety of primary and secondary pollutants, including trace metals (primary), secondary ionic species such as sulfate (SO_4^{2-}) and nitrate (NO_3^-), and carbonaceous material.[83] The carbonaceous fraction is composed of BC or EC, which are primary emissions from combustion, and organic matter (OM). Increased evidence suggests that $PM_{2.5}$ emissions from traffic may be harmful to human health, atmospheric visibility, traffic safety, construction, and ecosystems, as well as having complex interactions with climate.[4,33] Studies in urban Los Angeles, CA, have found associations between specific chemical components (*e.g.* PAHs and hopanes) from traffic emissions and biomarkers of systemic inflammation in older adults. Haze formation in China is largely due to the contributions of secondary aerosols (*e.g.* SO_4^{2-}, NO_3^-, and SOA) from traffic emissions,[84–86] which is expected to lead to an approximately 2% reduction in total rice production and an 8% reduction in total wheat production in China.[72] In addition, black crusts resulting from vehicular $PM_{2.5}$ components are the main features damaging cultural heritage buildings in urban areas.[87–90] In response to these side effects, target regulations and effective migration policies have been implemented in developed countries and have resulted in substantial reductions in exhaust emissions from road traffic.[12] Air pollutants from tailpipe exhaust are noticeably lower in these countries now than they were during earlier periods, but the tailpipe air pollutants in developing countries are rising significantly and are beginning to threaten public health.[91,92] Consequently, heavy regulations are being enacted in these developing countries (*e.g.* China and India) in order to effectively reduce vehicle emissions in their urban environments. If the regulations succeed, vehicle emissions should no longer pose serious health and air quality problems in the future.

Non-tailpipe emissions normally include the emissions from brake wear, road wear, tire wear, and the road dust resuspension.[4] Therefore, non-tailpipe emissions are typically characterized by trace metals (*e.g.* Cu, Zn, Ba, Sb, and Mn) in coarse PM.[3] However, emissions of trace metal markers are varied within the fleet composition, with greater emission factors being reported for some elements from heavy-duty vehicles.[40] The major factors affecting non-tailpipe emissions include traffic volume and pattern, vehicle fleet characteristics, driving and traffic pattern, and climate and geology of the region.[3,49] Since there are so many factors, the profile of trace metal concentrations in non-tailpipe PM is unique for every region of the world.[40] It is difficult to accurately calculate their contributions to ambient air quality when using the current SA approaches. In addition, non-tailpipe emissions from road transport have not been globally regulated and their contribution to total traffic emissions is expected to increase to 80–90% by the end of this decade in most developed countries.[33,93] Being able to understand non-exhaust emissions, including the key tracers of their source and their inherent toxicology, will help us to assess the relationship between emissions, concentrations, exposures, and health impacts and work out the effectiveness of potential remediation measures in the urban environment.

4 Impacts on Human Health

4.1 Health Impacts of Near-roadway Exposures and Urban Air Pollution from Traffic Emissions

Traffic-related air pollutants play an important role in the development of adverse health effects, as documented extensively in acute toxicity and epidemiologic studies.[94] People living near high traffic flows are more likely to develop lung cancer and leukemia, have a greater risk of asthma, and have a higher prevalence of most respiratory symptoms than those living further away or near lower traffic flows.[45,95–97] Epidemiological evidence of the effects of traffic sources on public health has been consistent on the basis of a large number of studies around the world (Table 5).

Table 5 Health impacts associated with near-road exposures.

References	Area	Exposure variable	Component	Outcomes
Ghosh et al. 2016[104]	Southern California, USA	Adults ≥45 years old	$PM_{2.5}$ and EC	Coronary heart disease mortality
Kheirbek et al. 2016[103]	New York City, USA	Adults ≥30 years old from chronic exposure, asthma from acute exposure among all age groups, and hospitalizations ≥20 years from acute exposure	$PM_{2.5}$ from on-road emissions	General mortality, hospitalizations, and emergency department visits
Thurston et al. 2016[102]	USA and Puerto Rico	ACS CPS-II volunteers	$PM_{2.5}$ and EC	Ischemic heart disease mortality
Lelieveld et al. 2015[94]	Global scale	Adults ≥30 years old and infants <5 years old	$PM_{2.5}$ and O_3	Premature mortality
Baumgartner et al. 2014[101]	Yunan, China	Healthy women	$PM_{2.5}$ and BC	Higher blood pressure
Heo et al. 2014[100]	Seoul, Korea	Healthy participants	$PM_{2.5}$, EC, and OC	Respiratory mortality and cardiovascular mortality
Beelen et al. 2014[99]	22 Cohorts in 13 countries across Europe	Healthy participants	$PM_{2.5}$	General mortality
Strak et al. 2012[96]	Utrecht, The Netherlands	Healthy, young, and non-smoking participants	Particle number concentrations, NO_2, and NO_x	Acute airway inflammation and impaired lung function
Nordling et al. 2008[70]	Stockholm, Sweden	Neonatal children	NO_2 and PM_{10}	Persistent wheezing
Vineis et al. 2007[98]	10 European countries	Healthy volunteers	NO_2	Lung cancer
McCreanor et al. 2007[97]	London, UK	Adults with asthma	UFPs and EC	Lung function in asthma

Vineis et al.[98] estimated exposures to environmental tobacco smoke and air pollution in non-smokers and ex-smokers in a large prospective study in ten European countries (European Prospective Investigation into Cancer and Nutrition; $n = 520\,000$) and found that 5–7% of lung cancers in European non-smokers and ex-smokers were attributable to high levels of air pollution, as expressed by NO_2 or proximity to roads with heavy traffic.

Nordling et al.[70] evaluated the health effects of children who were exposed to traffic-related NO_2 in a prospective birth cohort ($n = 4089$) in Sweden and concluded that exposure to moderate levels of locally emitted air pollution from traffic early in life appears to influence the development of airway disease and sensitization in preschool children.

In The Netherlands, Strak et al.[96] exposed healthy human volunteers to ambient PM at five locations (i.e. an underground train station, two traffic sites, a farm, and an urban background site) with various PM characteristics and observed associations between changes in particle number concentrations, NO_2, and NO_x and acute airway inflammation, as well as impaired lung function.

Beelen et al.[99] investigated associations between natural-cause mortality and long-term exposure to exhaust air pollutants in 367 251 participants from 22 European cohort studies and recorded a significantly increased hazard ratio for $PM_{2.5}$ of 1.07 (95% confidence interval: 1.02–1.13) per 5 $\mu g\,m^{-3}$.

Heo et al.[100] provided evidence that $PM_{2.5}$ derived from mobile sources is associated with both respiratory and cardiovascular mortality in Seoul, Korea. Baumgartner et al.[101] suggested that BC from combustion emissions is more strongly associated with blood pressure than PM mass and that BC's health effects may be more prominent in women living near a highway in Yunan, China, since they experience heavy exposure to motor vehicle emissions.

Based on epidemiological cohort studies that connect premature mortality to a wide range of causes, including the long-term health impacts of O_3 and $PM_{2.5}$, Lelieveld et al.[94] assessed the global burden of disease. They found that emissions from residential energy use, such as heating and cooking, have the largest impacts on premature mortality, whereas in much of the USA and in a few other countries, emissions from traffic and power generation are most important.

Thurston et al.[102] evaluated past associations between long-term (from 1982 to 2004) $PM_{2.5}$ concentrations and ischemic heart disease mortality, which were collected from 445 860 adults in 100 different US metropolitan areas. Diesel traffic-related EC was significantly associated with ischemic heart disease mortality (hazard ratio = 1.03 per 0.26 $\mu g\,m^{-3}$ EC increase; 95% confidence interval: 1.00–1.06). Kheirbek et al.[103] estimated that all on-road mobile sources in New York City (NYC) contribute to 320 (95% confidence interval: 220–420) deaths and 870 (95% confidence interval: 440–1280) hospitalizations and emergency department visits annually within NYC due to $PM_{2.5}$ exposures, accounting for a total of 5850 (95% confidence interval: 4020–7620) years of life lost. Ghosh et al.[104] investigated the coronary heart disease burden from near-roadway air pollution and compared it to the $PM_{2.5}$ burden in the California South Coast Air Basin for 2008, as well as a compact urban growth greenhouse gas reduction scenario for 2035. The results suggested that 1300 coronary heart disease deaths (6.8% of the total) were attributable to traffic density, 430 deaths (2.4%) were attributable to residential proximity to a major road, and 690 deaths (3.7%) were attributable to EC in 2008. The numbers of estimated deaths attributable to each of these exposures are anticipated to increase to 2500, 900, and 2900 in 2035, respectively, due to population aging.

In some epidemiologic studies, a common method characterizes residential locations relative to traffic proximity (*e.g.* within 20–500 m of a busy highway with traffic flow rates from 10 000–100 000 vehicles per day[70,101]). Researchers try to focus more on the associations between ambient concentrations of pollutants and their health effects. The proximity exposure paradigm simply assumes that personal exposure to outdoor concentrations of air pollutants is correlated with residential location, but people spend most of their time indoors, where concentrations of some air pollutants are lower than they are outdoors. In addition, the validity of this assumption may depend on the duration of the exposure of interest (weeks to decades), the nature of the health impact (acute *vs.* chronic), and the mobility of the subjects. This assumption may be problematic for school-age children and working adults because they do not stay in the residential location of interest for much of the day. Furthermore, many studies have been designed to emphasize the associations between criteria air pollutants from traffic and their related health effects. When a traffic-related pollutant also has major indoor sources (*e.g.* cooking or biomass burning), as do CO, NO_2, and $PM_{2.5}$, it may be difficult to distinguish between the health effects of indoor and outdoor exposures.

The solution is to apportion criteria air pollutants into different sources such that, although the resulting spatial map of the target pollutant is intended to represent that pollutant (alone), it actually includes all of the co-pollutants of source types involved. The personal exposure concentration that includes indoor, outdoor, and near-source concentrations is an accurate surrogate measure of exposure to all-day activities. Air pollution control strategies might thus be more effective when based on associations between specific types of emission sources and personal exposure concentrations, rather than those between individual pollutants and ambient concentrations.

4.2 The Contributions of Mobile Sources to PM and O_3 in Cities around the World

The preceding section discussed the chemical composition and health effects of traffic-related $PM_{2.5}$ and gaseous pollutants. This section will review the contributions from roadway emissions to $PM_{2.5}$ and O_3 in urban areas around the world.

The SA approach is usually adopted to quantify the contributions of mobile sources to $PM_{2.5}$, which is an important task in air pollution management, control, and policy options.[105,106] SA techniques that rely on statistical analysis of observations at monitor sites are called receptor models and include positive matrix factorization (PMF),[85] principal component analysis,[107] chemical mass balance (CMB),[108] factor analysis with multiple regression,[92] Unmix,[105] and back-trajectory analysis.[38] The CMB model requires the input of unique profiles for each major source,[108] while PMF is a multivariate model that identifies factors without needing source composition information.[85] PMF is convenient and practicable, which makes it the most popular method for $PM_{2.5}$ SA.

To achieve a direct view of recently reported mobile source contributions to $PM_{2.5}$, a comparative description is summarized in Table 6.[85,91,105-107,109-121] For these 20 cities around the world, mobile source contributions accounted for 7.8–44.6% of $PM_{2.5}$, with the mass ranging from 1.2–28.7 μg m^{-3}. From the results, it was evident that traffic was a common source among different cities in all six continents. In the megacities of the USA, Europe, and Australia, the average mass of mobile source emissions contributed 1.2–4.9 μg m^{-3} of the $PM_{2.5}$ mass concentration, while the mass of mobile source emissions was in the range of 5.9–28.7 μg m^{-3} in China, India, and Mexico, demonstrating that mobile source emission control is still a challenge for most megacities in

Table 6 Impacts from roadway emissions on $PM_{2.5}$ for cities and urban areas around the world. The data are reproduced from a selection of source apportionment studies.

Area	Mean concentration (μg m^{-3})	Contribution (%)	Time
New York City, USA[105]	15.2	26	2000–2001
Los Angeles, USA[106]	19.6	20	2002–2012
Mexico City, Mexico[109]	47	42	Mar. 2006
London, UK[110]	15.7	8	Jan. 2012 to Feb. 2012
Athens, Greece[111]	11.0	20	2013
Barcelona, Spain[111]	15.0	20	2013
Paris, France[112]	14.7	14	Sep. 2009 to Sep. 2010
Milan, Italy[113]	34.5	7.6	Dec. 2009 to Nov. 2010
Seoul, Korea[107]	42.6	23	Jun. 2009 to May 2010
Hong Kong, China[114]	55.5	29	Oct. 2014 to Sep. 2015
Beijing, China[115]	69	14.7	2010
Shanghai, China[85]	90.7	6.8	Jan. 2013
Guangzhou, China[85]	69.1	8.6	Jan. 2013
Xi'an, China[116]	142.6	14.2	2010
Jinan, China[117]	169	17	2010
New Delhi, India[91]	168	17	2013
Belen, Costa Rica[118]	36	7.8	Jun. 2010 to Jul. 2011
Nairobi, Kenya[119]	17	39	2008–2010
Recife, Brazil[120]	7.3	44.6 (with sea spray factor)	Jul. 2007 to Jun. 2008
Newcastle, Australia[121]	8.1	27	Feb. 1998 to Dec. 2013

developing countries. In some cities of South America and Africa, mobile source air pollution accounted for a large fraction of $PM_{2.5}$, which is likely attributable to the cities' outdated vehicle technologies. It is apparent that there is considerable spatial variability in mobile source contributions within cities around the world, mainly because of the diversity of human activities and economic development from city to city (Table 6).

As discussed in Section 3.3, NO_x, CO, and VOCs from road traffic emissions are responsible for ground-level O_3 formation in urban environments. According to a study that investigated the global impact of road traffic emissions in 1990 on ground-level O_3 formation by using a general circulation model coupled with a chemistry module,[122] total road traffic emissions contributed to the zonally averaged O_3 distribution by more than 12% near the surface in mid-latitudes and arctic latitudes of the Northern Hemisphere in July. In January of 1990, road traffic emissions contributed to the zonally averaged O_3 distribution by more than 8% near the surface in both the northern and southern extratropics. Increases in the surface O_3 concentration resulting from 1995 road traffic emissions are typically 5-15% at mid-latitudes of the Northern Hemisphere during summer.[123] If estimating the impacts of 1997 road traffic emissions,[124] automobile emissions enhance the summertime boundary layer O_3 by 20-40% in urbanized regions of the Northern Hemisphere, by 25% in the Mediterranean, by 40% in both Southeast Asia and California, and by 30% in Japan. Nevertheless, O_3 decreases by about 5-20% in the Northeastern USA, California, and East Asia during winter. According to the global average impact of road traffic emissions on O_3 changes in 2000,[125] road traffic emissions accounted for about 36.7% of O_3 increases. In the highly polluted industrialized and urbanized regions of China (Beijing-Tianjin-Heibei Region, China), a 30% reduction in industrial and transportation sources produced the most effective impacts on O_3 concentrations, with a maximum decrease of 20 ppbv[126] (Table 7). From these modeling studies, it could be concluded that the current fleet of road vehicles is moderately affecting the changes in tropospheric O_3, particularly in the Northern Hemisphere.

The previous section showed that mobile emissions have been well documented regarding their significant impacts on $PM_{2.5}$ mass and surface O_3 concentrations in all inhabited continents, including Asia, Europe, America (North and South), Africa, and Oceania. Considering the increased public health risk and potential impact of climate change associated with exhaust $PM_{2.5}$ and O_3, cost-effective control policies and advanced technology interventions that prioritize mobile source mitigation may also provide a greater benefit in protecting public health and climate change.

5 Impacts on the Natural and Built Environments

The negative impacts of air pollutants from traffic emissions on natural environments, built environments, and ecosystems have been well documented. Effects of air pollution on all ecosystem types and their threats to

Table 7 Impacts from roadway emissions on ozone for cities and urban areas around the world. The data are reproduced from a selection of modeling studies.

Area	Estimated contribution (%)	Time
Northern hemisphere[122]	+12	Jul. 1990
Northern and southern extratropics[122]	+8	Jan. 1990
Northern hemisphere[123]	+5–15	Jul. 1995
Northern hemisphere[124]	+20–40	
USA	+20	
California	+40	
Northeastern USA	+30	
Mediterranean	+25	
Europe	+20	Jul. 1997
Southeast Asia	+40	
Japan	+30	
Oceania	+5–10	
Northern hemisphere[124]	+20–40	
California	+20	
Mediterranean	+10–25	
East Asia	+15–20	Jan. 1997
Europe	+10–25	
Oceania	+5	
Global[125]	+36.7	2000
Beijing-Tianjin-Hebei Region, China[126]	A 30% reduction leads to a O_3 decrease of 20 ppbv	Aug. 2007

biodiversity conservation are appreciable, but not specific to vehicle emissions. In some historically important cities of the world, cultural heritage buildings as well as modern buildings present dark and soiled facades in spite of repetitive cleaning.[83,84,108] This building blackening is mainly caused by the anthropogenic air pollutants from industrial activities and traffic emissions.[83] Black crusts are recognized as one of the major forms of deterioration causing black soiling of these modern or cultural buildings. The formation of black crusts occurs mainly on carbonate building materials (*i.e.* marble and limestone), whose interaction with an SO_2-loaded atmosphere leads to the transformation of calcium carbonate (calcite) into calcium sulfate dihydrate (gypsum).[84]

Once SO_2 is emitted from its source sectors (*i.e.* traffic and industrial emissions), it may undergo a variety of complex reactions, including oxidation to sulfur trioxide and the formation of sulfuric acid in an aqueous form during its residence in the atmosphere.[88] Aerosols in the form of sulfuric acid can be transferred onto the surface of the building through either a wet or dry deposition process, resulting in gypsum formation.[127] During this process, soot particles embedded in black crusts are capable of catalyzing the destructive action because they contain transition metals (*e.g.* V, Ti, Fe, Mn, and Cu) for the oxidation of SO_2.[127] Apart from being aesthetically displeasing, such crusts contain various organic compounds and harbor micro-organisms. The latter can contribute to bio-deterioration of the calcareous substrata, which can produce copious quantities of hydrated calcium oxalate.[88] It is now widely accepted that these black crusts accumulating on buildings are comprised of both a carbonate and a non-carbonate fraction.[127] The non-carbonate fraction includes two different components: organic carbon (OC) of biogenic and anthropogenic origin and EC. Del Monte *et al.*[128] determined the existence of two types of such particles: (1) spherical shaped with an irregular, rough surface and high porosity derived from an oil combustion source; and (2) spherical shaped with a smooth surface derived from a coal combustion source. Saiz-Jimenez[129] analyzed the black crusts coating the surfaces of buildings in Dublin, Mechelen, and Seville using gas chromatography mass spectrometry and observed a variety of organic molecular markers from petroleum derivatives, including hopanes, *n*-alkanes, and fatty acids. Although there is a consensus on the composition of black, sulfated crusts in terms of their carbonate and non-carbonate fractions, quantitative data on the fractions of historical buildings and their environments have been reported only recently. Bonazza *et al.*[127] showed that sulfur and carbon were major components in the black crusts of European cultural heritage buildings. These specific buildings include the Milan Cathedral (Milan, Italy), the S. Maria del Fiore (Florence, Italy), the Corner Palace (Venice, Italy), the Vittoriano (Rome, Italy), St. Eustache Church (Paris, France), the Tuileries (Paris, France), the Seville Cathedral (Seville, Spain), the Tower of London (London, UK), and St. Rumbold's Cathedral (Mechelen, Belgium). The concentration of total carbon varied from 1.3% (S. Maria del Fiore) to 7.7% (Tower of London), while the OC/EC ratio ranged from 0.14 (Milan Cathedral) to 8.37 (Tower of London). Sulfates were found in all of the samples with their concentrations varying from 52 $mg\,g^{-1}$ (Tower of London) to 521 $mg\,g^{-1}$ (Milan Cathedral), which highlights the fact that sulfation due to SO_2–surface interactions was still an important factor in the damage process. Belfiore *et al.*[89] used laser ablation inductively coupled plasma mass spectrometry to quantify the trace elements (*e.g.* Fe, Zn, Pb, and V) of the black crusts from the inner parts to the external layers of European cultural buildings (*i.e.* the Corner Palace, St. Rumbold's Cathedral, and St. Eustache Church) and concluded that trace metals can be markers for evaluating the variation of fuels used over time. Bergin *et al.*[90] suggested that the deposition of light-absorbing PM (*e.g.* brown carbon) could influence the discoloration of the Taj Mahal, India, as well as natural and human-made surfaces of the buildings in regions of high aerosol loadings.

The contributing factors mentioned above have resulted in a need for cleaning of soiled historical buildings and monuments. This kind of cleaning can be accomplished by the use of lasers, micro-sandblasters, or chemical agents, but these methods could result in damage to the surface of the building. Abatement policies and regulations for traffic emissions should be key factors to focus on for improving air quality and decreasing building soil rates.

6 Impacts on Remote Sites

This section continues from Section 4.2, which was a discussion on the current understanding of how roadway emissions impact $PM_{2.5}$ and O_3 in remote sites around the world. Mobile source air pollution outweighs area source air pollution at urban sites, negatively affects human health, reduces visibility, and interacts with climate and ecosystems.[62] However, concerns about the contributions of local mobile source pollution to the deterioration of global air quality are increasing because these contributions could influence global meteorology, resulting in melting glaciers, frequent floods and droughts, and stronger hurricanes and other storms.[11] A quantitative establishment of temporal trends of mobile source emissions at remote (rural) sites is important for gaining perspectives on air quality control for remote zones in terms of import.[130–136]

The US Environmental Protection Agency has carried out substantial research on the chemical composition of natural aerosols at Interagency Monitoring of Protected Visual Environments (IMPROVE) monitoring network sites across the country.[130] Sufficient data are provided to quantify the background concentration of sea salt, biogenic OM, fugitive dust, and carbon from natural wildfires at these sites so that visibility in 156 national parks and wilderness areas would reach "natural conditions" by 2064.[131,137,138] This goal is very important in the implementation of the US Regional Haze Rule. In addition, the IMPROVE database is also available for estimating contributions from mobile source air pollution to natural-condition $PM_{2.5}$ and O_3, and such relevant research findings could present the new information on strategy planning needed for controlling regional haze.[132]

Chen et al.[137] applied the CMB method to speciated $PM_{2.5}$ measurements from the Chemical Speciation Network (CSN) and IMPROVE monitoring networks across the state of Minnesota. The results indicated that gasoline and diesel contributions to $PM_{2.5}$ at rural Minnesota IMPROVE sites were minor (<4%) and vice versa at urban sites, where gasoline and diesel contributed to 44% of the $PM_{2.5}$ mass concentration. Kim and Hopke[132] collected $PM_{2.5}$ samples at a rural monitoring site in Brigantine, NJ, using IMPROVE samplers and quantified source contributions to $PM_{2.5}$ by using the PMF method. The largest contributor to $PM_{2.5}$ during the monitoring period was sulfate-rich secondary aerosols (48%), followed by traffic emissions (16%). Potential source contribution function analysis showed that the influence of sulfate-rich secondary aerosols was regional, whereas traffic emission was a local source. The greatest mobile source impacts at this monitoring site were likely caused by a closely located highway carrying a large amount of traffic (Table 8).

The European Integrated Project on Aerosol Cloud Climate and Air Quality Interactions (EUCAARI) aimed to understand the interactions between air

Table 8 Impact of roadway emissions on $PM_{2.5}$ at remote sites around the world. The data are reproduced from a selection of source apportionment studies.

Area	Mean concentration (μg m^{-3})	Contribution (%)	Time
Voyageurs National Park, USA[137]	4.1 ($PM_{2.5}$)	2	2000–2006
Boundary Waters Canoe Area, USA[131]	4.4 ($PM_{2.5}$)	2	
Blue Mounds State Park, USA[131]	7.7 ($PM_{2.5}$)	1	2002–2005
Great River Bluffs State Park, USA[131]	8.4 ($PM_{2.5}$)	6	
Brigantine National Wildlife Refuge, USA[132]	9.7 ($PM_{2.5}$)	16	1992–2001
Finokalia, Greece[133]	5.5 (PM_1)	1	
Jungfraujoch, Switzerland[133]	1.5 (PM_1)	3	Spring 2008
Hyytiälä, Finland[133]	2.4 (PM_1)	2	
Puy de Dome, France[133]	4.1 (PM_1)	0.8	Fall 2008 and Spring 2009
Mace Head, Ireland[133]	2.1 (PM_1)	5	Spring 2008 and Spring 2009
Gosan, South Korea[134]	32 (0.56–2.5 μm)	7	Mar. 2009 to May 2009
Himalayas, South Asia[135]	26 ($PM_{2.5}$)	5	2006
Huaniao Island (~66 km east of Shanghai), China[136]	N/A	27% of PAH concentration	2011–2012

pollution and climate change in European countries.[82] Within this framework, PM_1 (an aerodynamic diameter less than 1 μm) SA using PMF with the multilinear engine on Aerodyne aerosol mass spectrometer data was performed at 17 different sites, including urban, rural, and remote sites, as well as high-altitude sites across Europe. Traffic emissions contributed to somewhere between 6% and 12% of PM_1 mass at urban sites, while the contributions to total mass varied from 0.8% to 5% at rural and remote sampling sites. Furthermore, the contributions of traffic emissions were not identified at some of the rural and remote sampling sites (Table 8).

Lefohn et al.[81] analyzed rural O_3 trends across the USA from 1994 to 2008 (15 years) and found no statistically significant decrease in O_3 concentration at most of the rural sites, but the emissions of NO_x and VOCs declined by 40% and 47%, respectively (Table 9). Wilson et al.[82] assessed the impacts of changing anthropogenic emission tracers on surface European rural

Table 9 Impact of roadway emissions on O_3 at remote sites around the world. The data are reproduced from a selection of field studies.

Area	Trends in emissions (%)	Response from ozone trends (% year^{-1})	Time
Denali NP, USA[81]		NS	
Lassen Volcanic, USA[81]		NS	
Anyonlands, USA[81]		NS	
Shenandoah, USA[81]	−40% and −47% in NO_x and VOCs during this period	−1.12	1994–2008
Mesa Verde, USA[81]		+0.74	
Whiteface, USA[81]		NS	
Great Smoky Mountains, USA[81]		NS	
Craters of the Moon, USA[81]		NS	
Great Britain[82]		NS	
Eastern Europe[82]		NS	
Central Europe[82]	−3.00% year^{-1} in NO_x and VOC emissions	NS	1996–2005
Northwestern Europe[82]		NS	
Austria and Hungary[82]		1.76 (Austria) −4.11 (Hungary)	

background O_3 concentrations from 1996 to 2005 using the European scale 3D chemistry-transport model (CTM) CHIMERE. They did not find that anthropogenic NO_x or VOC reductions had any substantial effect on observed annual mean rural background O_3 trends in most European countries, with the exception of Austria and Hungary. The European results are consistent with the US studies, suggesting that impacts from mobile sources on air quality at remote sites are not significant compared with urban sites.

Asian countries are known to emit large amounts of anthropogenic pollutants due to their high densities of industrial activities and increasingly high rates of energy consumption.[134] In order to estimate aerosol compositions and sources in this region, several international projects, including the Asia-Pacific Regional Aerosol Characterization Experiment (ACE-Asia),[139] the East Asia/North Pacific Regional Experiment (APARE),[140] the Pacific Exploratory Mission-West (PEM-WEST),[141] and the Atmospheric Brown Cloud (ABC)[135] program, were launched to formulate effective control strategies for ambient PM. Under the support of these grants, SA studies at Asian background sites were carried out in an effort to understand sources of $PM_{2.5}$ and their roles in regional air quality. Stone et al.[135] quantified the contributions of traffic emissions to $PM_{2.5}$ using the CMB receptor model in the Himalayas, which may serve as one of the background sites in South Asia. The average contribution of diesel engines to $PM_{2.5}$ was 5% of 26 μg m^{-3}, which is significantly lower than the average vehicular contribution (17% of 168 μg m^{-3}) in the megacities of India. Han et al.[134] conducted a size-resolved PM SA study in Gosan, South Korea, which is a representative background site of East Asia. Traffic contribution accounted for a small proportion of $PM_{0.56-2.5}$ mass (7% of 32 μg m^{-3}), which was much lower than the results

(23% of 42.6 µg m^{-3}) from one SA study conducted in Seoul, South Korea. Wang et al.[136] investigated the impacts of seasonal atmospheric transport of land-based PAHs on air quality in Huaniao Island, China, which is 66 km east of Shanghai and is also a downwind background site of East Asian continental outflow. The annual average concentration of PAHs in PM$_{2.5}$ is 5.24 ± 5.81 ng m^{-3}, with an estimated contribution of 27%, indicating that this background site might be influenced by exhaust emissions from the Yangtze River Delta, China (Table 7). The impacts of mobile sources on air pollution at remote locations in Asia are comparable to those in the USA and Europe, with the exception of China. PM$_{2.5}$ SA studies at the Chinese background sites have rarely been reported, so scientific conclusions cannot be made with such few data.

One should also recognize that these conclusions are based on a relatively limited review of the literature. Measurement methods might be biased, measurements maybe unrepresentative, and scientific hypotheses could change with additional information. Consequently, much of what is involved in this assessment has some degree of uncertainty. Continued efforts are vital to ensuring that the impacts of mobile sources on regional air quality in both developed and developing countries become well understood.

7 Global Trends in Emissions

This section analyzes differences in traffic-related air pollutants from various countries/regions around the world. An overview concerning region-specific air pollutants from transport emissions is given in the four maps of Figure 3a–d, based on the regional and global emission results from the Task Force on Hemispheric Transport of Air Pollution (TF HTAP).[142]

A comparison between the shares of national GDP (on the basis of purchasing power parity) in the global GDP (expressed in US dollars) revealed that the top three countries/regions are very close (i.e. the share in the world economy of the European Union, the USA, and China are 17%, 16%, and 15%, respectively).[142,143] Trailing these three are India (7%), the Russian Federation (3%), and Brazil (3%). However, when looking at their individual contributions to global economic growth over the last 10 years, which was 47% since 2002, China contributed 31%, India 11%, the USA 8%, the European Union 6%, the Russian Federation 4%, and Brazil and Indonesia each 3%.

This rapid economic development consumed a large amount of fossil fuels and produced significant quantities of relevant pollutant emissions, including PM$_{2.5}$, NO$_x$, VOCs, CO, and SO$_2$.[142] If observing the total emissions of 2010, the six countries/regions that emitted the most PM$_{2.5}$ were East and South Asia (22.2 Tg), Africa (4.6 Tg), the European Union (2.1 Tg), the USA (1.9 Tg), Latin America (1.4 Tg), and the Russian Federation (0.4 Tg). The country/region releasing the most NO$_x$ in 2010 was East and South Asia (48.4 Tg), followed by the USA (15.6 Tg), the European Union (11.2 Tg), Latin America (7.6 Tg), Middle Asia (6.0 Tg), Africa (4.6 Tg), and the Russian Federation (4.3 Tg). The top three emitting regions for non-methane VOC

(NMVOC) in 2010, together accounting for more than half (78%) of the total global emissions, were East and South Asia (63.1 Tg NMVOC, or 44%), Africa (32.8 Tg NMVOC, or 23%), and the USA (15.2 Tg NMVOC, or 11%). The CO emissions of countries/regions in 2010, in descending order, were East and South Asia (316.3 Tg), Africa (70.5 Tg), the USA (55.3 Tg), Latin America (48.8 Tg), the European Union (29.9 Tg), the Middle East (12.6 Tg), the Russian Federation (9.2 Tg), and Oceania (2.1 Tg). As for SO_2 emissions in 2010, East and South Asia led (45.7 Tg), followed by the USA (11.9 Tg), the European Union (8.8 Tg), the Middle East (7.9 Tg), the Russian Federation (7.7 Tg), Latin America (7.4 Tg), Africa (5.3 Tg), and Oceania (3.0 Tg).

As shown in Figure 3, ground transportation contributes ~7% of all $PM_{2.5}$ emissions in the world (excluding non-exhaust road abrasion dust and tire wear emissions). Road transport emissions in some developed regions represented only about 1.9% (European Union, ~1.0%; USA, ~0.9%) of the global emissions of $PM_{2.5}$ and its components (*i.e.* BC and OC) in 2010, while much higher contributions (~4.7% of global $PM_{2.5}$ mass) were derived from developing countries with large populations in Asia, such as China (~2.9%) and India (~1.4%). In some developing countries, large contributions to the global emissions of $PM_{2.5}$ were mainly due to industry and residential emissions, especially the latter, which accounted for ~88% of all African $PM_{2.5}$ mass. Approximately 42% of the global NO_x emissions originated from transport. The largest roadway NO_x contributor was China, accounting for ~13% of the global NO_x emissions, followed by the USA (~7%), the European Union (~6%), India (~4%), Latin America (~4%), Middle Asia (~3%), and Africa (~2%). In Central and South America, major NO_x emissions are attributable to the transportation sector (*e.g.* in Mexico, ~65% of the NO_x emissions originate from road transport), as well as the energy sector, to a minor extent. China's contribution to global NMVOC emissions was relatively high (~6.3%) and comparable to India's (~5.4%), with similar findings for $PM_{2.5}$ and NO_x. Other regions/countries, including Latin America, the USA, the Middle East, the European Union, the Russian Federation, Africa, and Oceania, contributed ~3.5%, ~3.1%, ~1.6%, ~1.5%, ~1.4%, ~0.9%, and ~0.1% of the global NMVOC emissions, respectively. In Brazil, particularly high usage of bio-gasoline is present, which is a factor in Latin America's ~3.5% contribution to global NMVOC emissions. The country that emitted the most vehicular CO in 2010 was China, producing ~7.2% of global CO emissions, with the second largest contributor being the USA at ~6.2%, and the third largest being Latin America at ~6.0%. Following these top three CO-emitting countries/regions in 2010 were the Middle East (~2.0%), the European Union (~1.9%), Africa (~1.2%), the Russian Federation (~0.6%), and Oceania (~0.07%). Remarkable trends were seen in the countries/regions emitting the most vehicular SO_2, which together only accounted for ~1.5% of total global SO_2 emissions. In China, vehicular SO_2 emissions accounted for ~1.0% of global SO_2 emissions, while India produced ~0.3% of the global SO_2 emissions in 2010. The European Union, as a whole, contributed 0.2% of global SO_2 emissions in 2010. These low

Local-acting Air Pollutant Emissions from Road Vehicles

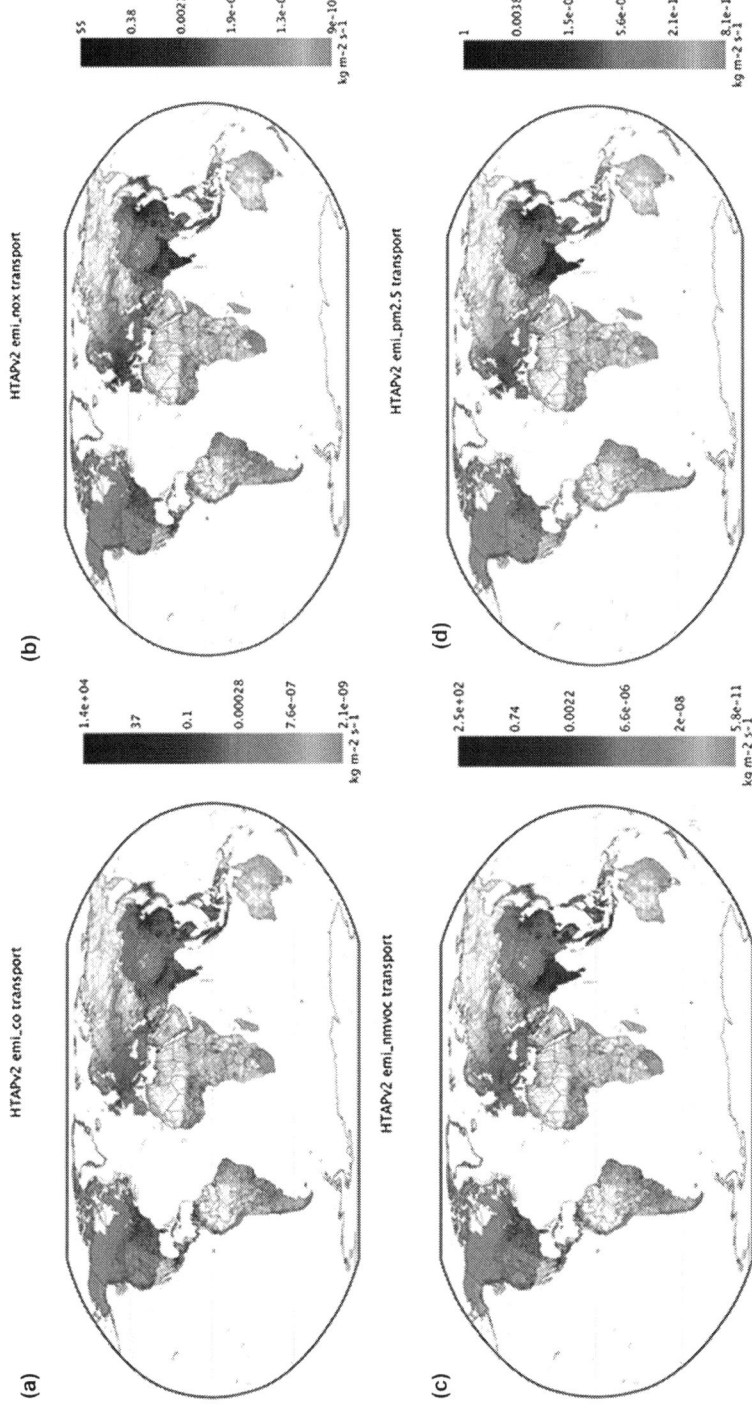

Figure 3 Global maps of region-specific air pollutants from transport emissions. The 2010 emission inventory is from http://edgar.jrc.ec.europa.eu/htap_v2/index.php?SECURE=123.

contributions of vehicular SO_2 emissions to the global total could be attributed to the advanced technology of sulfur removal in crude oil.

The large differences in the amounts of traffic-related air pollutants from different countries/regions reveal diversities in the types of vehicles, vehicle technologies, and fuel qualities around the world.[12] Different regions use individual modes of roadway transport to varying extents. Most developed countries rely on passenger light-duty vehicles (*e.g.* cars, sport utility vehicles, and minivans) much more than developing countries. Developing countries demonstrate far higher modal shares of diesel vehicles (for public transport) and motorized two-wheelers (*e.g.* scooters and motorcycles). In some regions with high population densities, the total amount of rail freight has been increasing, although the rates of increase vary widely between countries. Across the globe, total rail volumes are higher than total road volumes, but they are concentrated in a small number of countries.[144] In this case, reliance on diesel vehicles as a public transport tool results in relatively higher amounts of airborne pollutants. In addition, demand for biofuel for heavy-duty vehicles in some developing countries is increasing as a result of greater vehicular exhaust emissions.[135] Outdated vehicle emission standards in most developing counties may be to blame for their increased levels of air pollution. Taking diesel PM emission as an example, in the year 2010, Euro 5 regulated emissions such that the amount of PM from diesel vehicles in Europe was to be at or below 0.005 $g\,km^{-1}$. Meanwhile, some developing countries, including China, India, and Brazil, held their vehicle emission standards equivalent to Euro 3, which only aims to reduce diesel PM emissions to less than or equal to 0.05 $g\,km^{-1}$.[145,146] In some other developing countries of Middle Asia and Africa, the vehicle emission standards were set below Euro 3 or not required at all for after-treatment emission control devices, such as particulate filters and catalysts for on-road vehicles.[21] Also, in some developing countries, their fuel quality was not able to match their emission standards due to their technological inability to produce cleaner gasoline and diesel.[21] Once advanced technologies are available, developing countries should introduce stricter vehicle emission standards for both light- and heavy-duty vehicles that are 'technology forcing' and require cleaner vehicle technologies, including electric and fuel cell vehicles. The ultimate goal would be to provide state-of-the-art vehicle technologies that can significantly lower exhaust emissions.

8 Future Technologies and Projected Trends

Worldwide, transport sector energy use has been strongly linked to rising populations and incomes. Transportation continues to rely primarily on oil. Decoupling transport growth from income growth and shifting away from oil shall be a slow and difficult process. Despite steady global growth of energy use in the transport sector, different regions and countries show very different growth patterns.[143] The USA has a higher share of passenger travel by LDVs than any other country or region, but the production rate of passenger light-duty vehicles (LDVs) has declined in recent years. This reflects a slowing in population growth and is also a sign of car ownership saturation, which means that most families already own enough vehicles. Most developed countries in Europe also have a high share of passenger travel per family, and transport volumes are expected to remain relatively constant through to 2030.[147] However, in developing countries, rates of LDV ownership are growing rapidly. Many families purchase LDVs as soon as they can afford them. The emergence of low-cost LDVs will probably further accelerate LDV ownership rates.[147] Therefore, achieving large global reductions in traffic-related emissions in the coming decades will depend heavily on improvements in vehicle technology, fuel quality, and traffic emission standards.

In Europe, regulations for road transport were introduced for vehicles from Euro 1 in 1992 to Euro 6 in 2014.[146] Recent studies reported that European emissions of $PM_{2.5}$, NO_x, SO_2, BC, and OC in 2010 were 69%, 71%, 53%, 59%, and 32% lower, respectively, than the emissions that would have occurred in 2010 in the absence of legislative and technological measures. For this reason, the annual mean concentration in Europe of fine PM ($PM_{2.5}$) is 35% less than it would have been, along with there being 44% less sulfate, 56% less BC, and 23% less particulate OM.[12] These reductions in $PM_{2.5}$ emissions are expected to have prevented 80 000 premature deaths annually across the European Union, which equates to approximately $232 per beneficiary spent annually. Crippa et al.[12] also computed that a 50% decrease of global $PM_{2.5}$ road emissions was due to the implementation of the updated Euro standards on vehicles. Since 2000, many developing countries in Asia and Latin America have been adopting national legislation on mobile sources equivalent to Euro 3 (Figure 4), which has led to substantially lower growth rates in the emissions of NO_x and CO from the transport sector when compared to earlier studies. Cofala et al.[144] predicted that global NO_x emissions would increase by approximately 13% in 2030 compared to those of 2000 under the current legislation on mobile sources, despite a large increase in traffic volumes that is anticipated for the coming decades.

Regulations in the USA and Canada closely parallel those in Europe, but differ in their nomenclature. For example, Euro 4 was equivalent to US Tier 4, and Stage III equaled Tier 3. Euro 6 standards can be met with the technology that is currently available.[148] DPF will likely be included in every diesel engine in order to lower NO_x and PM emissions. Additionally, new powertrain technologies such as hybrids, fuel cells, and electric vehicles have started to penetrate the gasoline and diesel markets due to their strong energy efficiencies and zero emissions.[149] In developed countries, travel is shifting toward more efficient modes, but total travel growth will be restrained by better land use and the denser metropolitan areas of 2050.[148] If these policies are implemented, the need for fossil fuels in transport will be cut to nearly half of what it was in 2007, resulting in a large reduction in tailpipe exhaust emissions.

In the developing countries of Africa, Asia, Latin America, and Eastern Europe, major targets are producing low-sulfur fuels (<50 ppm) and making after-treatment technologies, such as DPF, more effective and efficient.[21]

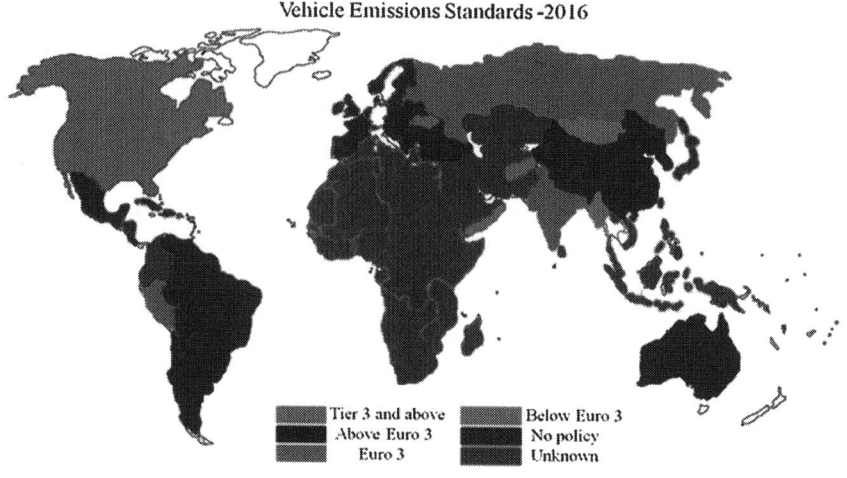

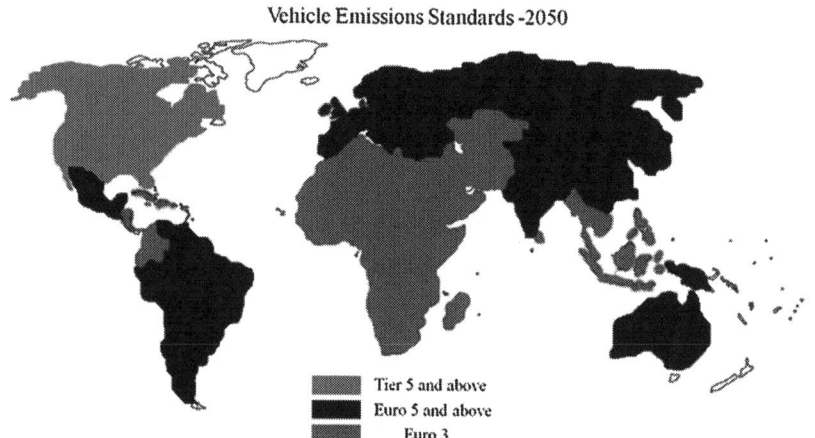

Figure 4 Global map of region-specific vehicle emissions standards for 2016 and 2050 (projected).
The data are adopted from https://www.dieselnet.com/standards/.

Once low-sulfur fuels are available, countries can introduce stricter vehicle emission standards (like Euro 3 and above) for both light- and heavy-duty vehicles. Unfortunately, it is not possible to achieve these improvements immediately. The introduction of advanced technologies will take time and be dependent on reductions in technology costs and increases in the capacities of businesses and consumers to afford these technologies in these countries. Some developing countries, like China, have already tightened their fuel economy standards and put themselves on an initial path to achieving significant reductions in traffic-related air pollutants by 2030.[148] This impact is expected to affect nearly 30% of transport air pollutant emissions in the world by 2050.

Acknowledgements

We would like to thank Professor Benjamin de Foy from Saint Louis University for his help with collecting data for the figures. We would like to thank Mr Marshall Schlick from the University of Wisconsin-Madison for his assistance with editing the chapter. Dr Qingyang Liu also thanks the China Scholarship Council (201608320139) and the China Postdoctoral Science Foundation (2016M600419) for supporting his research overseas.

References

1. A. J. Cohen, H. R. Anderson, B. Ostro, K. D. Pandey, M. Krzyzanowski, N. Kunzli, K. Gutschmidt, A. Pope, I. Romieu, J. M. Samet and K. Smith, *J. Toxicol. Environ. Health, Part A*, 2005, **68**, 1301–1307.
2. HEI, Health Effects of Outdoor Air Pollution in Developing Countries of Asia: A Literature Review, *Health Eff. Inst.*, 2004, 15.
3. P. Pant and R. M. Harrison, *Atmos. Environ.*, 2013, 77, 78–97.
4. R. M. Harrison, A. M. Jones, J. Gietl, J. Yin and D. C. Green, *Environ. Sci. Technol.*, 2012, **46**, 6523–6529.
5. J. J. Schauer, Source contributions to atmospheric organic compound concentrations: Emission measurements and model predictions. California Institute of Technology Thesis, Pasadena, California, 1998.
6. J. J. Schauer, M. J. Kleeman, G. R. Cass and B. R. T. Simoneit, *Environ. Sci. Technol.*, 1999, **33**, 1578–1583.
7. J. J. Schauer, M. J. Kleeman, G. R. Cass and B. R. T. Simoneit, *Environ. Sci. Technol.*, 2002, **36**, 1169–1180.
8. J. Chow, J. Bachmann, Y. C. Hsu, S. K. Chen, S. K. Chen, J. H. Tsai, H. L. Chiang, E. Fujita, B. Zielinska, D. Campbell, W. Arnott, L. J. Sagebiel, J. Chow, B. Zielinska, D. Campbell, W. Arnott, L. J. Sagebiel, J. Chow, P. Gabele, W. Crews, R. Snow, R. Snow, N. Clark, W. Wayne, W. Wayne, D. Lawson, E. Fujita, D. Campbell, W. Arnott, J. Chow, B. Zielinska, D. Campbell, W. Arnott, J. Chow, B. Zielinska, S. Brown, A. Frankel, S. Raffuse, P. Roberts, H. Hafner, A. Frankel, S. Raffuse, P. Roberts, H. Hafner, D. Anderson, B. Eklund, M. Simon, S. Regmi, M. Ongwandee, G. Morrison, M. Fitch, M. Ongwandee, G. Morrison, M. Fitch and R. Surampalli, *J. Air Waste Manage.*, 2007, **57**, 705–720.
9. T. W. Hesreberg, C. A. Lapin and W. B. Bunn, *Environ. Sci. Technol.*, 2008, **42**, 6438–6445.
10. U.S. Environmental Protecion Agency, EPA and NHTSA adopt standards to reduce greenhouse gas emissions and improve fuel efficiency of medium-and heavy-duty vehicles for model year 2018 and beyond. 2016.
11. J. J. West, A. M. Fiore, L. W. Horowitz and D. L. Mauzerall, *Proc. Natl. Acad. Sci. U. S. A.*, 2006, **103**, 3988–3993.
12. M. Crippa, G. Janssens-Maenhout, F. Dentener, D. Guizzardi, K. Sindelarova, M. Muntean, R. V. Dingenen and C. Granier, *Atmos. Chem. Phys.*, 2016, **16**, 3285–3841.

13. S. T. Turnock, E. W. Butt, T. B. Richardson, G. W. Mann, C. L. Reddington, P. M. Forster, J. Haywood, M. Crippa, G. Janssens-Maenhout, C. E. Johnson, N. Bellouin, K. S. Carslaw and D. V. Spracklen, *Environ. Res. Lett.*, 2016, **11**, 024010.
14. H. A. C. Denier van der Gon, M. E. Gerlofs-Nijland, R. Gehrig, M. Gustafsson, N. Janssen, R. M. Harrison, J. Hulskotte, C. Johansson, M. Jozwicka, M. Keuken, K. Krijgsheld, L. Ntziachristos, M. Riediker and F. R. Cassee, *J. Air Waste Manage.*, 2013, **63**, 136–149.
15. F. Amato, F. R. Cassee, H. A. Denier van der Gon, R. Gehrig, M. Gustafsson, W. Hafner, R. M. Harrison, M. Jozwicka, F. J. Kelly, T. Moreno, A. S. Prevot, M. Schaap, J. Sunyer and X. Querol, *J. Hazard. Mater.*, 2014, **275**, 31–36.
16. L. C. Marr, D. R. Black and R. A. Harley, *J. Geophys. Res.: Atmos.*, 2002, **107**, ACH 5-1–ACH 5-9.
17. L. C. Marr, G. S. Noblet and R. A. Harley, *J. Geophys. Res.: Atmos.*, 2002, **107**, ACH 6-1–ACH 6-11.
18. Z. G. Liu, D. R. Berg and J. J. Schauer, *SAE Int. J. Fuels Lubr.*, 2009, **1**, 184–191.
19. X. Fang, X. Hu, G. Janssens-Maenhout, J. Wu, J. Han, S. Su, J. Zhang and J. Hu, *Environ. Sci. Technol.*, 2013, **47**, 3848–3855.
20. Z. G. Liu, D. R. Berg, T. A. Swor and J. J. Schauer, *Environ. Sci. Technol.*, 2008, **42**, 6080–6085.
21. K. O. Blumberg, M. P. Walsh and C. Pera, Low-sulfur gasoline and diesel: the key to lower vehicle emissions, *The International Council on Clean Transportation*, 2003.
22. W. F. Rogge, L. M. Hildemann, M. A. Mazurek and G. R. Cass, *Environ. Sci. Technol.*, 1993, **27**, 636–651.
23. S. Biswas, V. Verma, J. J. Schauer and C. Sioutas, *Atmos. Environ.*, 2009, **43**, 1917–1925.
24. N. T. Kim Oanh, M. T. Thuy Phuong and D. A. Permadi, *Atmos. Environ.*, 2012, **59**, 438–448.
25. S. Geivanidis, P. Pistikopoulos and Z. Samaras, *Sci. Total Environ.*, 2003, **305**, 129–141.
26. A. L. Miller, C. B. Stipe, M. C. Habjan and G. G. Ahlstrand, *Environ. Sci. Technol.*, 2007, **41**, 6828–6835.
27. T. D. Durbin, J. R. Collins, J. M. Norbeck and M. R. Smith, *Environ. Sci. Technol.*, 2000, **34**, 349–355.
28. B. Zielinska, J. Sagebiel, J. D. McDonald, K. Whitney and D. R. Lawson, *J. Air Waste Manage.*, 2004, **54**, 1138–1150.
29. D. R. Gentner, G. Issacman, D. R. Worton, A. W. H. Chan, T. R. Dallmann, L. Davis, S. Liu, D. A. Day, L. M. Russell, K. R. Wilson, R. Weber, A. Guha, R. A. Harley and A. H. Goldstein, *Proc. Natl. Acad. Sci. U. S. A.*, 2012, **109**, 18318–18323.
30. D. R. Gentner, D. R. Worton, G. Isaacman, L. C. Davis, T. R. Dallmann, E. C. Wood, S. C. Herndon, A. H. Goldstein and R. A. Harley, *Environ. Sci. Technol.*, 2013, **47**, 11837–11848.

31. B. M. Graver, H. C. Frey and H. W. Choi, *Environ. Sci. Technol.*, 2011, **45**, 9044–9051.
32. W. Jing, Y. Yan, I. Kim and M. Sarvi, *Adv. Mech. Eng.*, 2016, **8**, 1–8.
33. A. Thorpe and R. M. Harrison, *Sci. Total Environ.*, 2008, **400**, 270–282.
34. H. Zhai, H. C. Frey, N. M. Rouphail, G. A. Gonçalves and T. L. Farias, *J. Air Waste Manage.*, 2009, **59**, 912–924.
35. I. J. George, M. D. Hays, J. S. Herrington, W. Preston, R. Snow, J. Faircloth, B. J. George, T. Long and R. W. Baldauf, *Environ. Sci. Technol.*, 2015, **49**, 13067–13074.
36. P. Geng, H. Zhang and S. Yang, *Fuel*, 2015, **145**, 221–227.
37. A. K. Chambers, M. Strosher, T. Wootton, J. Moncrieff and P. McCready, *J. Air Waste Manage.*, 2008, **58**, 1047–1056.
38. J. G. Watson, J. C. Chow and E. M. Fujita, *Atmos. Environ.*, 2001, **35**, 1567–1584.
39. E. M. Fujita, J. G. Watson and J. C. Chow, *Atmos. Environ.*, 1995, **29**, 3019–3035.
40. J. J. Schauer, G. C. Lough, M. M. Shafer, W. F. Christensen, M. F. Arndt, J. T. DeMinter and J. S. Park, *Health Eff. Inst.*, 2006, **133**, 21–48.
41. M. Ketzel, G. Omstedt, C. Johansson, I. Düring, M. Pohjola, D. Oettl, L. Gidhagen, P. Wåhlin, A. Lohmeyer, M. Haakana and R. Berkowicz, *Atmos. Environ.*, 2007, **41**, 9370–9385.
42. F. Amato, M. Pandolfi, A. Escrig, X. Querol, A. Alastuey, J. Pey, N. Perez and P. K. Hopke, *Atmos. Environ.*, 2009, **43**, 2770–2780.
43. N. Bukowiecki, P. Lienemann, M. Hill, M. Furger, A. Richard, F. Amato, A. S. H. Prévôt, U. Baltensperger, B. Buchmann and R. Gehrig, *Atmos. Environ.*, 2010, **44**, 2330–2340.
44. S. Lawrence, R. Sokhi, K. Ravindra, H. Mao, H. D. Prain and I. D. Bull, *Atmos. Environ.*, 2013, **77**, 548–557.
45. F. W. Lipfert and R. E. Wyzga, *J. Exposure Sci. Environ. Epidemiol.*, 2008, **18**, 588–599.
46. M. Pardo, M. M. Shafer, A. Rudich, J. J. Schauer and Y. Rudich, *Environ. Sci. Technol.*, 2015, **49**, 8777–8785.
47. Q. Liu, J. Baumgartner, Y. Zhang, Y. Liu, Y. Sun and M. Zhang, *Environ. Sci. Technol.*, 2014, **48**, 12920–12929.
48. R. M. Harrison and J. Yin, *Atmos. Environ.*, 2008, **42**, 1413–1423.
49. P. Pant, J. Yin and R. M. Harrison, *Atmos. Environ.*, 2014, **82**, 238–249.
50. J. Yin, S. A. Cumberland, R. M. Harrison, J. Allan, D. E. Young, P. I. Williams and H. Coe, *Atmos. Chem. Phys.*, 2015, **15**, 2139–2158.
51. Q. Liu and Y. Bei, *Atmos. Environ.*, 2016, **128**, 227–234.
52. G. S. W. Hagler, R. W. Baldauf, E. D. Thoma, T. R. Long, R. F. Snow, J. S. Kinsey, L. Oudejans and B. K. Gullett, *Atmos. Environ.*, 2009, **43**, 1229–1234.
53. G. D. Thurston, K. Ito and R. Lall, *Atmos. Environ.*, 2011, **45**, 3924–3936.
54. J. G. Su, J. S. Apte, J. Lipsitt, D. A. Garcia-Gonzales, B. S. Beckerman, A. de Nazelle, J. L. Texcalac-Sangrador and M. Jerrett, *Environ. Int.*, 2015, **78**, 82–89.

55. T. Cai, J. J. Schauer, W. Huang, D. Fang, J. Shang, Y. Wang and Y. Zhang, *Environ. Pollut.*, 2016, **219**, 821–828.
56. J. L. Durant, C. A. Ash, E. C. Wood, S. C. Herndon, J. T. Jayne, W. B. Knighton, M. R. Canagaratna, J. B. Trull, D. Brugge, W. Zamore and C. E. Kolb, *Atmos. Chem. Phys.*, 2010, **10**, 8341–8352.
57. A. L. Clements, Y. Jia, A. Denbleyker, E. McDonald-Buller, M. P. Fraser, D. T. Allen, D. R. Collins, E. Michel, J. Pudota, D. Sullivan and Y. Zhu, *Atmos. Environ.*, 2009, **43**, 4523–4534.
58. D. A. Olson, D. M. Hammond, R. L. Seila, J. M. Burke and G. A. Norris, *Atmos. Environ.*, 2009, **43**, 5647–5653.
59. Y. Zhu, J. Pudota, D. Collins, D. Allen, A. Clements, A. DenBleyker, M. Fraser, Y. Jia, E. McDonald-Buller and E. Michel, *Atmos. Environ.*, 2009, **43**, 4513–4522.
60. Y. Zhu, W. C. Hinds, S. Kim and C. Sioutas, *J. Air Waste Manage.*, 2002, **52**, 1032–1042.
61. Y. Zhu, W. C. Hinds, S. Shen and C. Sioutas, *Aerosol Sci. Technol.*, 2004, **38**, 5–13.
62. R. M. Harrison, A. M. Jones, D. C. S. Beddows, M. Dall'Osto and I. Nikolova, *Atmos. Environ.*, 2016, **125**, 1–7.
63. J. S. Wang, T. L. Chan, C. S. Cheung, C. W. Leung and W. T. Hung, *Atmos. Environ.*, 2006, **40**, 484–497.
64. S. Hu, S. Fruin, K. Kozawa, S. Mara, S. E. Paulson and A. M. Winer, *Atmos. Environ.*, 2009, **43**, 2541–2549.
65. T. Kuhn, M. Krudysz, Y. Zhu, P. M. Fine, W. C. Hinds, J. Froines and C. Sioutas, *J. Aerosol Sci.*, 2005, **36**, 291–302.
66. I. Nikolova, A. R. MacKenzie, X. Cai, M. S. Alam and R. M. Harrison, *Faraday Discuss.*, 2016, **189**, 529–546.
67. B. R. Bzdek, M. R. Pennington and M. V. Johnston, *J. Aerosol Sci.*, 2012, **52**, 109–120.
68. S. K. J. Suthawaree, K. Okuzawa, Y. Kanaya, P. Pochanart, H. Akimoto, Z. Wang and Y. Kajii, *Atmos. Chem. Phys.*, 2010, **10**, 1269–1285.
69. Y. Liu, M. Shao, L. Fu, S. Lu, L. Zeng and D. Tang, *Atmos. Environ.*, 2008, **42**, 6247–6260.
70. E. Nordling, N. Berglind, E. Melen, G. Emenius, J. Hallberg, F. Nyberg, G. Pershagen, M. Svartengren, M. Wickman and T. Bellander, *Epidemiology*, 2008, **19**, 401–408.
71. C. N. Hewitt, A. R. MacKenzie, P. Di Carlo, C. F. Di Marco, J. R. Dorsey, M. Evans, D. Fowler, M. W. Gallagher, J. R. Hopkins, C. E. Jones, B. Langford, J. D. Lee, A. C. Lewis, S. F. Lim, J. McQuaid, P. Misztal, S. J. Moller, P. S. Monks, E. Nemitz, D. E. Oram, S. M. Owen, G. J. Phillips, T. A. Pugh, J. A. Pyle, C. E. Reeves, J. Ryder, J. Siong, U. Skiba and D. J. Stewart, *Proc. Natl. Acad. Sci. U. S. A.*, 2009, **106**, 18447–18451.
72. X. Tie, R. J. Huang, W. Dai, J. Cao, X. Long, X. Su, S. Zhao, Q. Wang and G. Li, *Sci. Rep.*, 2016, **6**, 29612.

73. Y. J. Wang, A. DenBleyker, E. McDonald-Buller, D. Allen and K. M. Zhang, *Atmos. Environ.*, 2011, **45**, 43–52.
74. T. M. Barzyk, B. J. George, A. F. Vette, R. W. Williams, C. W. Croghan and C. D. Stevens, *Atmos. Environ.*, 2009, **43**, 787–797.
75. S. Hu, S. E. Paulson, S. Fruin, K. Kozawa, S. Mara and A. M. Winer, *Atmos. Environ.*, 2012, **51**, 311–319.
76. X. Huang, Y. Wang, Z. Xing and K. Du, *Sci. Total Environ.*, 2016, **565**, 698–705.
77. A. Vette, J. Burke, G. Norris, M. Landis, S. Batterman, M. Breen, V. Isakov, T. Lewis, M. I. Gilmour, A. Kamal, D. Hammond, R. Vedantham, S. Bereznicki, N. Tian, C. Croghan and C. Community Action Against Asthma Steering, *Sci. Total Environ.*, 2013, **448**, 38–47.
78. G. S. W. Hagler, E. D. Thoma and R. W. Baldauf, *J. Air Waste Manage.*, 2010, **60**, 328–336.
79. P. M. Baylon, D. A. Jaffe, R. B. Pierce and M. S. Gustin, *Environ. Sci. Technol.*, 2016, **50**, 2994–3001.
80. E. Chan and R. J. Vet, *Atmos. Chem. Phys.*, 2010, **10**, 8629–8647.
81. A. S. Lefohn, D. Shadwick and S. J. Oltmans, *Atmos. Environ.*, 2010, **44**, 5199–5210.
82. R. C. Wilson, Z. L. Fleming, P. S. Monks, G. Clain, S. Henne, I. B. Konovalov, S. Szopa and L. Menut, *Atmos. Chem. Phys.*, 2012, **12**, 437–454.
83. J. J. Schauer, *J. Exposure Anal. Environ. Epidemiol.*, 2003, **13**, 443–453.
84. Q. Liu, J. Baumgartner, Y. Zhang and J. J. Schauer, *Atmos. Environ.*, 2016, **126**, 28–35.
85. R. J. Huang, Y. Zhang, C. Bozzetti, K. F. Ho, J. J. Cao, Y. Han, K. R. Daellenbach, J. G. Slowik, S. M. Platt, F. Canonaco, P. Zotter, R. Wolf, S. M. Pieber, E. A. Bruns, M. Crippa, G. Ciarelli, A. Piazzalunga, M. Schwikowski, G. Abbaszade, J. Schnelle-Kreis, R. Zimmermann, Z. An, S. Szidat, U. Baltensperger, I. El Haddad and A. S. Prevot, *Nature*, 2014, **514**, 218–222.
86. S. Guo, M. Hu, M. L. Zamora, J. Peng, D. Shang, J. Zheng, Z. Du, Z. Wu, M. Shao, L. Zeng, M. J. Molina and R. Zhang, *Proc. Natl. Acad. Sci. U. S. A.*, 2014, **111**, 17373–17378.
87. F. Cappitelli, L. Toniolo, A. Sansonetti, D. Gulotta, G. Ranalli, E. Zanardini and C. Sorlini, *Appl. Environ. Microbiol.*, 2007, **73**, 5671–5675.
88. S. A. Ruffolo, V. Comite, M. F. La Russa, C. M. Belfiore, D. Barca, A. Bonazza, G. M. Crisci, A. Pezzino and C. Sabbioni, *Sci. Total Environ.*, 2015, **502**, 157–166.
89. C. M. Belfiore, D. Barca, A. Bonazza, V. Comite, M. F. La Russa, A. Pezzino, S. A. Ruffolo and C. Sabbioni, *Environ. Sci. Pollut. Res.*, 2013, **20**, 8848–8859.
90. M. H. Bergin, S. N. Tripathi, J. Jai Devi, T. Gupta, M. McKenzie, K. S. Rana, M. M. Shafer, A. M. Villalobos and J. J. Schauer, *Environ. Sci. Technol.*, 2015, **49**, 808–812.

91. P. Pant, A. Shukla, S. D. Kohl, J. C. Chow, J. G. Watson and R. M. Harrison, *Atmos. Environ.*, 2015, **109**, 178–189.
92. C. S. Liang, F. K. Duan, K. B. He and Y. L. Ma, *Environ. Int.*, 2016, **86**, 150–170.
93. C. Holman, R. Harrison and X. Querol, *Atmos. Environ.*, 2015, **111**, 161–169.
94. J. Lelieveld, J. S. Evans, M. Fnais, D. Giannadaki and A. Pozzer, *Nature*, 2015, **525**, 367–371.
95. R. R. Patil, S. K. Chetlapally and M. Bagavandas, *Int. J. Occup. Med. Environ. Health*, 2014, **27**, 523–535.
96. M. Strak, N. A. Janssen, K. J. Godri, I. Gosens, I. S. Mudway, F. R. Cassee, E. Lebret, F. J. Kelly, R. M. Harrison, B. Brunekreef, M. Steenhof and G. Hoek, *Environ. Health Perspect.*, 2012, **120**, 1183–1189.
97. J. McCreanor, P. Cullinan, M. J. Nieuwenhuijsen, J. Stewart-Evans, E. Malliarou, L. Jarup, R. Harrington, M. Svartengren, I. K. Han, P. Ohman-Strickland, K. F. Chung and J. F. Zhang, *N. Engl. J. Med.*, 2007, **357**, 2348–2358.
98. P. Vineis, G. Hoek, M. Krzyzanowski, F. Vigna-Taglianti, F. Veglia, L. Airoldi, K. Overvad, O. Raaschou-Nielsen, F. Clavel-Chapelon, J. Linseisen, H. Boeing, A. Trichopoulou, D. Palli, V. Krogh, R. Tumino, S. Panico, H. B. Bueno-De-Mesquita, P. H. Peeters, E. E. Lund, A. Agudo, C. Martinez, M. Dorronsoro, A. Barricarte, L. Cirera, J. R. Quiros, G. Berglund, J. Manjer, B. Forsberg, N. E. Day, T. J. Key, R. Kaaks, R. Saracci and E. Riboli, *Environ. Health*, 2007, **6**, 7.
99. R. Beelen, O. Raaschou-Nielsen, M. Stafoggia, Z. J. Andersen, G. Weinmayr, B. Hoffmann, K. Wolf, E. Samoli, P. Fischer, M. Nieuwenhuijsen, P. Vineis, W. W. Xun, K. Katsouyanni, K. Dimakopoulou, A. Oudin, B. Forsberg, L. Modig, A. S. Havulinna, T. Lanki, A. Turunen, B. Oftedal, W. Nystad, P. Nafstad, U. De Faire, N. L. Pedersen, C.-G. Östenson, L. Fratiglioni, J. Penell, M. Korek, G. Pershagen, K. T. Eriksen, K. Overvad, T. Ellermann, M. Eeftens, P. H. Peeters, K. Meliefste, M. Wang, B. Bueno-de-Mesquita, D. Sugiri, U. Krämer, J. Heinrich, K. de Hoogh, T. Key, A. Peters, R. Hampel, H. Concin, G. Nagel, A. Ineichen, E. Schaffner, N. Probst-Hensch, N. Künzli, C. Schindler, T. Schikowski, M. Adam, H. Phuleria, A. Vilier, F. Clavel-Chapelon, C. Declercq, S. Grioni, V. Krogh, M.-Y. Tsai, F. Ricceri, C. Sacerdote, C. Galassi, E. Migliore, A. Ranzi, G. Cesaroni, C. Badaloni, F. Forastiere, I. Tamayo, P. Amiano, M. Dorronsoro, M. Katsoulis, A. Trichopoulou, B. Brunekreef and G. Hoek, *Lancet*, 2014, **383**, 785–795.
100. J. Heo, J. J. Schauer, O. Yi, D. Paek, H. Kim and S. M. Yi, *Epidemiology*, 2014, **25**, 379–388.
101. J. Baumgartner, Y. Zhang, J. J. Schauer, W. Huang, Y. Wang and M. Ezzati, *Proc. Natl. Acad. Sci. U. S. A.*, 2014, **111**, 13229–13234.

102. G. D. Thurston, R. T. Burnett, M. C. Turner, Y. Shi, D. Krewski, R. Lall, K. Ito, M. Jerrett, S. M. Gapstur, W. R. Diver and C. A. Pope, *Environ. Health Perspect.*, 2016, **124**, 785–794.
103. I. Kheirbek, J. Haney, S. Douglas, K. Ito and T. Matte, *Environ. Health*, 2016, **15**, 89.
104. R. Ghosh, F. Lurmann, L. Perez, B. Penfold, S. Brandt, J. Wilson, M. Milet, N. Kunzli and R. McConnell, *Environ. Health Perspect.*, 2016, **124**, 193–200.
105. K. Ito, N. Xue and G. Thurston, *Atmos. Environ.*, 2004, **38**, 5269–5282.
106. S. Hasheminassab, N. Daher, B. D. Ostro and C. Sioutas, *Environ. Pollut.*, 2014, **193**, 54–64.
107. J. K. Choi, J. B. Heo, S. J. Ban, S. M. Yi and K. D. Zoh, *Sci. Total Environ.*, 2013, **447**, 370–380.
108. K. R. Bullock, R. M. Duvall, G. A. Norris, S. R. McDow and M. D. Hays, *Atmos. Environ.*, 2008, **42**, 6897–6904.
109. V. Mugica, E. Ortiz, L. Molina, A. De Vizcaya-Ruiz, A. Nebot, R. Quintana, J. Aguilar and E. Alcántara, *Atmos. Environ.*, 2009, **43**, 5068–5074.
110. J. Yin, S. A. Cumberland, R. M. Harrison, J. Allan, D. E. Young, P. I. Williams and H. Coe, *Atmos. Chem. Phys.*, 2015, **15**, 2139–2158.
111. F. Amato, A. Alastuey, A. Karanasiou, F. Lucarelli, S. Nava, G. Calzolai, M. Severi, S. Becagli, V. L. Gianelle, C. Colombi, C. Alves, D. Custódio, T. Nunes, M. Cerqueira, C. Pio, K. Eleftheriadis, E. Diapouli, C. Reche, M. C. Minguillón, M.-I. Manousakas, T. Maggos, S. Vratolis, R. M. Harrison and X. Querol, *Atmos. Chem. Phys.*, 2016, **16**, 3289–3309.
112. M. Bressi, J. Sciare, V. Ghersi, N. Bonnaire, J. B. Nicolas, J. E. Petit, S. Moukhtar, A. Rosso, N. Mihalopoulos and A. Féron, *Atmos. Chem. Phys.*, 2013, **13**, 7825–7844.
113. N. Daher, A. Ruprecht, G. Invernizzi, C. De Marco, J. Miller-Schulze, J. B. Heo, M. M. Shafer, B. R. Shelton, J. J. Schauer and C. Sioutas, *Atmos. Environ.*, 2012, **49**, 130–141.
114. Y. Cheng, S. Lee, Z. Gu, K. Ho, Y. Zhang, Y. Huang, J. C. Chow, J. G. Watson, J. Cao and R. Zhang, *Particuology*, 2015, **18**, 96–104.
115. S. Wu, F. Deng, H. Wei, J. Huang, X. Wang, Y. Hao, C. Zheng, Y. Qin, H. Lv, M. Shima and X. Guo, *Environ. Sci. Technol.*, 2014, **48**, 3438–3448.
116. P. Wang, J. J. Cao, Z. X. Shen, Y. M. Han, S. C. Lee, Y. Huang, C. S. Zhu, Q. Y. Wang, H. M. Xu and R. J. Huang, *Sci. Total Environ.*, 2015, **508**, 477–487.
117. J. Gu, S. Du, D. Han, L. Hou, J. Yi, J. Xu, G. Liu, B. Han, G. Yang and Z.-P. Bai, *Air Qual., Atmos. Health*, 2014, **7**, 251–262.
118. J. H. Murillo, S. R. Roman, J. F. Rojas Marin, A. C. Ramos, S. B. Jimenez, B. C. Gonzalez and D. G. Baumgardner, *Atmos. Pollut. Res.*, 2013, **4**, 181–190.
119. S. M. Gaita, J. Boman, M. J. Gatari, J. B. C. Pettersson and S. Janhäll, *Atmos. Chem. Phys.*, 2014, **14**, 9977–9991.

120. L. H. M. dos Santos, A. A. F. S. Kerr, T. G. Veríssimo, M. d. F. Andrade, R. M. de Miranda, A. Fornaro and P. Saldiva, *J. Air Waste Manage. Assoc.*, 2014, **64**, 519–528.
121. E. Stelcer, D. D. Cohen and A. J. Atanacio, *Environ. Chem.*, 2014, **11**, 644.
122. S. Matthes, V. Grewe, R. Sausen and G. J. Roelofs, *Atmos. Chem. Phys.*, 2007, **7**, 1707–1718.
123. C. Granier and G. P. Brasseur, *Geophys. Res. Lett.*, 2003, **30**, 1086.
124. U. Niemeier, C. Granier, L. Kornblueh, S. Walters and G. P. Brasseur, *J. Geophys. Res.*, 2006, **111**, D09301.
125. J. B.-K. P. Hoor, D. Caro, O. Dessens, O. Endresen, M. Gauss, V. Grewe, D. Hauglustaine, I. S. A. Isaksen, P. Jockel, J. Lelieveld, G. Myhre, E. Meijer, D. Olivie, M. Prather, C. Schnadt Poberaj, K. P. Shine, J. Staehelin, Q. Tang, J. van Aardenne, P. van Velthoven and R. Sausen, *Atmos. Chem. Phys.*, 2009, **9**, 3113–3136.
126. Y. Qu, J. An, J. Li, Y. Chen, Y. Li, X. Liu and M. Hu, *Adv. Atmos. Sci.*, 2014, **31**, 787–800.
127. A. Bonazza, C. Sabbioni and N. Ghedini, *Atmos. Environ.*, 2005, **39**, 2607–2618.
128. M. Del Monte, C. Sabbioni and O. Vittori, *Atmos. Environ.*, 1981, **15**, 645–652.
129. C. Saiz-Jimenez, *Atmos. Environ.*, 1993, **278**, 77–85.
130. L. W. A. Chen, H. Moosmuller, W. P. Arnott, J. C. Chow, J. G. Waston, R. A. Susott, R. E. Babbitt, C. E. Wold, E. N. Lincoln and W. M. Hao, *Environ. Sci. Technol.*, 2007, **41**, 4317–4326.
131. J. C. Chow, J. G. Watson, D. H. Lowenthal, L. W. Antony Chen and N. Motallebi, *Atmos. Environ.*, 2011, **45**, 5407–5414.
132. E. Kim and P. K. Hopke, *J. Air Waste Manage.*, 2005, **55**, 1456–1463.
133. M. Crippa, F. Canonaco, V. A. Lanz, M. Äijälä, J. D. Allan, S. Carbone, G. Capes, D. Ceburnis, M. Dall'Osto, D. A. Day, P. F. DeCarlo, M. Ehn, A. Eriksson, E. Freney, L. Hildebrandt Ruiz, R. Hillamo, J. L. Jimenez, H. Junninen, A. Kiendler-Scharr, A. M. Kortelainen, M. Kulmala, A. Laaksonen, A. A. Mensah, C. Mohr, E. Nemitz, C. O'Dowd, J. Ovadnevaite, S. N. Pandis, T. Petäjä, L. Poulain, S. Saarikoski, K. Sellegri, E. Swietlicki, P. Tiitta, D. R. Worsnop, U. Baltensperger and A. S. H. Prévôt, *Atmos. Chem. Phys.*, 2014, **14**, 6159–6176.
134. J. S. Han, K. J. Moon, S. J. Lee, Y. J. Kim, S. Y. Ryu, S. S. Cliff and S. M. Yi, *Atmos. Chem. Phys.*, 2006, **6**, 211–213.
135. E. A. Stone, J. J. Schauer, B. B. Pradhan, P. M. Dangol, G. Habib, C. Venkataraman and V. Ramanathan, *J. Geophys. Res.*, 2010, **115**, D06031.
136. F. Wang, T. Lin, Y. Li, T. Ji, C. Ma and Z. Guo, *Atmos. Environ.*, 2014, **92**, 484–492.
137. L. W. A. Chen, J. G. Waston, J. C. Chow, D. W. DuBois and L. Herschberger, *J. Air Waste Manage.*, 2011, **61**, 1204–1217.
138. E. Kim and P. K. Hokpe, *J. Geophys. Res.*, 2004, **109**, D09204.

139. A. Hakami, D. K. Henze, J. H. Seinfeld, T. Chai, Y. Tang, G. R. Carmichael and A. Sandu, *J. Geophys. Res.: Atmos.*, 2005, **110**, D14031.
140. Y. Sawa, H. Tanimoto, S. Yonemura, H. Matsueda, A. Wada, S. Taguchi, T. Hayasaka, H. Tsuruta, Y. Tohjima, H. Mukai, N. Kikuchi, S. Katagiri and K. Tsuboi, *J. Geophys. Res.*, 2007, **112**, D22S26.
141. J. M. Hoell, D. D. Davis, S. C. Liu, R. Newell, M. Shipham, H. Akimoto, R. J. McNeal, R. J. Bendura and J. W. Drewry, *J. Geophys. Res.: Atmos.*, 1996, **101**, 1641–1653.
142. G. Janssens-Maenhout, M. Crippa, D. Guizzardi, F. Dentener, M. Muntean, G. Pouliot, T. Keating, Q. Zhang, J. Kurokawa, R. Wankmüller, H. Denier van der Gon, J. J. P. Kuenen, Z. Klimont, G. Frost, S. Darras, B. Koffi and M. Li, *Atmos. Chem. Phys.*, 2015, **15**, 11411–11432.
143. G. J. Frost, P. Middleton, L. Tarrasón, C. Granier, A. Guenther, B. Cardenas, H. Denier van der Gon, G. Janssens-Maenhout, J. W. Kaiser, T. Keating, Z. Klimont, J.-F. Lamarque, C. Liousse, S. Nickovic, T. Ohara, M. G. Schultz, U. Skiba, J. van Aardenne and Y. Wang, *Atmos. Environ.*, 2013, **81**, 710–712.
144. J. Cofala, M. Amann, Z. Klimont, K. Kupiainen and L. Höglund-Isaksson, *Atmos. Environ.*, 2007, **41**, 8486–8499.
145. C. Guido, C. Beatrice and P. Napolitano, *Appl. Energy*, 2013, **102**, 13–23.
146. G. Fontaras, V. Franco, P. Dilara, G. Martini and U. Manfredi, *Sci. Total Environ.*, 2014, **468–469**, 1034–1042.
147. J. Dargay, D. Gately and M. Sommer, *Energy*, 2007, **28**, 143–170.
148. M. M. Roy, J. Calder, W. Wang, A. Mangad and F. C. M. Diniz, *Appl. Energy*, 2016, **180**, 52–65.
149. V. R. J. H. Timmers and P. A. J. Achten, *Atmos. Environ.*, 2016, **134**, 10–17.

Water and Soil Pollution Implications of Road Traffic

ASHANTHA GOONETILLEKE,* BUDDHI WIJESIRI AND ERICK R. BANDALA

ABSTRACT

Rising vehicle ownership around the world has resulted in greater traffic volumes on roads and associated traffic-generated pollutants. Vehicular traffic and, by implication, road surfaces are among the most important sources of pollutants to the environment, particularly the urban environment, and can have significant human and ecosystem health impacts. Research studies have confirmed the vulnerability of the population living near high-traffic areas, particularly the elderly. An equal concern is the uptake of trace metals by plants near road environments. Road surfaces also act as repositories for pollutants from other sources that can react with traffic-generated pollutants, leading to changes to pollutant characteristics, influencing their solubility, reactivity, bioavailability, mobility, toxicity and persistence. Pollutants deposited on road surfaces are incorporated in road dust. The primary sources of traffic pollutants are exhaust emissions, fuel and lubrication system leakages, component wear such as tyres, brakes and chassis and road surface wear. Traffic and road surface factors play significant roles in influencing the type, amount and rate of pollutant generation. It has been found that traffic congestion is more significant than traffic volume in terms of pollutant generation. The pollutant emissions from heavy-duty vehicles that use diesel are different compared to light-duty vehicles that use gasoline. Very wide ranges of trace

*Corresponding author.

metals and hydrocarbons are emitted into the environment as a result of fuel combustion and vehicle component and road surface wear. The metals most commonly originating from traffic activities include Cd, Cu, Cr, Ni and Zn. Pb is added to road surfaces by the accumulation in soils due to past usage of leaded fuels, as well as ongoing contributions from the wear of vehicle components and road paint. Among the hydrocarbons, polycyclic aromatic hydrocarbons are of most concern, as these compounds represent a significant class of suspected carcinogens. Used lubrication oil and incomplete combustion of fossil fuels, particularly diesel, are significant contributors of polycyclic aromatic hydrocarbons to the environment. It is hypothesised that climate change-driven alterations to climate characteristics will influence changes to traffic pollutant build-up on ground surfaces and wash-off with stormwater runoff.

1 Introduction

Vehicle ownership is rising around the world for a range of reasons. Firstly, it can be attributed to the increase in population in most parts of the world. Secondly, the increase in urbanisation has resulted in the conversion of outlying agricultural areas into human settlements. This in turn has resulted in greater distances between areas of employment and recreation and human habitation, leading to greater dependency on motor vehicles. Thirdly, improvements in living standards in most parts of the world and the corresponding increases in expectations have resulted in greater vehicle ownership, as this is considered to be an indicator of affluence. The consequences include greater traffic volume on roads, congestion and increased traffic-generated emissions.

Vehicular traffic and by association road surfaces are among the most important sources of pollutants to the environment. These two sources act synergistically to generate traffic-related pollutants. As illustrated in Figure 1, road surfaces receive pollutants from a range of sources in addition to traffic. Although the focus of this chapter is on traffic pollutants, pollutants from the other sources also need to be taken into consideration, as these can react with traffic pollutants. The resulting pollutant transformations can change the characteristics of the traffic pollutants. These issues are discussed in further detail below. Therefore, it is important to keep in mind that it is not an easy task to separate a pollutant group into solely traffic-generated pollutants, as their physical and chemical characteristics are subjected to a range of influential factors.

As Figure 1 has further illustrated, the generation of traffic-related pollutants and the processes involved in the transmission, deposition on ground surfaces and incorporation in stormwater runoff lead to soil and water pollution. This chapter discusses the key pollutants, the factors that influence pollutant generation, the inherent processes and the implications of traffic pollutants. Gaseous emissions from road traffic do not form a part of

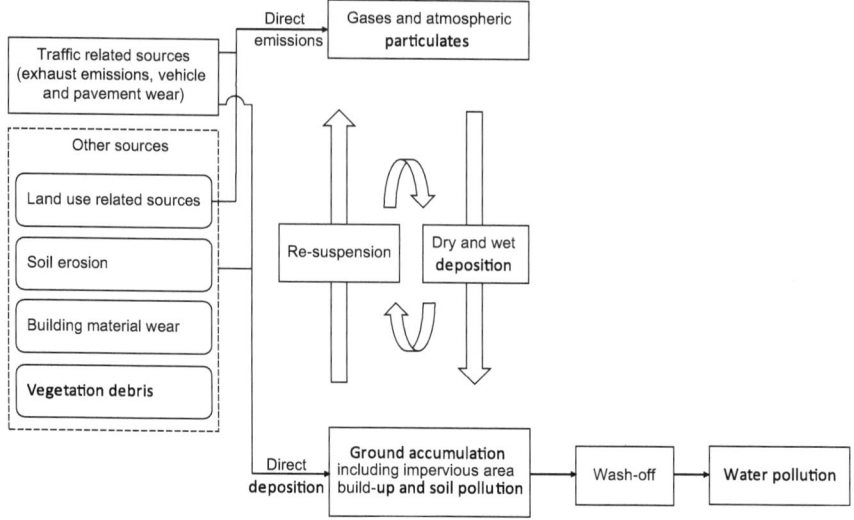

Figure 1 Schematic of traffic pollutant emissions, build-up and wash-off.

the discussion in this chapter. Furthermore, the role of atmospheric particulates are discussed only to the extent that they influence pollutant characteristics and behaviour on ground surfaces.

In relation to traffic pollutants, road dust plays an important role and has been the focus of significant research. Essentially, road dust incorporates traffic pollutants as well as pollutants from other sources, as illustrated in Figure 1. The accumulation of traffic pollutants in road dust is a temporary sink, the ultimate fate being deposition in water or soil due to transport by turbulent airstreams and stormwater runoff. Road dust particles can migrate *via* saltation, creep (>100 µm), suspension (<100 µm) or incorporation into aerosols (>10 µm), causing hazardous impacts and endangering environmental health.[1] These issues are further discussed below.

2 Primary Pollutants from Road Traffic

2.1 Pollutant Sources

Exhaust emissions and abrasion wear products are the most important traffic-related pollutant sources. These sources have distinct characteristics in generating different pollutant species and different particle sizes in relation to atmospheric emission and ground deposition phases. The primary sources of traffic-related pollutants can be categorised as follows:

- Exhaust emissions
- Fuel and lubrication system leakages
- Component wear, such as tyres, brakes and chassis
- Road surface wear

The exhaust emissions can be either gaseous or particulate matter. These initially build up in the atmosphere and contribute to atmospheric pollution. As illustrated in Figure 1, these pollutants eventually return to the ground as dry or wet depositions. Observations by Gunawardena et al.[2] of relatively high pollutant loads in the atmosphere during weekdays compared to weekends confirm that traffic-related emissions are most prominent in the urban environment. The processes of wet and dry deposition are discussed later in the chapter. Patches of fuels and lubricants due to leakage are common features in traffic areas. These are primarily found in car parks and near traffic intersections. Vehicle component wear, such as the abrasion of tyres and brakes, is a key contributor of metals to the urban environment. Road surface wear is considered to be an integral part of traffic-related pollutants as the pollutant loads generated are directly influenced by traffic characteristics. This is discussed in detail below.

2.2 Influential Factors in Pollutant Generation

2.2.1 Traffic Characteristics. A range of traffic and road surface factors play a significant role in influencing the type, amount and rate of pollutant generation. The following traffic factors were identified by Gunawardena et al.[3,4] as specifically relevant for characterising traffic pollutants:

- Traffic volume, as represented by average daily traffic (ADT);
- Traffic congestion during peak hours, as represented by the ratio of traffic volume to the design traffic capacity of the road (V/C);
- Traffic mix, as represented by the proportion of heavy-duty traffic compared to the total traffic.

These characteristics influence fuel combustion and the wear of vehicle components. It is to be expected that a relatively high traffic volume will lead to greater emission of combustion- and abrasion-related pollutants. Furthermore, high traffic volumes lead to greater redistributions of traffic-generated pollutants due to the increase in turbulent airstreams on road surfaces.[5] Additionally, the abrasion by vehicle tyres of particulates already deposited on road surface can result in changes to pollutant characteristics, such as particle size.

As illustrated in Figure 2, the pollutant load accumulated on road surfaces increases with an increase in traffic congestion. This is attributed to reduced vehicle speeds and increased stop–start activity. As illustrated in Figure 3, which shows the relationship between trace metal generation, ADT and V/C, it is evident that the impact of traffic congestion on pollutant generation is more significant than traffic volume. The negative correlation between ADT and V/C can be explained by the fact that in a traffic lane with fixed capacity, V/C will increase with the slow movement of vehicles, which in turn results in a decrease in ADT.[6] Based on an extensive study of dry and wet atmospheric deposition, Gunawardena et al.[3] noted that reducing traffic

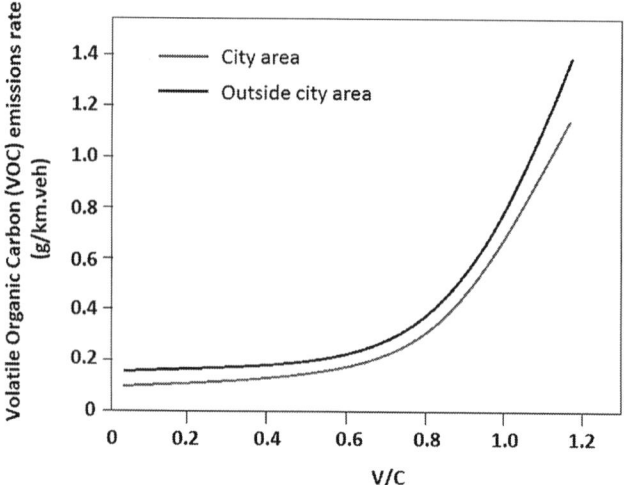

Figure 2 Impact of traffic congestion (V/C ratio) on volatile hydrocarbon emissions on a dual carriageway arterial road (adapted from *Urban Air Pollution in Australia*, AATSE, 1997[7]).

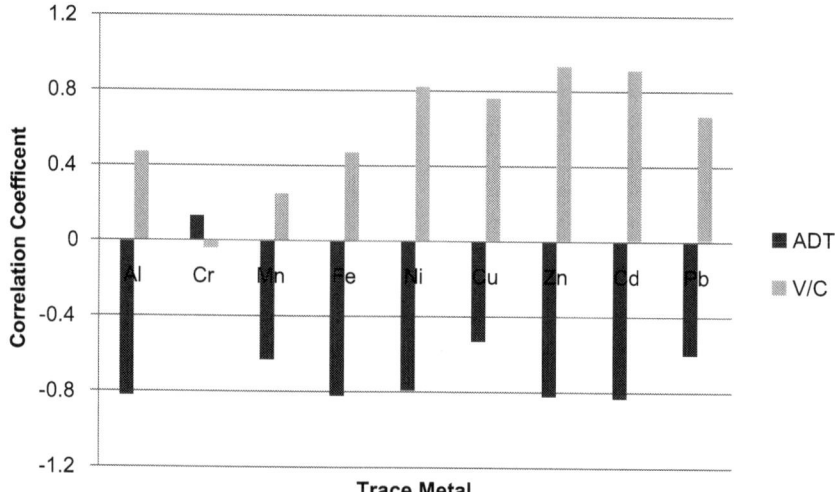

Figure 3 Correlations between trace metal load and ADT and V/C (data obtained from Mahbub[8]).

congestion would be more effective than reducing traffic volume for improving air quality in relation to Pb, Cd, Ni and Cu.

The pollutant emissions from heavy-duty vehicles that use diesel are different compared to light-duty vehicles that use gasoline. An extensive study on traffic-generated pollutants undertaken by Gunawardena[9] found that hydrocarbons (particularly polycyclic aromatic hydrocarbons [PAHs]) in both the atmospheric phase and during build-up show strong correlations with heavy-duty traffic volume and traffic congestion, suggesting that these pollutants predominately originate from diesel combustion and in congested traffic environments due to incomplete combustion. A similar result was found for metals in the atmospheric phase showing strong affinity to traffic characteristics such as total traffic volume, heavy-duty traffic volume and traffic congestion. This suggests that these metals mostly originate from traffic-related sources, such as exhaust emissions and abrasion products, and traffic-influenced sources, such as resuspension, with heavy-duty traffic playing a key role.

2.2.2 Road Characteristics. The road characteristics that play an influential role in pollutant generation include:

- Road layout, including the location of signalised and other intersections, pedestrian crossings, longitudinal slope and the presence of parking bays;
- Pavement, including surface roughness, construction materials used, condition, age and maintenance practices.[8,10]

These characteristics play important roles in the contribution to the total pollutant load, such as from tyre and brake wear. Tyre wear is generated due to the abrasion resulting from contact between tyres and road surfaces. In addition to the tyre material, size, pressure and contact area, the intensity of abrasion depends on the road pavement material and surface roughness, which can be defined in terms of surface texture depth.

Figure 4 shows the correlation between the generation of trace metals commonly associated with traffic and surface texture depth. It is evident that

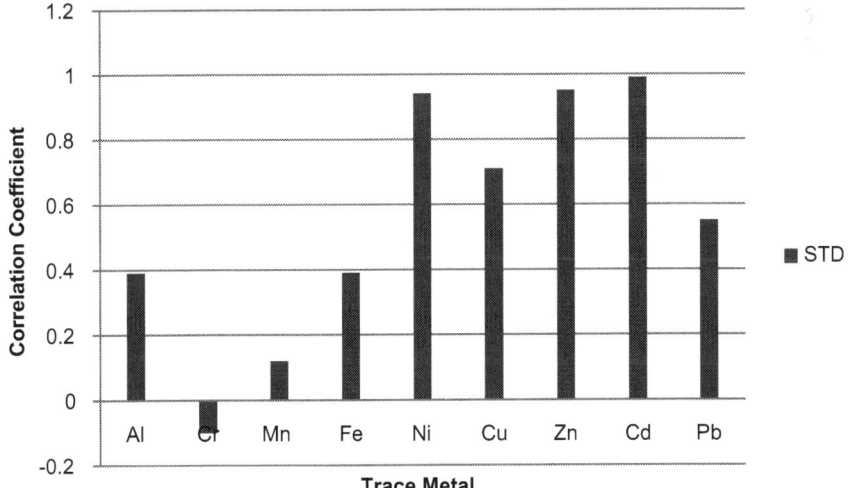

Figure 4 Correlations between trace metals and road surface texture depths (STDs) (data obtained from Mahbub[8]).

roads with higher texture depths can influence the abrasion of vehicle tyres. De Silva et al.[11] observed that the concentration of different species of particle-bound trace metals can be related to the age of a road where traffic characteristics are comparatively similar. For example, with similar traffic characteristics (high traffic volume ≥15 000 vehicles day^{-1}), the highest Mn load per unit mass of roadside soil is found at roads of age 2–5 years, while roads of age ≥30 years generate the highest loads of Pb in roadside soil. This is attributed to the use of Mn- and Pb-related products in the automobile industry at different time periods.

The road characteristics also play a significant role in the braking action of vehicles. For example, road slope influences the generation of brake wear, as the steeper the slope, the greater the engine power required, resulting in higher loads of brake wear. Furthermore, the load of brake wear generation can also vary at different locations, such as intersections and pedestrian crossings.[12]

2.3 Primary Pollutants

Vehicular traffic contributes solid, liquid and gaseous pollutants that exert a significant qualitative impact on water and soil environments. In the context of environmental pollution, the most noteworthy include organic carbon, trace metals and hydrocarbons. Nutrients such as nitrogen and phosphorus do not form part of the discussion. Although nitrogen compounds can originate from fuel combustion, past research has shown that transport activities contribute less than 5% to overall nitrogen emissions.[13] Phosphorus, though present on road surfaces, is primarily contributed by soil sources[14] and fertiliser application.

Organic carbon can be considered to be a pollutant in its own right. However, the more important role of organic carbon is in relation to interactions with metals and hydrocarbons and the resulting changes to the chemical characteristics of these pollutants. Consequently, the primary focus of the discussion will be on traffic-generated metals and hydrocarbons. This focus also needs to be viewed in the context that the presence of metals and hydrocarbons in the environment is significant due to their potential toxicity to human and to ecosystem health.

A very wide range of trace metal species are emitted into the environment as a result of fuel combustion and vehicle component and road surface wear. Over time, the species and pollutant loads vary due to the introduction of new technologies and processes. The metals most commonly present on road surfaces are Cd, Cu, Cr, Ni, Zn, Pb, Al, Mn and Fe.

Research has shown that Cd, Cu, Cr, Ni and Zn primarily originate from road traffic activities,[10,15] whilst Al, Mn and Fe are primarily contributed by roadside soil, although they could also originate from vehicle component wear, but to a lesser extent.[16,17] Pb builds up on road surfaces, stemming from accumulation in soils from past usage of leaded fuel, as well as ongoing contributions from the wear of vehicle components and road paint.[16,18]

Additionally, a large number of other metals in trace quantities originate from road traffic. For example, the presence of Mn can also be due to the introduction of methylcyclopentadienyl manganese tricarbonyl as a gasoline additive.[19] Similarly, Pd, Rh and Pt are released from catalytic converters in vehicles,[15,20] and Sb is used as a colorant (antimony pentasulphide) in tyre production.[21] Table 1 provides a list of the trace metals that are commonly associated with vehicular traffic. However, this list may not be exhaustive, as new products and processes may introduce other metal species.

However, the detrimental impact of metals in the water environment is dependent on bioavailability. For example, the partitioning of metal species between dissolved and particulate-bound fractions is a dynamic process that can change at various stages along the continuum from the point of deposition to the receiving environment due to bonding mechanisms. These mechanisms vary between different metals.[25-29]

A number of different classifications can be used for categorising hydrocarbon pollutants originating from road traffic, none of which are very precise. A common classification is total petroleum hydrocarbons (TPHs),

Table 1 Trace metals associated with vehicular traffic (adapted from De Silva et al.[11] and other references therein; Mummullage et al.[17] and other references therein; Adachi and Tainosho;[18] Huber et al.[22] and other references therein; Fatemi et al.[23] and other references therein; and Tiarks et al.[24] and other references therein).

	Fuel	Engine oil	Tyre wear	Brake wear	Catalytic converters	Chassis	Road surface wear	Road paint
Ag	X			X				
Al		X				X	X	
As				X				
Ba			X	X			X	X
Cd	X	X	X	X				
Co		X					X	
Cr	X	X	X	X		X	X	X
Cu			X	X			X	
Fe			X	X		X		
Hg	X							
Mn	X		X	X		X		
Ni	X	X	X	X			X	
Pb	X		X	X			X	X
Pd					X			
Pt					X			
Rh					X			
Se	X		X					
Sb				X				
Ti			X	X			X	X
Zn	X	X	X	X			X	
Zr				X				
V							X	
W		X						

which includes several hundred chemical compounds. For simplicity, TPHs can be categorised into gasoline-range and diesel-range organics. The gasoline-range organics are also identified as volatile TPHs and the diesel-range organics are also identified as semi- and non-volatile hydrocarbons.[8] These originate primarily from exhaust emissions and leakages from vehicles.

In the case of hydrocarbons, PAHs deserve special mention as these compounds are pervasive in the urban environment and represent a significant class of suspected carcinogens. The US Environmental Protection Agency has classified 16 PAHs as priority pollutants based on their prevalence and toxicity. These are naphthalene, acenaphthene, acenaphthylene, flourene, anthracene, phenanthrene, flouranthene, benzo[*a*]anthracene, benzo[*b*]flouranthene, benzo[*k*]flouranthene, chrysene, pyrene, benzo[*a*]pyrene, dibenzo[*a,h*]anthracene, benzo[*ghi*]perylene and indeno[1,2,3-*cd*]pyrene.[30]

Research undertaken by Brown and Peake[31] and Latimer et al.[32] has shown that while unused crankcase oil has no PAHs, used oil contains appreciable concentrations of PAHs with two (light molecular weight) to six (heavy molecular weight) benzene rings, with their presence increasing with usage. The greater the number of benzene rings, the greater the toxicity. Used lubrication oil is a major contributor of PAHs to the environment. Additionally, PAHs are typically by-products of the incomplete combustion of fossil fuels. Road traffic is estimated to contribute between 46% and 90% of total PAHs ambient in urban areas.[33] Mummullage et al.[34] have noted that non-combusted lubrication oil, non-combusted diesel fuel and tyre and road surface wear as the three most critical hydrocarbon sources in relation to traffic pollutants.

3 Pollutant Processes

As illustrated in Figure 1, traffic-generated pollutants, similarly to other pollutants, are deposited on ground surfaces *via* two primary pathways, contributing to soil and water pollution. These pathways are direct deposition of pollutants on ground surfaces and contribution to atmospheric pollutant load and subsequent build-up on ground surfaces as wet or dry deposition. Figure 5 provides a more detailed overview of the linkages and influential factors in relation to pollutant processes.

The pollutant load that will build up on impervious surfaces such as roads, roofs and driveways over the antecedent dry period is removed by stormwater runoff, as illustrated in Figure 6. The deposited particles undergo a number of intermediate processes that can be classified as redistribution (Figures 1 and 5) under the influence of traffic, periodic street sweeping and wind-driven forces. Particle redistribution occurs as resuspension, aggregation and redeposition.

Stormwater runoff will eventually contribute these pollutants to receiving waters. The amount of build-up pollutants removed by a single rainfall event is dependent on the rainfall characteristics. Traffic-generated pollutants

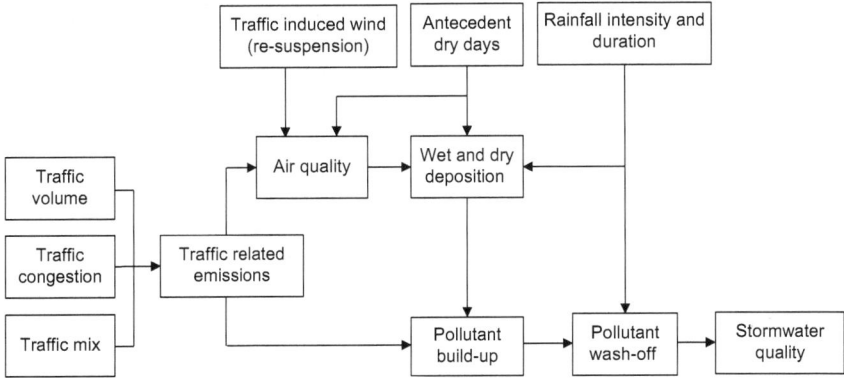

Figure 5 Pollutant processes, linkages and influential factors.

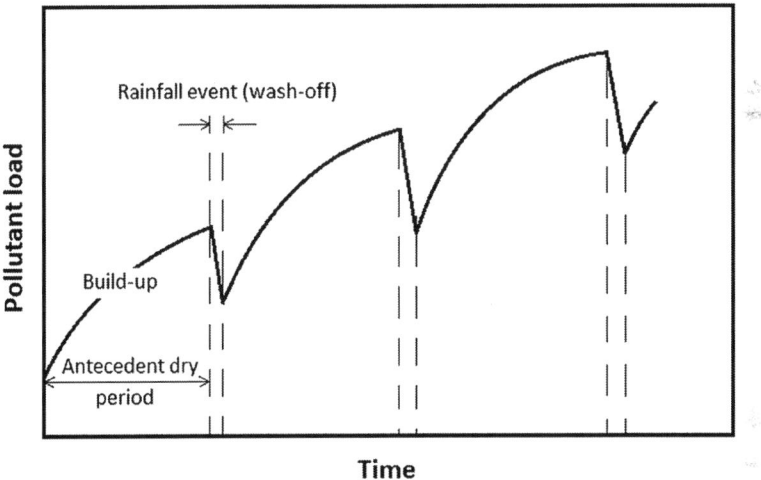

Figure 6 Hypothetical representation of pollutant build-up and wash-off processes (adapted from Sartor and Boyd[35]).

such as metals and hydrocarbons are commonly found in association with particulate solids. Whilst undergoing the processes of build-up and wash-off, the particle-bound nature of these pollutants can influence pollutant mobility, reactivity and toxicity.[36]

3.1 Pollutant Build-up

In the case of pollutant build-up, there is a dynamic equilibrium between pollutant deposition and removal. The load of pollutant build-up depends on the length of the antecedent dry period, the rate of deposition and the effects of redistribution over the antecedent dry period. The depositional and redistributional processes are influenced by both natural and anthropogenic factors, including climate factors such as the length of the antecedent dry period and wind speeds, traffic characteristics such as traffic volume and congestion, speed, traffic mix between diesel- and gasoline-powered vehicles and land use. In fact, traffic characteristics are related to land use. For example, diesel-powered heavy-duty vehicles are common in industrial areas, whereas gasoline-powered vehicular activities are typical to residential and commercial areas.

Pollutant load accumulated on road surfaces is a composite of a number of sources and can influence the build-up of traffic-related pollutants that are also present on the surface. Traffic-generated pollutants will interact with pollutants released from other sources (*e.g.* particulate solids, nutrients, organic matter from residential areas and metals and hydrocarbons from industrial activities) during build-up.

This can result in changes to the characteristics of the traffic pollutants. For example, Cu and Zn can combine with ligands containing nitrogen to form weak metal–nitrogen complexes,[37] which are relatively more soluble in water. The presence of organic matter from sources such as leaf litter and oxides of Fe, Mn and Al from geogenic sources such as roadside soil can influence chemisorption reactions between traffic-generated metals and soil particulates.[38] Similarly, the presence of finer particulates and clay-forming minerals contributed by roadside soils will influence the association of metals to particulates *via* ion exchange reactions.[28] These phenomena are discussed in further detail below. These changes, in turn, will define the bioavailability of pollutants in stormwater runoff, as the partitioning between dissolved and particulate fractions is based on the nature of the interactions with solid particles during adsorption.

3.2 Pollutant Wash-off

The mobilisation of pollutants adhering to an impervious surface is influenced by the kinetic energy of raindrops and the turbulence created by stormwater runoff. Although rainfall characteristics such as intensity, duration and runoff volume and flow velocity play significant roles in influencing the mechanisms of pollutant wash-off, the load and the characteristics of the pollutant built-up on the surface prior to a storm event primarily influence the pollutant load in the wash-off.[39] It is important to note that a rainfall event has the capacity to wash off only a fraction of the initially available pollutant load, and the fraction of wash-off would include a majority of fine particles in the initial build-up.[40] Therefore, changes to pollutant characteristics during build-up, as discussed above, will influence stormwater runoff quality. Consequently, the pollutant characteristics in wash-off would be different to the pollutant characteristics in build-up.

In relation to pollutant wash-off, the phenomenon of 'first flush' requires important consideration. The first flush has been identified as a significant and distinctive occurrence within the pollutant wash-off process. Essentially, the first flush produces relatively high pollutant concentrations at the early stages of a stormwater runoff event and a concentration peak prior to the stormwater flow peak.[41,42] This can result in 'shock' loads of pollutants to receiving water bodies, which can exceed their assimilative capacities. Therefore, in assessing the impact of traffic pollutants on receiving water bodies, it is not necessarily the average pollutant concentration that merits attention, but rather the first flush concentration.

3.3 Impact of Climate Change on Pollutant Processes

Climate change is predicted to exert significant impacts on rainfall patterns in most parts of the world.[43] Based on predicted changes to three key rainfall-related characteristics (as illustrated in Figure 7), the potential impacts of climate change on water and soil pollution due to traffic-generated pollutants can be hypothesised. The influential rainfall characteristics are:

- Increase in frequency of droughts, which would translate to greater antecedent dry periods between rainfall events;
- Increase in intensity of typical rainfall events in a particular area, which would translate to greater raindrop impacts on build-up pollutants and stormwater runoff turbulence;
- Decrease in rainfall duration, which would translate to opposite impacts to those listed due to the increase in rainfall intensity.[44,45]

An increased antecedent dry period would mean that there is greater opportunity for pollutants to build up on urban surfaces such as roads.

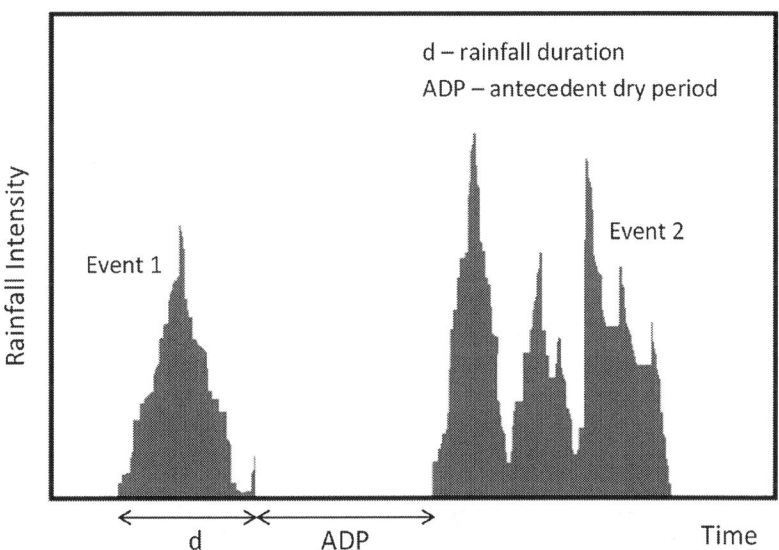

Figure 7 Changes in rainfall characteristics due to climate change.

Particulate solids carrying trace metals and hydrocarbons generated from automobile use activities will continue to accumulate on urban surfaces. At the same time, the impacts of vehicle movement and wind-driven forces on the pollutants accumulated on road surfaces will increase, resulting in increased pollutant redistribution and enhancing the pollution of roadside soil. Furthermore, the greater opportunity for redistribution and the abrasion by vehicle tyres will change pollutant characteristics, particularly the characteristics of particulate solids such as particle size distribution. This, in turn, will influence the adsorption of pollutants such as metals and hydrocarbons to particulates. Research has confirmed the greater preference of metals and hydrocarbons for adsorption to finer particles.[28,29,46] This is discussed in detail below.

Additionally, the predicted increase in ambient temperature due to climate change, compounded by longer residence times as build-up on urban surfaces, may influence various reactions between pollutants, such as adsorption, photolysis, oxidation–reduction and hydrolysis. These reactions can lead to the transformation of pollutants and to changes to the composition of the pollutant built-up on road surfaces available for wash-off with stormwater runoff. In this regard, photolysis of PAHs merits specific mention.[47–49] As PAHs are exposed to different forms of light (e.g. infrared, visible and ultraviolet) over a longer period of time, the molecules adsorb light energy and can transform into different species. Moreover, the excited molecules (molecules with elevated energy) of a particular species can react with the molecular forms of other pollutants and transform into different species.[50–52]

These changes to pollutant characteristics due to enhanced exposure to the influence of a number of natural and anthropogenic phenomena during build-up can result in potential changes to pollutant solubility, reactivity, bioavailability, toxicity and persistence. The changes, particularly to pollutant mobility, will influence the temporal variations in pollutant load and composition during build-up, which in turn will affect the wash-off of pollutants during rainfall events.[53,54]

The predicted increase in high-intensity rainfall events could increase the wash-off load of pollutants, primarily due to the increase in raindrop kinetic energy. This could be in the form of an increased first flush effect, as the impact of raindrop kinetic energy is influential at the initial stages. Later pollutant wash-off is more due to runoff turbulence as stormwater forms a sheet flow on the impervious surfaces and raindrop impact is cushioned. The changes to built-up pollutant characteristics as discussed above will dictate the composition of the pollutant load in the wash-off. However, it is important to note that the predicted decrease in the duration of rainfall events could play a counteracting role. It may influence the wash-off load due to the reduced contribution of coarser particles. This is due to the fact that coarser particles are primarily mobilised by the turbulent streams created by runoff. Therefore, the majority of coarser particles are washed off when the runoff develops with time.[40] Therefore, it is hypothesised that the compounding impact of an increase in rainfall intensity and a reduction in rainfall duration will mean an enhanced first flush effect, greater wash-off of finer particulates and reduced wash-off of coarser particulates.

3.4 Pollutant–Particulate Relationships and Mobility of Particle-bound Pollutants

Pollutant-particulate relationships essentially define the mobility of pollutants deposited on impervious surfaces. Traffic-generated metals and hydrocarbons are adsorbed by particulate solids through surface complexation. This occurs when ionic and molecular forms of metals and hydrocarbons interact with surface functional groups, which are chemically reactive molecular units protruding from particle surfaces.[55,56] The surface charge density of a particle increases with decreasing particle size due to the increase in particle-specific surface area.[57,58] Consequently, the concentration of trace metals and PAHs in road dust increases with decreasing particle size.

Another particle characteristic that influences pollutant adsorption is surface coatings such as hydrous metal oxides and organic matter. These coatings generate different electrical charges on the particle surface depending on the surface electrochemical properties (e.g. point of zero charge) and the particle mineralogical composition.[59] Thus, inorganic oxides such as iron and aluminium oxides have the highest impact on varying both positive and negative surface charge, while organic matter that produces negative charges specifically influences the adsorption of cationic forms of pollutants.

The physicochemical description of the relationship between particles and metals/hydrocarbons is different among different pollutants. For example, Cr, Cu, Zn, Ni, Hg and Pb loads have been reported to be different between different particle size fractions of road dust.[60,61] Therefore, variations of the mobility of these pollutants are expected during build-up and wash-off.

Moreover, strong relationships can be found between the mobility and amounts of particle-bound pollutants and how they are geochemically bound to particulate solids. For example, the geochemical origin of traffic-generated metals can be distinguished between exchangeable, bound to carbonates, bound to Fe–Mn oxides, bound to organic matter and residual fractions.[62] The order of mobility of these metals has been characterised as Cd > Zn > Pb > Co > Mn > Ni > Cu > Cr.[62–65] High mobility means that these metals are weakly bound to particulates.

Similarly to metals, different PAH concentrations and specific patterns of PAH distributions have been reported for different particle size fractions during build-up and wash-off.[66] For example, a particle size fraction of <0.7 μm has been found to contain the highest concentration of carcinogenic high-molecular-weight PAHs comprising four to six aromatic rings.[67]

4 Impacts of Traffic Pollutants

Road traffic releases a wide variety of pollutants that contaminate soil and water. Organic compounds and trace metals originating from fuel combustion, vehicle component wear and road surface wear are frequently found in various environmental compartments of air, soil, water and sediments.[11,68,69] Pollutants such as trace metals and PAHs are of special concern due to their ubiquity, potential toxicity, bioavailability and persistence in the environment. The study by Suman et al.[70] in Dhanabad City, India, found that soil in urban areas with high traffic densities had PAH concentrations that were more than six-times higher when compared to a control site in a rural area. Furthermore, in urban areas, more than 50% of the total PAHs consisted of higher-molecular-weight four- and five-ring species, whereas in rural areas, there was a predominance of low-molecular-weight three-ring PAHs. The higher-molecular-weight PAHs are relatively more toxic than lower-molecular-weight species. The presence of elevated concentrations of traffic-related metals in soils in urban areas was also reported by De Silva et al.[11] in a study in Melbourne, Australia.

Once entrained in stormwater runoff, pollutants associated with road dust become mobile and are the primary non-point source of pollution to receiving water bodies. The study undertaken by Van Metre et al.[71] provides a good case study of the role of stormwater runoff in the transport of traffic pollutants. The study outcomes revealed that hydrocarbon pollution in a number of water bodies was not solely due to urbanisation, as increased concentrations were also noted in areas where urbanisation was stable, but where there was an increase in traffic activity. The study by Liu et al.[69] found that intensive land use and associated traffic activities play significant roles in influencing the presence of high-molecular-weight PAHs in the sediments of Brisbane River in Australia. The sediment bed is a primary habitat for flora and fauna and disturbances such as flooding can release pollutants back into the water environment.

In the context of environmental pollution, dry and wet deposition are also significant. These processes are closely related with the pollutant build-up and wash-off processes described earlier. Both processes are very significant in cleaning the atmosphere: wet deposition as an episodic and orders of magnitude faster process, while dry deposition as a continuous and dependable process.[72] Another important effect of deposition is the transference of pollutants from air to water and/or soil. Atmospheric deposition is a significant source of potentially toxic substances in areas with anthropogenic perturbations of the natural biogeochemical cycles.

Traffic-generated pollutants are primarily confined to the region used by vehicles, including the road pavement and associated parking bays and a narrow roadside area on either side. Studies have found that there is a very rapid decline in pollutant deposition with distance from the road centre.[73–75] The distance of soil pollution from the road edge is influenced by a range of factors such as wind direction, wind speeds, vehicular speed, road age and traffic density,[11] with research studies reporting widely varying outcomes. Hewitt and Rashed[68] found that traffic pollutants are deposited within a distance of 50 m, whilst Sabin et al.[75] noted a distance of 150 m downwind from a freeway. In a review, Trombulak and Frissell[74] noted that soil pollution declines within about 20 m, but that elevated levels can occur 200 m or more from the road.

However, what is of most concern is the potential health risks associated with traffic pollutants due to the continuous flow of pollutants from the air to water and/or soil and the consequent impacts on the exposed population and ecosystems. Traffic-generated pollutants can harm human health as they are directly adsorbed into human lung tissues or through the consumption of products from terrestrial or aquatic ecosystems impacted by pollutant inputs *via* dry or wet deposition.[76] For example, the concentrations of Cu, Zn, Pb, Cd and Cr associated with road dust in Changsha City, Xiandao District, China, were found to represent an environmental risk, particularly in relation to Cd, with a mean concentration over 15-times higher than local environmental standards.[1] The significance of this finding needs to be viewed from the outcomes of the study by Murphy and Hutchinson,[77] showing that high Cd concentration in road dust can be linked to increased risk of rheumatoid arthritis in people living within 50 m of a highway. The presence of Pt, Pd and Rh (commonly used in catalytic converters in vehicles) in road dust in major cities in India has been found to be associated with high amounts of these metals being present in blood samples from the elderly population exposed to extreme traffic conditions.[78]

Furthermore, the uptake or deposition of traffic-generated pollutants in trees, grasses, crops, mosses and lichens, particularly near road environments, has been noted in research studies.[79–81] High concentrations of traffic-generated metals in roadside soil can be toxic to roadside flora and fauna. High concentrations of metals in mammals have been found to increase with proximity to roads, with the consequent accumulation in fatty tissue affecting the function of organs, nerves and endocrine systems.[82,83] Therefore, growing food crops in urban areas, particularly close to roads, could be a cause for concern. This issue is important in view of the fact that urban agriculture is being widely promoted in many countries for strengthening urban food security and as a sustainability initiative.

5 Conclusions

Vehicle ownership is expanding significantly across the world, resulting in an increase in road trips and traffic congestion and an ever-greater dependency on motor vehicles. This trend is expected to continue into the future due to the improvements in living standards in most parts of the world. Although the ownership of a vehicle can be considered to be a convenience, road traffic is associated with significant adverse human and ecosystem health impacts.

Vehicular traffic and road surfaces act synergistically to emit a range of pollutants into the environment. Road surfaces also act as reaction surfaces, enabling the deposition of traffic-generated pollutants as well as pollutants generated by anthropogenic and natural processes, which can react, forming new types of pollutants. These reactions can change the characteristics of traffic pollutants.

The primary sources of traffic pollutants are exhaust emissions, fuel and lubrication system leakages, component wear such as tyres, brakes and chassis and road surface wear. The pollutants are emitted into the atmosphere or directly deposited on the ground. The atmospheric pollutants eventually return as dry or wet deposition. Pollutants deposited on road surfaces are incorporated into road dust.

A range of traffic and road surface factors play a significant role in influencing the type, amount and rate of pollutant generation. The primary traffic factors include traffic volume, traffic congestion and traffic mix. It has been found that traffic congestion is more significant than traffic volume in terms of pollutant generation. The pollutant emissions from heavy-duty vehicles that use diesel are different compared to light-duty vehicles that use gasoline. The emission of hydrocarbons, particularly PAHs, in both the atmospheric phase and during build-up shows a strong correlation with heavy-duty traffic volume and traffic congestion. The road characteristics that play an influential role in pollutant generation are road layout including the locations of intersections, pedestrian crossings and profile, road surface characteristics, construction material used, condition, age and maintenance practices.

A very wide range of trace metals and hydrocarbons are emitted into the environment as a result of fuel combustion and vehicle component and road surface wear. The metals most commonly present on road surfaces are Cd, Cu, Cr, Ni, Zn, Pb, Al, Mn and Fe. Among these, Cd, Cu, Cr, Ni and Zn primarily originate from road traffic activities, whilst Al, Mn and Fe are primarily contributed by roadside soil, although they can also originate from vehicle component wear to a lesser extent. Pb builds up on road surfaces, stemming from accumulation in soils from past usage of leaded fuel, as well as ongoing contributions from the wear of vehicle components and road paint. Among hydrocarbons, PAHs are of most concern, as these compounds represent a significant class of suspected carcinogens. Used lubrication oil and incomplete combustion of fossil fuels, particularly diesel, are significant contributors of PAHs to the environment.

Traffic-generated pollutants are deposited on ground surfaces, contributing to soil and water pollution *via* direct deposition or dry or wet deposition and contributing to pollutant build-up during the dry period and wash-off with stormwater runoff. The pollutants accumulated on road surfaces are composites of a number of sources. Traffic-generated pollutants will interact with pollutants from other sources, which can result in changes to the characteristics of the pollutants, influencing their solubility, reactivity, bioavailability, mobility, toxicity and persistence. The load and the characteristics of the pollutants that have built up on the surface primarily influence the pollutant load in wash-off. Therefore, changes to pollutant characteristics during build-up will influence stormwater runoff quality. Additionally, the phenomenon of 'first flush' requires important consideration in relation to pollutant wash-off. It can result in the discharge of fluxes of high pollutant concentrations at the initial stage of a stormwater runoff event, which can exceed the assimilation capacity of the water body.

Climate change-driven alterations to climate characteristics will lead to changes to pollutant build-up and wash-off processes. Changes to the intervening dry period or antecedent dry period between successive rainfall events will influence the pollutant build-up load and composition. Changes to rainfall characteristics such as rainfall intensity and duration will influence the load and the characteristics of the pollutants transported by stormwater runoff.

As traffic-generated trace metals and hydrocarbons are adsorbed by particulates through surface complexation, the pollutant–particulate relationships essentially define the mobility of the pollutants deposited on impervious surfaces. The surface charge density, as influenced by particle size and surface coatings such as hydrous metal oxides and organic matter, are the key factors that dictate pollutant adsorption to solids. Similarly, different PAH concentrations and patterns of PAH distributions have been reported for different particle size fractions during build-up and wash-off.

Road traffic releases a wide range of trace metals and hydrocarbons that contaminate soil and water. Traffic-generated pollutants are primarily confined to the traffic area and a narrow roadside area on either side. Studies have found that there is a very rapid decline in pollutant deposition with distance from the road centre. The distance of soil pollution from the road edge is influenced by factors such as wind direction, wind speeds, vehicular speed, road age and traffic density. The human health implications of traffic pollutants are of significant concern. Research studies have confirmed the vulnerability of the population living near high-traffic areas, particularly the elderly. An equal concern is the uptake of trace metals by plants near road environments. Therefore, growing food crops in urban areas, particularly close to roads, needs to be viewed with caution.

References

1. J. Huang, F. Li, G. Zeng, W. Liu, X. Huang, Z. Xiao, H. Wu, Y. Gu, X. Li, X. He and Y. He, *Sci. Total Environ.*, 2016, **541**, 969–976.
2. J. Gunawardena, P. Egodawatta, G. A. Ayoko and A. Goonetilleke, *Atmos. Environ.*, 2012, **54**, 502–510.
3. J. Gunawardena, P. Egodawatta, G. A. Ayoko and A. Goonetilleke, *Atmos. Environ.*, 2013, **68**, 235–242.
4. J. Gunawardena, A. M. Ziyath, P. Egodawatta, G. A. Ayoko and A. Goonetilleke, *Chemosphere*, 2014, **99**, 267–271.
5. A. K. Namdeo, J. J. Colls and C. J. Baker, *Sci. Total Environ.*, 1999, **235**, 3–13.

6. P. Mahbub, A. Goonetilleke, P. K. Egodawatta, T. Yigitcanlar and G. A. Ayoko, *Water Sci. Technol.*, 2011, **63**, 2077–2085.
7. AATSE, *Urban Air Pollution in Australia*, Australian Academy of Technological Sciences and Engineering, Victoria, Australia, 1997.
8. S. M. P. B. Mahbub, Queensland University of Technology, 2011.
9. J. Gunawardena, Queensland University of Technology, 2012.
10. A. Thorpe and R. M. Harrison, *Sci. Total Environ.*, 2008, **400**, 270–282.
11. S. De Silva, A. S. Ball, T. Huynh and S. M. Reichman, *Environ. Pollut.*, 2016, **208, Part A**, 102–109.
12. S. W. N. Mummullage, Queensland University of Technology, 2015.
13. M. Sutton, U. Dragosits, Y. Tang and D. Fowler, *Atmos. Environ.*, 2000, **34**, 855–869.
14. D. Grobler and M. Silberbauer, *Water Res.*, 1985, **19**, 975–981.
15. M. C. Lim, G. A. Ayoko, L. Morawska, Z. D. Ristovski, E. R. Jayaratne and S. Kokot, *Atmos. Environ.*, 2006, **40**, 3111–3122.
16. P. Egodawatta, A. M. Ziyath and A. Goonetilleke, *Environ. Pollut.*, 2013, **176**, 87–91.
17. S. Mummullage, P. Egodawatta, G. A. Ayoko and A. Goonetilleke, *Sci. Total Environ.*, 2016, **541**, 1303–1309.
18. K. Adachi and Y. Tainosho, *Environ. Int.*, 2004, **30**, 1009–1017.
19. NICNAS, *Methylcyclopentadienyl Manganese Tricarbonl (MMT), Priority Existing Chemical Assesment* 24, National Industrial Chemical Notification and Assesment Scheme, Sydney, NSW, Australia, 2003.
20. M. A. Palacios, M. Gómez, M. Moldovan and B. Gómez, *Microchem. J.*, 2000, **67**, 105–113.
21. P. Kennedy and J. Gadd, *Prepared for Ministry of Transport, New Zealand, Infrastructure Auckland*, 2003.
22. M. Huber, A. Welker and B. Helmreich, *Sci. Total Environ.*, 2016, **541**, 895–919.
23. S. Fatemi, M. K. Varkani, Z. Ranjbar and S. Bastani, *Prog. Org. Coat.*, 2006, **55**, 337–344.
24. F. Tiarks, T. Frechen, S. Kirsch, J. Leuninger, M. Melan, A. Pfau, F. Richter, B. Schuler and C. L. Zhao, *Prog. Org. Coat.*, 2003, **48**, 140–152.
25. J. Marsalek and P. Marsalek, *Water Sci. Technol.*, 1997, **36**, 117–122.
26. G. Morrison, D. Revitt and J. Ellis, *Water Sci. Technol.*, 1990, **22**, 53–60.
27. D. Revitt, R. Hamilton and R. Warren, *Sci. Total Environ.*, 1990, **93**, 359–373.
28. C. Gunawardana, P. Egodawatta and A. Goonetilleke, *Environ. Pollut.*, 2014, **184**, 44–53.
29. C. Gunawardana, P. Egodawatta and A. Goonetilleke, *Chemosphere*, 2015, **119**, 1391–1398.
30. E. Manoli and C. Samara, *TrAC, Trends Anal. Chem.*, 1999, **18**, 417–428.
31. J. N. Brown and B. M. Peake, *Sci. Total Environ.*, 2006, **359**, 145–155.
32. J. S. Latimer, E. J. Hoffman, G. Hoffman, J. L. Fasching and J. G. Quinn, *Water, Air, Soil Pollut.*, 1990, **52**, 1–21.
33. I. J. Keyte, A. Albinet and R. M. Harrison, *Sci. Total Environ.*, 2016, **566**, 1131–1142.

34. S. Mummullage, P. Egodawatta, G. A. Ayoko and A. Goonetilleke, *Environ. Pollut.*, 2016, **216**, 80–85.
35. J. D. Sartor and G. B. Boyd, *Water Pollution Aspects of street Surface Contaminants* U.S. Environmental Protection Agency, Washington, D.C., 1972.
36. B. A. Dempsey, Y.-L. Tai and S. Harrison, *Water Sci. Technol.*, 1993, **28**, 225–230.
37. A. M. Ziyath, P. Egodawatta and A. Goonetilleke, *Ecotoxicol. Environ. Saf.*, 2016, **127**, 193–198.
38. C. Gunawardana, P. Egodawatta and A. Goonetilleke, *Chemosphere*, 2015, **119**, 1391–1398.
39. H. P. Duncan, *A Review of Urban Stormwater Quality Processes*, Coorporative Research Centre for Catchment Hydrology, 1995.
40. H. Zhao, X. Chen, S. Hao, Y. Jiang, J. Zhao, C. Zou and W. Xie, *Sci. Total Environ.*, 2016, **563–564**, 62–70.
41. N. Alias, A. Liu, A. Goonetilleke and P. Egodawatta, *Ecol. Eng.*, 2014, **64**, 301–305.
42. J. H. Lee, K. W. Bang, L. H. Ketchum Jr, J. S. Choe and M. J. Yu, *Sci. Total Environ.*, 2002, **293**, 163–175.
43. IPCC, *Climate Change and Water, IPCC Technical Paper VI*, Intergovernmental Panel on Climate Change, Geneva, 2008.
44. AGO, *Climate Change: An Australian Guide to the Science and Potential Impacts*, Australian Greenhouse Office, Canberra, Australia, 2003.
45. I. Delpla, A. V. Jung, E. Baures, M. Clement and O. Thomas, *Environ. Int.*, 2009, **35**, 1225–1233.
46. L. F. Herngren, Queensland University of Technology, 2005.
47. H. Jia, J. Zhao, L. Li, X. Li and C. Wang, *Appl. Catal., B*, 2014, **154**, 238–245.
48. H. Jia, L. Li, H. Chen, Y. Zhao, X. Li and C. Wang, *J. Hazard. Mater.*, 2015, **287**, 16–23.
49. J. Niu, P. Sun and K.-W. Schramm, *J. Photochem. Photobiol., A*, 2007, **186**, 93–98.
50. T. D. Behymer and R. A. Hites, *Environ. Sci. Technol.*, 1985, **19**, 1004–1006.
51. J. S. Miller and D. Olejnik, *Water Res.*, 2001, **35**, 233–243.
52. L. Pirjola, T. Lähde, J. V. Niemi, A. Kousa, T. Rönkkö, P. Karjalainen, J. Keskinen, A. Frey and R. Hillamo, *Atmos. Environ.*, 2012, **63**, 156–167.
53. B. Wijesiri, P. Egodawatta, J. McGree and A. Goonetilleke, *Sci. Total Environ.*, 2015, **518–519**, 434–440.
54. B. Wijesiri, P. Egodawatta, J. McGree and A. Goonetilleke, *Sci. Total Environ.*, 2015, **527–528**, 344–350.
55. D. L. Sparks, *Environmental Soil Chemistry*, Access Online via Elsevier, 2003.
56. G. Sposito, *The Chemistry of Soils*, Oxford University Press, 2008.
57. C. Cristina, J. Tramonte and J. Sansalone, *Water, Air, Soil Pollut.*, 2002, **136**, 33–53.
58. C. Gunawardana, A. Goonetilleke, P. Egodawatta, L. Dawes and S. Kokot, *J. Environ. Eng.*, 2012, **138**, 490–498.
59. X. N. Zhang and A. Z. Zhao, in *Chemistry of Variable Charge Soils*, ed. T. R. Yu, Oxford University Press, 1997.

60. L. Herngren, A. Goonetilleke and G. A. Ayoko, *Anal. Chim. Acta*, 2006, **571**, 270–278.
61. M. Murakami, F. Nakajima and H. Furumai, *Chemosphere*, 2005, **61**, 783–791.
62. B. Wijesiri, P. Egodawatta, J. McGree and A. Goonetilleke, *Water Res.*, 2016, **101**, 582–596.
63. X. Li, C.-s. Poon and P. S. Liu, *Appl. Geochem.*, 2001, **16**, 1361–1368.
64. E. Manno, D. Varrica and G. Dongarrà, *Atmos. Environ.*, 2006, **40**, 5929–5941.
65. Ş. Tokalıoğlu and Ş. Kartal, *Atmos. Environ.*, 2006, **40**, 2797–2805.
66. M. Murakami, F. Nakajima and H. Furumai, *Water Res.*, 2004, **38**, 4475–4483.
67. K. Nielsen, Y. Kalmykova, A.-M. Strömvall, A. Baun and E. Eriksson, *Sci. Total Environ.*, 2015, **532**, 103–111.
68. C. N. Hewitt and M. B. Rashed, *Sci. Total Environ.*, 1990, **93**, 375–384.
69. A. Liu, G. O. Duodu, S. Mummullage, G. A. Ayoko and A. Goonetilleke, *Environ. Pollut.*, 2017, **223**, 81–89.
70. S. Suman, A. Sinha and A. Tarafdar, *Sci. Total Environ.*, 2016, **545**, 353–360.
71. P. C. Van Metre, B. J. Mahler and E. T. Furlong, *Environ. Sci. Technol.*, 2000, **34**, 4064–4070.
72. Y. Pan and Y. Wang, *Atmos. Chem. Phys.*, 2015, **15**, 951–972.
73. H. Zhang, Z. Wang, Y. Zhang, M. Ding and L. Li, *Sci. Total Environ.*, 2015, **521**, 160–172.
74. S. C. Trombulak and C. A. Frissell, *Conserv. Biol.*, 2000, **14**, 18–30.
75. L. D. Sabin, J. Hee Lim, M. Teresa Venezia, A. M. Winer, K. C. Schiff and K. D. Stolzenbach, *Atmos. Environ.*, 2006, **40**, 7528–7538.
76. S. M. Mohan, *Int. J. Environ. Sci. Technol.*, 2016, **13**, 387–402.
77. D. Murphy and D. Hutchinson, *J. Inflamm.*, 2015, **12**, 58.
78. V. Balaram, C. Kamala, A. S. Rao, M. Satyanarayanan, K. Subramanyam and S. Sawant, *Geostatistical and Geospatial Approaches for the Characterization of Natural Resources in the Environment*, Springer, 2016, pp. 415–420.
79. H. Zhang, Y. Zhang, Z. Wang, M. Ding, Y. Jiang and Z. Xie, *Sci. Total Environ.*, 2016, **573**, 915–923.
80. C. L. Wiseman, F. Zereini and W. Püttmann, *Sci. Total Environ.*, 2013, **442**, 86–95.
81. S. Vingiani, F. De Nicola, W. Purvis, E. Concha-Graña, S. Muniategui-Lorenzo, P. López-Mahía, S. Giordano and P. Adamo, *Water, Air, Soil Pollut.*, 2015, **226**, 240.
82. B. Bocca, A. Alimonti, F. Petrucci, N. Violante, G. Sancesario, G. Forte and O. Senofonte, *Spectrochim. Acta, Part B*, 2004, **59**, 559–566.
83. T. Hamers, L. A. M. Smit, A. T. C. Bosveld, J. H. J. van den Berg, J. H. Koeman, F. J. van Schooten and A. J. Murk, *Arch. Environ. Contam. Toxicol.*, 2002, **43**, 0345–0355.

Cardiovascular Health Effects of Road Traffic Noise

ANNA HANSELL,* YUTONG SAMUEL CAI AND JOHN GULLIVER

ABSTRACT

The most common source of environmental noise is road traffic. Environmental noise is being increasingly recognised as being associated with long-term impacts on health, in addition to causing annoyance and sleep disturbance. In recent years, a lot of progress has been made in estimating population-level exposures to road traffic noise that permit large-scale epidemiological studies. While studies are not fully consistent, a relatively small number have been conducted to date. For hypertension and ischaemic heart disease, where enough studies have been conducted for combination in a meta-analysis, small but statistically significant associations showing increasing risks with increasing exposure to noise have been demonstrated. Further evidence from longitudinal studies is needed in order to refine estimates and also to examine potential mechanisms and intermediate outcomes such as inflammatory biomarkers and atherosclerosis.

1 Introduction

Environmental noise is well recognised as causing annoyance and disturbing sleep, but there is increasing awareness of the possibility that there may be long-term impacts on health.[1] The major source of environmental noise is from transport sources (aircraft, rail and road), with road noise

*Corresponding author.

being the most ubiquitous. In the European Union, about 56 million people (54%) in areas with >250 000 inhabitants are exposed to road traffic noise of >L_{den} 55 dB per year[2] (where L_{den} is an average noise level with 5 dB penalty weightings for evening noise and 10 dB penalty weightings for night noise). The day–night level of 55 dB is generally held to be the level at which chronic health effects are likely to become apparent,[1] although this may also be because accurate estimation of community noise exposures below this level become more difficult, as discussed in Section 2 below. A maximum recommended exposure level of 55 dB for night-time average noise measured outside ($L_{Aeq,outside}$) has also been identified by the World Health Organization (WHO),[3] with a long-term goal for this to be 40 dB, as nocturnal noise has been recognised as being more important in terms of potential impacts on health and well-being.[4]

The evidence on health effects of noise is limited, especially by comparison with environmental exposures to air pollutants, but there have been increasing numbers of publications in this area in recent years. The WHO published in 2009 'Night Noise Guidelines for Europe'[3] and in 2011 'Burden of Disease from Environmental Noise'.[4] A regional update of the 1999 WHO Community Noise Guidelines is currently in progress, aiming to provide WHO Environmental Noise Guidelines for the European Region, expected in 2017. These will be based on systematic reviews of the literature and meta-analyses where possible, related to a range of outcomes, including cognitive development, sleep and annoyance, and on different types of noise, including transport and also that from wind farms. The majority of evidence that is currently available on the health effects of noise relates to cardiovascular disease in adults in relation to road traffic noise, and that is the focus of this narrative review.

1.1 Biological Mechanisms

There is good evidence for acute effects of noise on the body, especially the cardiovascular system. These are mediated through direct impacts on the autonomic nervous system and endocrine system and indirect effects on these systems *via* sleep disturbance and annoyance (Figure 1). Experimentally, this is associated with increases in adrenaline and noradrenaline (the 'fight–flight' mechanism) and plasma cortisol (the 'defeat' mechanism).[5] These result in short-term increases in heart rate, blood pressure and cardiac output, as well as changes in blood lipids, inflammatory markers and other factors. These reactions appear to be independent of cognitive perception of noise involving subcortical parts of the brain. Noise occurring during sleep is recognised and evaluated at subcortical levels, with a gating function undertaken by the thalamus. If information is passed onto the cortex, it causes arousals that may not result in full awakening, but may nevertheless cause increases in heart rate and blood pressure, disturbing normal circadian rhythms and fragmenting sleep.[6] There is some habituation to noise impacts on sleep if exposure occurs on a regular basis, but the

Cardiovascular Health Effects of Road Traffic Noise

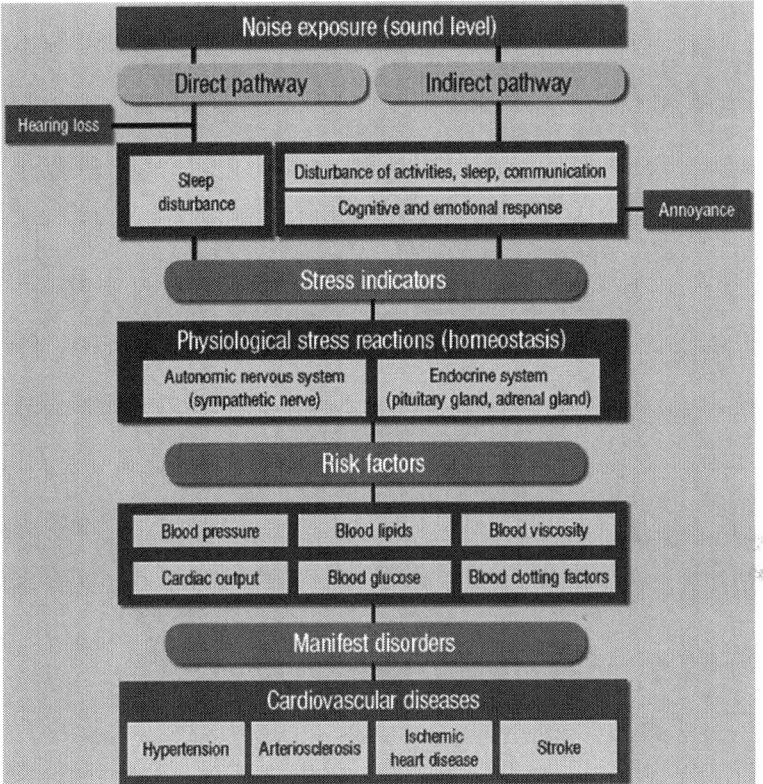

Figure 1 Noise effects reaction scheme from Munzel et al. Adapted with permission from Babisch, 2014.
Reproduced with permission from Oxford University Press on behalf of T. Munzel, T. Gori, W. Babisch and M. Basner, Cardiovascular effects of environmental noise exposure, *Eur. Heart J.*, 2014, **35**, 829–836 and Professor W. Babisch, Updated exposure-response relationship between road traffic noise and coronary heart diseases: a meta-analysis, *Noise Health*, 2014, **16**(68), 1–9.

extent to which this occurs is likely to vary between individuals.[5,7] Also, autonomic arousals habituate to a lesser degree than cortical arousals.[8] Different types of transport noise appear to have different impacts on sleep—in a laboratory study of 72 individuals, road noise was found to have the largest impact on sleep structure and continuity, but rail and aircraft noise were perceived as most annoying.[8] This may be because the traffic noise events were shorter than the aircraft and rail noise events, and therefore not consciously perceived by the individuals in the study.[6]

The biological mechanisms by which chronic exposure to noise results in physiological impacts are likely to be *via* the same mechanisms as those for acute reactions. However, habituation and also the psychological processing of noise (*e.g.* annoyance reactions) make experimental studies more difficult to perform. A very small number of studies are available on the intermediate outcomes and biological pathways of noise, and these are discussed further below.

2 Assessment of Traffic Noise Exposure in Epidemiological Studies

Epidemiological studies typically involve very large numbers of individuals, and it is not currently practical to measure personal noise exposures, particularly over long periods, in order to examine chronic impacts on health. Unlike air pollution, there are no national networks available for monitoring noise exposure. Therefore, most epidemiological studies of environmental exposures such as noise rely on modelled exposure, typically given as daytime and night-time average sound levels (*e.g.* L_{day} and L_{night}) or weighted day–night averages (*e.g.* L_{den}). An advantage of using a modelling approach is that it is possible to model noise from a particular source (*e.g.* road traffic, railway or aircraft noise), which may be important both biologically and from a policy perspective. An understanding of the way in which noise is modelled is important for the design and interpretation of health studies.

Models of road traffic noise have been in existence for several decades, but their application to exposure assessment has mostly taken place in the last decade as interest in the health effects of noise has grown, imbued by advancements in computer processing power that have allowed exposure studies to be undertaken over large geographical areas.[9,10] Noise modelling broadly comprises four steps in order to estimate, from one or more roads sources, the sound pressure level (SPL) in decibels at a receiver (*e.g.* an address in exposure assessments):

(1) Definition of the source emission by characterisation of the vehicle fleet (type and speed) in terms of noise frequency (Hz) in order to determine the 'basic noise level' (SPL) and how this in turn is modulated by road gradient and road surface (*i.e.* pervious or impervious);
(2) Noise propagation, which is variably attenuated depending on land cover, terrain and the maximum building height and width (diffraction) along the line of sight between the source (road section) and receiver (point location close to a building facade);
(3) An ordered series (first order: first reflection; second order: reflection from a second surface, and so on) of reflections from surfaces intersecting each line of sight between the source and receiver;
(4) The integrated SPL at a receiver from the combination (by incoherent sum) of noise propagated along line(s) of sight between each source and the receiver, as well as the received reflected noise.

Noise models use the SPL from the above steps to estimate A-weighted noise, the noise metric covering the audio range related to human hearing

(20 Hz to 20 kHz). Other metrics such as C-weighting (to reveal the peak SPL) and Z-weighting (no weighting of the initial SPL, similar to that of L for 'linear weighting') are potentially of interest, but are seldom used in epidemiological studies. Various noise metrics can be produced from A-weighted values of L (SPL), often expressed as the 'A-weighted equivalent value' (L_{Aeq}) for each hour of the day ($L_{Aeq1\,h}$) and then averaged to provide, for example, L_{DAY} (07.00–19.00), $L_{EVENING}$ (19.00–23.00), L_{NIGHT} (23.00–07.00) and $L_{Aeq16\,h}$ (07.00–23.00).

Proprietary noise models (*e.g.* SoundPLAN[11] and CadnaA[12]), mostly designed for engineering/planning assessments and later to produce strategic noise maps (*e.g.* as part of the European Noise Directive (END)[13]), provide an advanced level of noise modelling to account for the abovementioned steps (often in a very detailed manner), and these models have been applied in some epidemiological studies.[9,14] Current versions of most proprietary noise models include a range of national methods that share some similarities and sometimes more substantial differences. Most methods are, for example, similar in terms of the geographical inputs that are used to run models and the calculation of the basic noise level (source emission), whereas the methods for propagation tend to vary; for example, the UK Calculation of Road Traffic Noise (CoRTN) method[15] accounts only for single reflections due to buildings opposite each receiver and does not account for meteorology, compared to the Nord2000 (Scandinavia) method, which includes a solution for multiple-order reflections, and wind and temperature gradient are used to approximate the vertical effective sound speed profile by a log-linear relationship. A more complex model does not, however, always improve accuracy; it depends on the characteristics of the study area. A detailed comparison between eight major national/international methods (ASJ RTN [Japan], CoRTN [UK], FHWA [USA], HARMONOISE [predecessor to CNOSSOS-EU], NMBP-Routes [France], Nord2000 [Scandinavia], RLS90 [Germany] and SonRoad [Switzerland]) is provided in Garg and Maji.[16]

Different noise models may not be directly comparable and, if combining geographic areas where noise has been estimated using different models, there may be apparent step changes in sound levels where maps join. Recently, CNOSSOS-EU[17] has been developed in order to have a common methodology for noise modelling across Europe (formerly HARMONOISE), allowing results from different countries to be directly comparable, rather than having individual countries producing assessments with differently parameterised models. This harmonisation of methods is a requirement of the European Directive on the Assessment and Management of Environmental Noise (2002/49/EC) (END).

Notwithstanding substantial improvements in computer processing power and the increasing availability of spatially resolved geographical data, proprietary models that apply one or more of the national/international methods are usually limited in an exposure context to cities/regions, which may still require months of computer processing in order to achieve noise estimation at between hundreds of thousands to several million receivers.

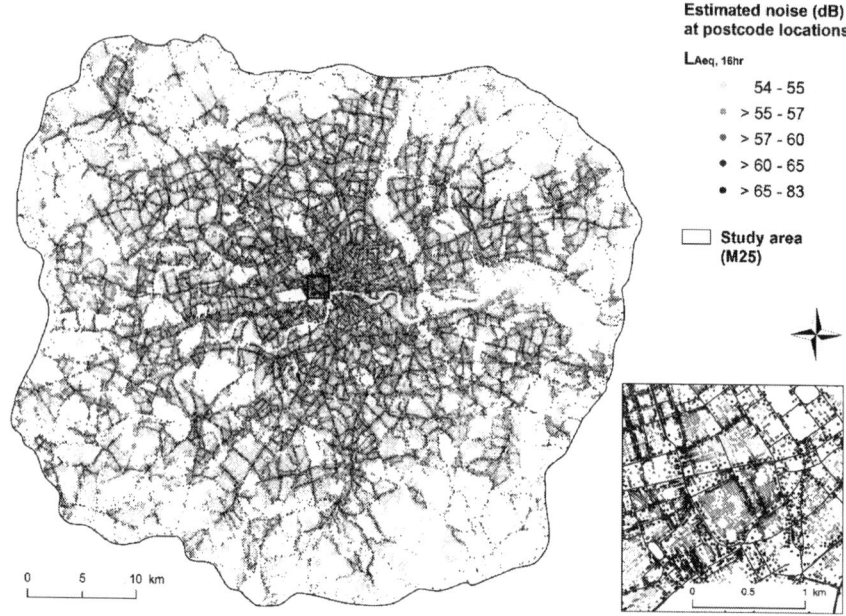

Figure 2 Predicted postcode noise levels from TRAffic Noise EXposure (TRANEX). Reproduced with permission from the Elsevier Publishing and J. Gulliver, D. Morley, D. Vienneau, F. Fabbri, M. Bell, P. Goodman et al., Development of an open-source road traffic noise model for exposure assessment, Environ. Model. Softw., 2015, **74**, 183–193. Sharing and adaptation of this figure is covered under a creative commons license. https://creativecommons.org/licenses/by/4.0/.

Some noise models can be applied to larger areas/numbers of receivers by lowering the geographical resolution of inputs and reducing the number of propagation terms by, for example, excluding reflections, but this is sometimes at the expense of reduced accuracy in noise predictions. Due to its relatively simple treatment of reflections, a full implementation of CoRTN was developed (TRANEX: TRAffic Noise EXposure) specifically with the aim of modelling noise exposures at up to several million address locations in London, UK,[18] which has been subsequently used (Figure 2) for epidemiologic analysis.[19–21] The software was developed in an open-source geographical information system and is freely available.[22]

There is a growing interest in estimating noise exposures in national, 'administrative' cohorts as well as international studies undertaking noise assessment across different cohorts (*e.g.* pooled analysis). This is also of interest in air pollution studies where noise is treated as a potential confounder. The interest in this type of analysis creates challenges in modifying elements of noise models so that they can be run over very large geographical areas whilst retaining the accuracy of noise predictions. One such example is a version of the CNOSSOS-EU methodology (Figure 3) that was

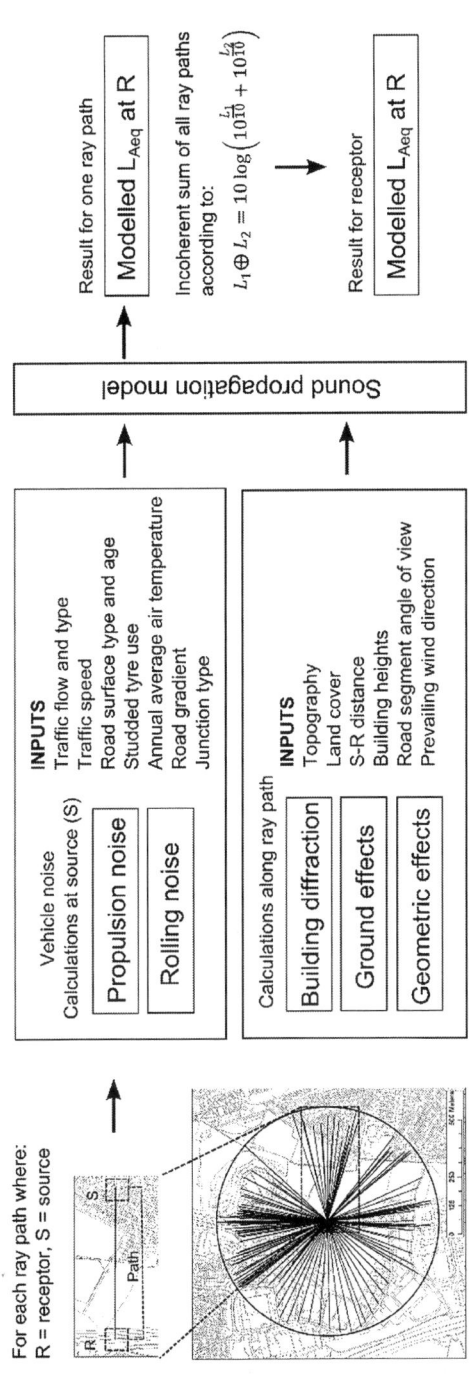

Figure 3 Flow diagram from CNOSSOS-EU showing the steps in noise modelling. Reproduced from D. W. Morley, K. de Hoogh, D. Fecht, F. Fabbri, M. Bell, P. S. Goodman, P. Elliott, S. Hodgson, A. L. Hansell and J. Gulliver, International scale implementation of the CNOSSOS-EU road traffic noise prediction model for epidemiological studies, *Environ. Pollut.*, **206**, 332–341. Copyright (2015) with permission from Elsevier.

applied with a lower resolution of geographical inputs (mostly a degradation of detail in land cover whilst retaining a relatively accurate representation of the road geography and traffic flows) in order to be able to produce an international-scale (UK, The Netherlands and Norway) noise exposure assessment for the EU BioSHaRe project.[10] This has subsequently contributed noise exposures to a number of epidemiological studies.[23,24] A detailed noise model might, for example, provide average noise level errors in the range 1–2 dB(A) with an international-scale noise model having errors in the range 3–5 dB(A); however, this is based on a very limited number of model evaluation studies that have been undertaken in exposure studies,[9,10,25–27] as noise, unlike air pollution, is not routinely measured.

Noise modelling is not confined to deterministic models. In recent years, there have been a number of noise models developed from (stochastic) land use regression (LUR) modelling,[28,29] a method that is widely adopted for air pollution exposure studies.[25,30] Whilst being relatively simple to implement, LUR modelling requires bespoke monitoring of noise level measurement campaigns at tens to hundreds of sites, depending on the size of the study area, in order to derive and evaluate models. Emerging low-cost monitors designed for creating networks may facilitate the further use of noise LUR modelling in exposure studies.[31]

Despite efforts to run a detailed noise model with highly resolved traffic information and geographical data, there is a great deal of potential for exposure misclassification in epidemiological studies. Many studies have to rely on using a single receiver per dwelling (address location) due to the time taken to process large numbers of receivers over large geographical areas. This may lead to variable under- and over-estimation of exposures where there is variability in noise levels between building facades at the same dwelling; this is particularly noticeable for dwellings close to main roads, where noise level gradients tend to be greatest between the fronts and rears of dwellings, especially in street canyons. Where it is computationally possible to produce a noise exposure based on noise estimates at multiple facades, the average exposure may be better determined, but determining which facade noise level is the most relevant exposure provides an even greater challenge and requires insight into the configuration of the dwelling (*e.g.* on which facade is the bedroom if night-time noise is the focus). A further issue is that estimates of noise at the facade do not directly relate to the outdoor noise levels that become variably attenuated indoors (*e.g.* due to number and type [*e.g.* double glazing] of windows, orientation of dwellings, building materials, absorption properties including furnishings and insulation, *etc.*). Proprietary noise models can provide 'building modules' in order to determine levels of indoor noise, but this is currently impractical to apply to epidemiological studies, given the potential for variability in the factors that determine indoor noise levels. Other sources of uncertainty include poorly captured diurnal and seasonal traffic patterns and the effects of meteorology (wind direction). Misclassification of individual noise exposures where random and not systematically linked to health outcomes is likely to reduce the precision of estimates, therefore making detection of small increases in relative risk (RR) difficult in epidemiological analyses.

An equally important consideration is misclassification due to the lack of traffic on minor roads, which does not necessarily limit missing traffic information at low flows (*e.g.* <500 vehicles in terms of annual average daily traffic). Many roads with more substantial traffic flows are missing from the traffic data sets used to run noise models, particularly moderately trafficked roads (*e.g.* 1000–5000 vehicles in terms of annual average daily traffic). This problem is especially acute in national-scale exposure studies, where traffic information tends to be sparser. The result is often noise distributions where the majority of the address locations are found in the same noise category (*e.g.* 50 to <55 dB). The impact of this on epidemiological analyses would be to constrain estimation of linear exposure response (due to differential misclassification at the lower end of the exposure range). However, many epidemiological studies also include analyses by predefined noise category, often with the lowest categories being <55 dB or <50 dB, as well as linear estimation, which helps to address this. Methods to estimate traffic on minor roads and to improve the characterisation of exposure related to minor roads are, however, being developed.[32]

Despite the potential sources of misclassification, the capacity for noise modelling in epidemiological studies continues to improve, and international-scale studies to allow for harmonised exposure assessment are possible, which allows very large studies to be conducted.[23,24]

3 Health Studies of Cardiovascular Disease in Adults

3.1 Hypertension

Acute exposure to road traffic noise is linked to transient increases of both systolic blood pressure (SBP) and diastolic blood pressure (DBP),[33] which may potentially lead to hypertension in a chronic timeframe of noise exposure. It is therefore not surprising that the largest number of studies to date on associations between noise and cardiovascular health relate to hypertension. An influential systematic review was published by van Kempen and Babisch in 2012, in which 45 effect estimates from 24 cross-sectional studies published up to the year 2010 were included for a meta-analysis.[34] The authors reported an odds ratio (OR) of 1.034 (95% confidence interval [CI]: 1.011–1.056)[†] for each 5 dB(A) increase of the 16 hour average road traffic noise ($L_{Aeq16\,h}$) (Figure 1). This pooled estimate was based on results from studies in which noise levels ranged from 45 to 75 dB(A). For each 5 dB(A) increase of $L_{Aeq16\,h}$, a positive significant association was seen in men (OR: 1.056, 95% CI: 1.012–1.103), but not in women, and in those aged less than 60 years (OR: 1.062, 95% CI: 1.033–1.092), but not in those aged 60 years and older. However, heterogeneity across the studies was also

---

[†]An odds ratio or hazard ratio of 1.034 implies a 3.4% increase in risk relative to the control (low-exposure group). If the lower 95% CI exceeds 1.00, the association is regarded as significant.

revealed in this meta-analysis. Important sources of heterogeneity were age, sex and assessment approaches to road noise exposure. Studies using model-based noise estimates also showed a pooled positive significant association with prevalent hypertension (OR: 1.032, 95% CI: 1.009–1.055), compared to the null association based on studies using measured noise estimates. The authors speculated that studies using model-based noise estimates may have better captured the chronic exposure to road traffic noise than limited measured noise that may include noise from other sources and introduce exposure misclassification.

In the KORA (Cooperative Health Research in the Augsburg Region) study of Augsburg areas, Germany, a 10 dB(A) increase of L_{dn} (weighted day–night average noise level) was positively associated with prevalent hypertension (OR: 1.16, 95% CI: 1.00–1.35) in the city of Augsburg population, but not in the Greater Augsburg area.[35] This may in part be due to the fact that the quality of noise assessment was more robust in the city of Augsburg. However, this association was slightly attenuated (OR: 1.11, 95% CI: 0.94–1.30) after further adjustment for $PM_{2.5}$ (particulate matter with a diameter of less than 2.5 µm). In contrast to this, the association between L_{dn} and isolated systolic hypertension (defined as SBP $\geq$ 140 mmHg and DBP <90 mmHg and not using antihypertensive medication) was robust to adjustment for $PM_{2.5}$ (OR: 1.43, 95% CI: 1.10–1.86) in the city of Augsburg populations. In a study in the city of Berlin, Germany, a 10 dB(A) increase of L_{den} (weighted day–evening–night average noise level) was significantly associated with prevalent hypertension (OR: 1.112, 95% CI: 1.004–1.231).[36] This association was independent of background air pollution. Interestingly, this study reported a stronger association with exposure in the living room during daytime (OR: 1.237, 95% CI: 1.083–1.413), but there was a null association with exposure in the bedroom during night-time; the authors speculated that activities during daytime may also play a role in the pathway between road traffic noise and hypertension, which needs further investigation.

Most of the studies included in the van Kempen and Babisch review did not take into account the potential confounding role of air pollution, which has also been associated with hypertension.[37] A few more recent noise studies have simultaneously investigated the effects of both exposures on hypertension or blood pressure levels, which in general support the notion that both road traffic noise and air pollution play independent roles in hypertension.[38]

The association between long-term road traffic noise exposures and prevalent hypertension is clearer than that with actual blood pressure levels, potentially because of problems accounting for antihypertensive medication use. In the Heinz Nixdorf Recall Study, Germany, long-term exposure to L_{den} > 60 dB(A) was linked to non-significantly elevated SBP and DBP compared to those exposed to a level of <55 dB(A), adjusting for both short- and long-term PM exposure.[39] An analysis of three European cohorts with a total of 88 336 individuals found no associations between L_{den} road noise and SBP and an unexpected small protective association with DBP, but the results differed by cohort, with adverse effects in one cohort, no effects in a second and protective effects in a third.[23] Results were adjusted for air pollution exposures.

The Hypertension and Exposure to Noise near Airports (HYENA) study included an examination of average road traffic noise ($L_{Aeq24\ h}$) in relation to the use of antihypertensive medication and found a null overall association.[40] When stratifying the association by sex, a positive but non-significant association was seen for men and a negative association was seen for women. However, this more pronounced effect in men has been not observed in some other studies.[41,42]

All of the aforementioned studies were cross-sectional in design,‡ which warrants caution in interpretation of the findings (*i.e.* the temporal direction of the association is uncertain). However, it seems reasonable to hypothesise that exposure to road traffic noise is linked to hypertension, but not *vice versa*, because it is unlikely that subjects with hypertension will prefer to move to areas with higher traffic noise levels. Longitudinal studies are nevertheless needed in order to strengthen a possible link between road traffic noise and incident hypertension. In the Danish Diet, Cancer and Health (DCH) cohort of 44 083 participants, positive significant cross-sectional associations were found between 1 year mean L_{den} and SBP in men and older participants (aged >60 years); however, no association was found between road traffic noise and incident self-reported hypertension.[43] This is the first report in a prospective study design and more studies with rigorously validated incident hypertension outcomes are needed in order to allow for a proper conclusion to be drawn. In a recent study in London of 169 849 adults, although incident hypertension was ascertained through general practice or hospital records between 2005 and 2011, no associations were observed with night-time road traffic noise.[20] In the Women's Health Initiative (WHI) study in the USA, women living ≤50 m from the nearest major roadway had a hazard ratio (HR) of 1.13 (95% CI: 1.00–1.28) for incident hypertension compared to those living >1000 m from the nearest major roadway.[44] Incident hypertension was ascertained based on blood pressure measurements at baseline and during follow-up visits. Residential proximity to major roadways, however, is only a surrogate marker for both traffic-related air and noise pollution. To better disentangle the effects of each, high-resolution spatial and temporal individual-level estimates for both air pollution and road traffic noise will be needed in future studies.

In summary, most current evidence supports a link between long-term road traffic noise exposure and prevalent hypertension in adult populations; the association seems to be independent of ambient air pollution effects. However, the role of road traffic noise on incident hypertension is not clear and needs further investigation.

---

‡Cross-sectional studies consider gradients in exposure and health outcomes across different population groups measured over the same time period. Longitudinal studies consider how health impacts in a given population change with time as exposures change.

3.2 Cardiovascular Disease Incidence, Morbidity and Mortality

3.2.1 Cardiovascular Mortality. The short-term effects of road traffic noise exposure on cardiovascular deaths have not been extensively investigated. Only two reports have been published to date. In the city of Madrid, Spain, each 1 dB(A) increase of diurnal noise (08.00–22.00) at lag 1 day was significantly associated with a 4.5% (95% CI: 0.6–8.7%) increase in mortality from all cardiovascular causes at all ages; corresponding increment for night-time noise (22.00–08.00) was 3.9% (95% CI: 0.6–7.3%) and for daily noise exposure was 6.2% (95% CI: 2.1%–10.6%)[45] When stratifying the associations by age, the significant associations were only seen among those aged ≥65 years. Cause-specific cardiovascular mortality was later examined in relation to night-time noise exposures, defined as the maximum hourly equivalent noise level (L_{nmax}, nocturnal: 00.00–08.00), at the same day and up to 4 days after the exposure.[46] The study found that, at lag 0, each 1 dB(A) increase of L_{nmax} was associated with a 2.9% (95% CI: 1.0–4.8%) increase of ischaemic heart disease (IHD) mortality and a 2.4% (95% CI: 0.1–4.8%) increase of cerebrovascular disease mortality in those aged ≥65 years. For those aged <65 years, significant associations were only seen for IHD and myocardial infarction (MI) mortality. Interestingly, this study also observed that L_{nmax} exposures at lag 1 were significantly associated with mortality from chronic obstructive pulmonary disease and pneumonia.

A small number of studies have examined the association between long-term road traffic noise exposures and cardiovascular mortality. In the Netherlands Cohort Study on Diet and Cancer (NLCS), a significant higher risk (adjusted relative risk (ARR): 1.25, 95% CI: 1.01–1.53) for cardiovascular mortality was seen among those exposed to an annual average traffic noise levels of ≥65 dB(A) compared to those exposed to a level of <50 dB(A).[47] When further adjusted for background black smoke exposures and traffic intensity of the nearest road, this association became non-significant (ARR: 1.17, 95% CI: 0.94–1.45). By accounting for traffic intensity in the adjusted model, it is possible that traffic noise effects on cardiovascular mortality may have been over-adjusted. In the Danish DCH cohort, residential road traffic noise (L_{den}) at enrolment was not associated with stroke mortality (incident rate ratio: 1.09, 95% CI: 0.86–1.39, per 10 dB[A]).[48] Similarly, in a small-area study of 8.6 million residents in Greater London, whilst daytime (07.00–23.00) noise was significantly associated with all-cause mortality in adults (RR: 1.04, 95% CI: 1.00–1.07, in areas with >60 *vs.* <55 dB),[19] the associations with mortality from IHD or stroke were positive but non-significant. In another London-based study, a 5 dB increase of road traffic noise (L_{Aeq16h}) was associated with all-cause mortality (HR: 1.02, 95% CI: 0.99–1.06) among MI survivors, independent of $PM_{2.5}$ air pollution.[49] In a large population-based study of 445 868 residents in Vancouver, Canada, where annual average noise estimates were assessed as community noise at residence after taking into account noise sources from roads, railways and aircrafts, a 10 dB(A) increase in noise levels was significantly associated with a

9% (95% CI: 1–18%) increase in mortality from coronary heart disease.[9] This association was independent of age, sex, comorbidity, socioeconomic status and air pollution exposures.

In summary, there is emerging evidence that both short- and long-term exposure to road traffic noise may have effects on both overall cardiovascular mortality and cause-specific mortality, particularly for IHD. Given only the small number of studies that have been conducted to date, more prospective cohort studies are needed in this area of research in order to better quantify the exposure–response relationship between road traffic noise and cardiovascular mortality in all group and subgroup populations.

3.2.2 Ischaemic Heart Disease Incidence. Findings of road traffic noise effects on IHD incidence before the year 2005 were mainly drawn from those of the Berlin study[50] and the Caerphilly–Speedwell study;[51,52] the latter mainly reported a positive non-significant association in these men-only cohorts. Babisch *et al.* retested the hypothesis in the Berlin study using samples of both men and women, in which they confirmed the previously suggestive findings in men, but found no clear dose–response relationship in women.[53] The Caerphilly–Speedwell and Berlin studies were used to derive an exposure–response function for the 2011 WHO global burden of disease from environmental noise,[4] and this, combined with the exposure distribution of road noise, gave an estimated risk of 1.8% of all MIs being attributable to road traffic noise in western European countries.

For the Berlin and Caerphilly–Speedwell studies, noise exposure at the home address was calculated from a coarse gridded noise map, which is likely to have had greater exposure misclassification compared to later model-based noise studies. In addition, these studies did not consider the possible confounding effects of air pollution. Studies using more sophisticated model-derived noise exposures began to emerge from 2005. A case–control study of 3666 participants in Stockholm County, Sweden, reported a positive non-significant association between address-level 24 hour A-weighted equivalent noise level (L_{Aeq}) and MI incidence (OR: 1.12, 95% CI: 0.95–1.33) when comparing addresses with noise levels of ≥50 or <50 dB(A).[54] A significant association was only found among those without hearing impairment and those with noise exposures other than road traffic noise. A recent study of 13 512 individuals in Skåne, Sweden, found mainly null associations between 24 hour noise (L_{den}) and MI incidence.[55] In contrast, in the Danish DCH cohort of 50 614 participants, a statistically significant association with incident MI (RR: 1.12, 95% CI: 1.02–1.22, per 10 dB[A] L_{den}) was found.[56] These studies published since 2005 adjusted for traffic-related air pollution (nitrogen oxides or black smoke) and suggested no confounding effects.

A meta-analysis of studies of road traffic noise effects on both prevalent and incident IHD was published by Babisch in 2014.[57] In this updated review including 11 studies of men and women, a significant pooled estimate of the risk for IHD (OR: 1.08, 95% CI: 1.04–1.13) per 10 dB(A) increase of weighted day–night road noise level within the range of 52–77 dB(A) was reported (see Figure 1). Vienneau et al. later updated this review by including eight incidence studies only and reported a pooled effect of 1.04 (95% CI: 1.00–1.10) per 10 dB(A) increase of L_{den} on IHD incidence.[58] The Vienneau et al. review suggested a possible threshold of 50 dB, above which a linear dose–response relationship was seen. In agreement with Babisch's review, men and those over 65 years of age were possibly at higher risk.

Two very large ecological studies and a longitudinal study using general practice data have been published since Vienneau et al.'s review in 2015, with mixed findings. A London-wide small-area study covering a population of 8.6 million found that daytime road traffic noise was not associated with IHD hospital admissions when comparing areas with an annual mean noise level of >60 vs. <55 dB(A) in both adult (aged ≥25 years) and elderly populations (aged ≥75 years).[19] In contrast, a study in the Rhine-Main region of Germany identified 19 632 incident MI cases and 834 734 controls through health insurance databases and reported a statistically significant association between 24 hour road noise level and incident MI (2.8% increase per 10 dB increase, 95% CI: 1.2–4.5%).[59] Both studies were only adjusted for age, sex and a range of area-level variables, hence they are not directly comparable to individual-level cohort studies. In a recent longitudinal study of 211 016 adults aged 40–79 years who were registered in general practices across London, neither incident coronary heart disease nor incident MI were associated with night-time road traffic noise.[20]

There have also been mixed results reported with respect to subgroup analyses in studies. Some previous studies either found a stronger association with IHD in men[53,56] or no effect modification by sex.[47,54] It remains inconclusive whether there exists a sex difference with respect to noise exposure on IHD. Notably, the possible confounding effects of the intake of sex hormones or the menopausal status of women participants were not considered in most of the previous studies, which may plausibly influence the stress response either positively or negatively.[53] Also, it was reported in a UK study that women were more sensitive to noise,[38] but it is unclear whether noise sensitivity would modify the associations with IHD morbidity and mortality. Two studies reported a stronger association among the elderly,[19,56] but one study did not.[54] Inconsistency was also observed across smoking status, with studies reporting effects in ex-smokers only,[54] or no effect modifications by smoking.[47]

In summary, therefore, studies are not fully consistent in terms of finding associations between noise and IHD. Systematic reviews were conducted in 2014 and 2015 and did suggest statistically significant associations, but these need to be updated as further evidence accumulates in order to be able to revise the estimate of 1.8% of MIs being attributable to environmental noise from road traffic provided in the 2011 WHO global burden of disease.[4]

3.2.3 Cerebrovascular Disease Incidence.

Very few studies have investigated road traffic noise effects on incident stroke. The Danish DCH study was the first to report a positive significant association between L_{den} and incident stroke (HR: 1.14, 95% CI: 1.03–1.25, per 10 dB[A] increase) among 57 053 participants.[48] The same research group later reported that this association was mainly confined to ischaemic stroke, but not haemorrhagic stroke.[60] The small-area study in London discussed in Section 3.2.2 reported significantly elevated risks for stroke hospital admissions in areas exposed to annual mean daytime noise levels of >60 vs. <55 dB(A) in both adults (RR: 1.05, 95% CI: 1.02–1.09) and the elderly (RR: 1.09, 95% CI: 1.04–1.14).[19] However, no associations were found for night-time noise effects. Another London-based study based on general practice registry data, also discussed in Section 3.2.2, did not report an association between night-time road traffic noise and incident stroke, accounting for both individual-level covariates and area-level deprivation index.[20]

Two other studies investigating stroke incidence in combination with other heart diseases have reported either null associations[61] or that the observed significant association was confounded by air pollution.[62] Heterogeneity was found across these studies for stroke outcomes in a recent review, in which the authors suspected that road traffic noise effects on stroke may follow a non-linear trend.[63] The nature of any exposure–response relationship, however, remains to be established.

3.2.4 Other Cardiovascular Disease.

A small number of studies have investigated cardiovascular outcomes other than IHD and stroke, and these have been in relation to heart failure and atrial fibrillation, but have had inconsistent findings.

Carey *et al.* recently reported that those exposed to a night-time road traffic noise level of ≥60 dB(A) in London had a higher risk of incident heart failure (HR: 1.09, 95% CI: 0.94–1.26) compared to those exposed to a level of <55 dB(A).[20] In fact, heart failure was the only cardiovascular outcome that had a positive association with road traffic noise in this study. However, this association was not co-adjusted for traffic-related $PM_{2.5}$, which itself has a significant positive association with heart failure in the same population. In the Dutch NLCS study, road traffic noise was significantly associated with mortality from heart failure (RR: 1.99, 95% CI: 1.05–3.79 for >65 vs. ≤50 dB[A]).[47] When further adjusting for both background black smoke and traffic intensity on the nearest road, this association was slightly attenuated and became non-significant (RR: 1.90, 95% CI: 0.96–3.78). While both reports suggested that a higher exposure level of road traffic noise at residence may be associated with a higher risk of heart failure, it remains difficult to separate such effects from exposure to air pollution, and there is a possibility that collinearity of exposures resulted in some over-adjustments. Also, as heart failure usually represents the end stage of heart disease, it is possible that the effects of road traffic noise or air pollution may only be exacerbatory or short term in nature, which would not be possible to detect with the long-term average noise exposure levels used.[20]

Carey et al. found a null association between road traffic noise and incident atrial fibrillation;[20] in contrast, the Danish DCH cohort study reported that for each 10 dB increase in 5 year time-weighted mean exposure to road traffic noise, the risk of incident atrial fibrillation increased by 6% (95% CI: 0–12%).[64] This is the first study to report a positive association between road traffic noise and atrial fibrillation. Again, when nitrogen dioxide (NO_2) air pollution was added into the main adjusted model, there was no significant association (RR: 1.01, 95% CI: 0.94–1.09), suggesting some possible confounding effects by air pollution.

3.3 Cardiovascular Risk Factors

As evidence has begun to emerge regarding the associations between road traffic noise exposures and cardiovascular disease, attention is being paid to associations between noise and cardiovascular risk factors in order to assess coherence of effect and to help strengthen the evidence base relating to causality and biological mechanisms. Currently, the data are still very limited, and a greater number of rigorous studies are clearly needed in this area of research to improve mechanistic insights into noise-induced cardiovascular disease events.

3.3.1 Atherosclerosis. The first examination of the association of noise with atherosclerosis came from the German Heinz Nixdorf Recall Study.[65] This cross-sectional analysis included 4238 participants aged 45–75 years for whom measurement of thoracic aortic calcification (TAC) was available at baseline recruitment. It was found that for 5 dB increases in night-time (22.00–06.00) noise, the TAC score increased by 4.6% (95% CI: 0.7–8.7%) in the main model without adjustment for $PM_{2.5}$ and by 3.9% (95% CI: 0–8.0%) after $PM_{2.5}$ was added into the model. L_{den}, however, was not associated with TAC score. This study was the first to suggest that both road traffic noise and air pollution play independent roles in atherosclerosis. A further cross-sectional pooled analysis of 2592 participants from two London cohorts found a positive but non-significant association between both daytime and night-time noise levels and carotid artery intimal media thickness (cIMT), which was not affected by air pollution adjustment (increasing cIMT reflects changes in the vascular walls due to plaque formation as part of the atherosclerotic process).[21]

Future studies with a longitudinal design that look at the progression of markers of atherosclerosis such as TAC and cIMT in relation to long-term noise exposure are needed before firm conclusions can be drawn regarding associations of noise with atherosclerosis.

3.3.2 Heart Rate. Zijlema et al. conducted a pooled cross-sectional analysis of 88 336 participants of three European cohorts using the

CNOSSOS-EU exposure model (see Section 2) and found that each 10 dB(A) increase in L_{den} noise was associated with a 0.93 (95% CI: 0.76–1.11) beats per minute increase in resting heart rate, reducing to 0.64 (95% CI: 0.44–0.80) after adjustment for NO_2 air pollution.[23] Effect estimates were slightly higher in males and in those not on antihypertensive medication (which may include drugs that modify heart rate).

3.3.3 Adiposity Markers. It is reasonable to hypothesise that continued exposure to noise may result in impacts on adiposity through stress and cortisol release, with most likely result in increased fat deposition around the waist, affecting the waist/hip ratio (WHR) and waist circumference (WC). The small number of studies on traffic noise exposures and adiposity markers to date have been conducted in Scandinavian populations.

In a cross-sectional analysis of the Oslo Health Study (HUBRO) of 15 085 adults living in Oslo, Norway, no overall significant associations were seen between road traffic noise and any obesity makers.[66] However, in women with high noise sensitivity (assessed by a single question of 'How sensitive are you towards noise?'), a positive significant association was observed between road traffic noise and both body mass index (BMI) and WC. In another cross-sectional study of 5075 adults in Stockholm County, Sweden, central obesity as indicated by WC and WHR was associated with traffic noise exposures, whilst no association was found with BMI as a general adiposity marker.[67] Traffic noise levels from road, rail and air were each significantly associated with WC with 0.21 cm (95% CI: 0.01–0.41), 0.46 cm (95% CI: 0.03–0.89) and 0.99 cm (95% CI: 0.62–1.37) increases, respectively, per 5 dB(A) increase. The risk of central obesity was 18% (95% CI: 3–34%) higher among those exposed to road traffic noise levels of ≥45 *vs.* <45 dB(A). Importantly, when combining noise sources from road, rail and air traffic, the risk became higher (OR: 1.95, 95% CI: 1.24–3.05).

As with the Stockholm study, the Danish DCH cohort study also reported a significant association between road traffic noise and WC (0.35 cm, 95% CI: 0.21–0.50 per 10 dB increase in L_{den}).[68] Small but significant associations were also observed for body fat mass and lean mass. Unlike the Stockholm study, this Danish study also reported a significant positive association between road traffic noise and BMI (0.18, 95% CI: 0.12–0.23 per 10 dB increase in L_{den}). However, these cross-sectional results were not confirmed in a longitudinal analysis of this DCH cohort, in which 5 year time-weighted mean L_{den} exposures preceding follow-up visits was not associated with changes in weight or WC over a mean follow-up time of 5.3 years in the fully adjusted model.[69] However, the study did show that the risk of gaining weight by more than 5 kg during the follow-up period was 7% (95% CI: 2–13%) higher per 10 dB increase in 5 year exposure preceding follow-up. Another interesting study of the Danish National Birth Cohort reported that exposure to 10 dB higher road traffic noise (L_{den}) during pregnancy (OR: 1.06, 95% CI: 1.00–1.12) and childhood (OR: 1.06, 95% CI: 0.99–1.12) was associated with overweight at 7 years of age.[70]

It is difficult to draw conclusions from the studies to date as they are not fully consistent, but they do raise the possibility of a link between traffic noise exposures and adiposity, although replication studies in other populations as well as more longitudinal studies will be needed in order to strengthen the link.

3.3.4 Type 2 Diabetes Mellitus and Blood Glucose. As with adiposity, links have also been suggested between road traffic noise exposure and diabetes, again through potential impacts on stress and endocrine disruption. The largest study to date was from the Danish DCH cohort, in which a 10 dB increase in road traffic noise at diagnosis and during the 5 years preceding diagnosis was associated with a higher risk of incident diabetes, with IRRs of 1.08 (95% CI: 1.02–1.14) and 1.11 (95% CI: 1.05–1.18), respectively.[71] These associations were adjusted for potential confounders, including air pollution exposures. In a small cross-sectional study of 513 participants in Bulgaria, self-reported doctor-diagnosed type 2 diabetes mellitus was significantly associated with exposure to a L_{den} level of 71–80 vs. 51–70 dB(A).[72] Short-term noise exposure at night was also significantly associated with diabetes mortality (11%, 95% CI: 4.0–19% per 1 dB increase) in a study in Madrid.[46]

The role of noise in glucose metabolism has rarely been studied. We have only located one published study relating to road traffic noise, conducted in the early 1990s. The study reported a positive trend (p-value < 0.05) between higher daytime noise levels and increased glucose levels in men.[51] The authors of the current chapter are preparing a manuscript at the time of writing this chapter. In our cross-sectional analysis of 62 765 Lifelines participants from the north of The Netherlands, an increase of 4.2 dB[A] daytime road traffic noise (the interquartile range) was significantly associated with a 0.013 mmol L^{-1} (95% CI: 0.006–0.019) increase in fasting blood glucose. Further adjusting for ambient air pollution did not materially change the association. Our study, the largest to date, suggests a link between road traffic noise exposure and increased fasting glucose independent of air pollution adjustment. However, the findings were not fully consistent, as we did not see a clear corresponding association with glycated haemoglobin (HbA1c). Replication of our study is therefore needed.

3.3.5 Blood Lipids. As for glucose metabolism, evidence of road traffic noise effects on lipid abnormalities in population-based studies is very limited. One previous analysis of the Danish DCH cohort in the Copenhagen area reported that the positive non-significant association (0.48 mg dL^{-1}, 95% CI: −0.19 to 1.15) with total cholesterol per 9.3 dB(A) increase in L_{den} became null after adjustment for air pollution.[73] Analyses by the current authors described in Section 3.3.4 showed that daytime road traffic noise was significantly associated with both higher triglycerides and high-density lipoprotein cholesterol levels, although the association with triglycerides became null after air pollution was further adjusted for (manuscript in preparation).

3.4 Further Factors to Consider in the Interpretation of Epidemiological Studies: Confounding and Effect-modifying Factors

Most studies, especially the more recently published ones, take careful account of potential confounders, including socioeconomic status and air pollution. Despite similar major sources of road vehicles, an analysis of noise and air pollution in London, UK, at differing spatial scales generally found only moderate correlations between noise and air pollution exposures, with low correlations existing near roads due to differing propagation rates of the different pollutants.[74] This suggests that it should be possible to detect independent effects of noise and air pollution. This was supported by a review by Stansfeld and Shipley in 2015, who reviewed the epidemiological evidence for associations with the health effects of noise given air pollution exposures, and they concluded that effects of noise were independent from those of air pollution.[38]

A limitation of many epidemiological studies is that, generally, exposure estimates relate to outdoor noise and not indoor noise, and the tacit assumption is that variability in outdoor noise is reflected in variability in indoor exposures. While this assumption has been reasonable for air pollution, it needs further testing for noise where factors such as housing type, window opening and double glazing may cause important exposure misclassification and/or effect modifications. Very few studies have attempted to investigate this in an epidemiological setting. Babisch investigated cross-sectional associations of road noise and hypertension in relation to various effect modifiers in 4861 individuals in the HYENA study.[2] Fully adjusted analyses (including for education as a measure of socioeconomic status) showed associations between road noise and hypertension in those living in terraced houses, apartment blocks or parts of converted buildings, but not those in detached or semi-detached houses. Associations were also seen with noise for those with living rooms at the rear of the house, but there was no impact for location of the bedroom (street side or quiet side). Remedies to reduce noise, such as shutting windows or window shutters, were found to be indicators of perceived exposure, rather than effect modifiers.

Annoyance presents a potential biological pathway by which road traffic noise may have impacts on health, and it may also function as an effect modifier, intensifying the biological impacts of noise on those who are also annoyed by it. Again, there are few studies directly investigating this. Babisch *et al.* used the same HYENA data to investigate this hypothesis.[75] However, no clear interaction was seen with annoyance in cross-sectional analyses of road traffic noise and hypertension, although one was described for aircraft noise exposure, with stronger associations in more annoyed subjects. More studies are needed in order to investigate this further, but it raises the possibility that different types of noise exposure have impacts on health through different biological mechanisms.

Noise sensitivity is a recognised and stable response to noise exposure that has associations with the anxiety trait.[76] It has high prevalence in populations (*e.g.* 48% in a study of English civil servants[38] and 40% in a study of Finnish civil servants[77]). Noise sensitivity is recognised as being an independent predictor of the annoyance response to environmental noise.[78] A longitudinal analysis of the Whitehall II study by Stansfeld and Shipley found that noise sensitivity was predictive of angina pectoris in lower employment grades, but not of cardiovascular mortality or morbidity.[38] Noise sensitivity has the potential to modify the effects of environmental noise, with potential for enhanced effects in such individuals. Very few studies have included this in their analyses, but the HUBRO study described above did find stronger evidence for associations with obesity markers in noise-sensitive women.[66]

4 Conclusions

Most current evidence suggests a role for road traffic noise in hypertension and the development of IHD, for which a linear exposure–response relationship may possibly start at a level of around 50 dB(A). The evidence base is still small in comparison with air pollution, and more work is needed in order to strengthen this and to provide a robust basis for the quantification of the effect size. The evidence for other incident cardiovascular disease outcomes including stroke, heart failure and atrial fibrillation is less consistent across studies, partly due to the small number of studies published to date. A small number of studies have been conducted in relation to the impacts of road traffic noise on cardiovascular risk factors and intermediate outcomes, including atherosclerosis, adiposity and diabetes, which help us to explore potential mechanisms. Further issues that need to be carefully investigated in future studies include combining different sources of traffic noise, separating the effects from air pollution exposure, the effects in subgroups with different characteristics, in particular the role of noise annoyance and noise sensitivity, and the impacts of physical factors such as building and window type on outdoor–indoor noise ingress.

Acknowledgements

The work of the UK Small Area Health Statistics Unit is funded by Public Health England as part of the MRC-PHE Centre for Environment and Health, funded also by the UK Medical Research Council. Dr Anna Hansell is also supported by the National Institute for Health Research Health Protection Research Unit (NIHR HPRU) in Health Impacts of Environmental Hazards at King's College London in partnership with Public Health England.

References

1. M. Basner, W. Babisch, A. Davis, M. Brink, C. Clark, S. Janssen and S. Stansfeld, Auditory and non-auditory effects of noise on health, *Lancet*, 2014, **383**, 1325–1332.
2. W. Babisch, *Exposure to environmental noise: risks for health and the environment. Workshop on "sound level of motor vehicles". Directorate General for Internal Policies of the European Parliament; Brussels.* 2012.
3. World Health Organisation Europe. Night Noise Guidelines for Europe. 2009; Available from: http://www.euro.who.int/__data/assets/pdf_file/0017/43316/E92845.pdf.
4. N. Nashashibi, E. Cardamakis, G. Bolbos and V. Tzingounis, Investigation of kinetic of lead during pregnancy and lactation, *Gynecol. Obstet. Invest.*, 1999, **48**, 158–162.
5. T. Munzel, T. Gori, W. Babisch and M. Basner, Cardiovascular effects of environmental noise exposure, *Eur. Heart J.*, 2014, **35**, 829–836.
6. K. I. Hume, M. Brink and M. Basner, Effects of environmental noise on sleep, *Noise Health*, 2012, **14**, 297–302.
7. T. Munzel, M. Sorensen, T. Gori, F. P. Schmidt, X. Rao, F. R. Brook, L. C. Chen, R. D. Brook and S. Rajagopalan, Environmental stressors and cardio-metabolic disease: part II-mechanistic insights, *Eur. Heart J.*, 2017, **38**(8), 557–564.
8. M. Basner, U. Muller and E. M. Elmenhorst, Single and combined effects of air, road, and rail traffic noise on sleep and recuperation, *Sleep*, 2011, **34**, 11–23.
9. W. Q. Gan, H. W. Davies, M. Koehoorn and M. Brauer, Association of long-term exposure to community noise and traffic-related air pollution with coronary heart disease mortality, *Am. J. Epidemiol.*, 2012, **175**, 898–906.
10. D. W. Morley, K. de Hoogh, D. Fecht, F. Fabbri, M. Bell, P. S. Goodman, P. Elliott, S. Hodgson, A. L. Hansell and J. Gulliver, International scale implementation of the CNOSSOS-EU road traffic noise prediction model for epidemiological studies, *Environ. Pollut.*, 2015, **206**, 332–341.
11. *SoundPLAN Acoustics*, http://www.soundplan.eu/english/soundplan-acoustics/, *(accessed October 2016)*.
12. *CadnaA - State of the art Noise Prediction Software*, http://www.datakustik.com/en/products/cadnaa, *(accessed October 2016)*.
13. *Environmental Noise Directive European Commission*, http://ec.europa.eu/environment/noise/directive_en.htm, *(accessed October 2016)*.
14. N. Roswall, V. Hogh, P. Envold-Bidstrup, O. Raaschou-Nielsen, M. Ketzel, K. Overvad, A. Olsen and M. Sorensen, Residential exposure to traffic noise and health-related quality of life–a population-based study, *PLoS One*, 2015, **10**, e0120199.
15. Department of Transport, Welsh Office (1988) *Calculation of Road Traffic Noise*, HMSO, London.
16. N. Garg and S. Maji, A critical review of principal traffic noise models: Strategies and implications, *Environ. Impact Assess. Rev.*, 2014, **46**, 68–81.

17. S. Kephalopoulos, M. Paviotti, F. Anfosso-Ledee, D. Van Maercke, S. Shilton and N. Jones, Advances in the development of common noise assessment methods in Europe: The CNOSSOS-EU framework for strategic environmental noise mapping, *Sci. Total Environ.*, 2014, **482–483**, 400–410.
18. J. Gulliver, D. Morley, D. Vienneau, F. Fabbri, M. Bell, P. Goodman, S. Beevers, D. Dajnak and D. Fecht, Development of an open-source road traffic noise model for exposure assessment, *Environ. Model. Softw.*, 2015, **74**, 183–193.
19. J. I. Halonen, A. L. Hansell, J. Gulliver, D. Morley, M. Blangiardo, D. Fecht, M. B. Toledano, S. D. Beevers, H. R. Anderson, F. J. Kelly and C. Tonne, Road traffic noise is associated with increased cardiovascular morbidity and mortality and all-cause mortality in London, *Eur. Heart J.*, 2015, **36**, 2653–2661.
20. I. M. Carey, H. R. Anderson, R. W. Atkinson, S. Beevers, D. G. Cook, D. Dajnak, J. Gulliver and F. J. Kelly, Traffic pollution and the incidence of cardiorespiratory outcomes in an adult cohort in London, *Occup. Environ. Med.*, 2016, **73**(12), 849–856.
21. J. Halonen, H. Dehbi, A. Hansell, J. Gulliver, D. Fecht, M. Blangiardo, F. Kelly, N. Chaturvedi, M. Kivimäki and C. Tonne, Associations of nighttime road traffic noise with carotid intima-media thickness and blood pressure: the Whitehall II and SABRE study cohorts, *Environ. Int.*, 2017, **98**, 54–61.
22. *Small Area Health Statistics Unit Data Download by Environmental Exposures Group*, http://www.sahsu.org/content/data-download, *(accessed October 2016)*.
23. W. Zijlema, Y. Cai, D. Doiron, S. Mbatchou, I. Fortier, J. Gulliver, K. de Hoogh, D. Morley, S. Hodgson, P. Elliott, T. Key, H. Kongsgard, K. Hveem, A. Gaye, P. Burton, A. Hansell, R. Stolk and J. Rosmalen, Road traffic noise, blood pressure and heart rate: Pooled analyses of harmonized data from 88,336 participants, *Environ. Res.*, 2016, **151**, 804–813.
24. Y. Cai, W. Zijlema, D. Doiron, M. Blangiardo, P. Burton, I. Fortier, A. Gaye, J. Gulliver, K. de Hoogh, K. Hveem, S. Mbatchou, D. Morley, R. Stolk, P. Elliott, A. Hansell and S. Hodgson, Ambient air pollution, traffic noise and adult prevalent asthma: a BioSHaRE approach, *Eur. Respir. J.*, 2017, **49**(1), 1502127.
25. J. Gulliver and K. de Hoogh, *Environmental Exposure Assessment: Modelling Air Pollution Concentrations*, The 6th Oxford Book of Public Health, Oxford University Press, 2015.
26. J. Ko, S. I. Chang and B. Lee, Noise impact assessment by utilizing noise map and GIS: A case study in the city of Chungju, Republic of Korea, *Appl. Acoustics*, 2011, **72**, 544–550.
27. E. Suarez and J. L. Barros, Traffic noise mapping of the city of Santiago de Chile, *Sci. Total Environ.*, 2014, **466–467**, 539–546.
28. M. S. Ragettli, S. Goudreau, C. Plante, M. Fournier, M. Hatzopoulou, S. Perron and A. Smargiassi, Statistical modeling of the spatial variability

28. of environmental noise levels in Montreal, Canada, using noise measurements and land use characteristics, *J. Exposure Sci. Environ. Epidemiol.*, 2016, **26**(6), 597–605.
29. D. Xie, Y. Liu and J. Chen, Mapping urban environmental noise: a land use regression method, *Environ. Sci. Technol.*, 2011, **45**, 7358–7364.
30. G. Hoek, R. Beelen, K. de Hoogh, D. Vienneau, J. Gulliver, P. Fischer and D. Briggs, A review of land-use regression models to assess spatial variation of outdoor air pollution, *Atmos. Environ.*, 2008, **42**, 7561–7578.
31. M. Bell and F. Galatioto, Novel wireless pervasive sensor network to improve the understanding of noise in street canyons, *Appl. Acoustics*, 2012, **74**, 169–180.
32. D. W. Morley and J. Gulliver, Methods to improve traffic flow and noise exposure estimation on minor roads, *Environ. Pollut.*, 2016, **216**, 746–754.
33. L. T. Chang, K. J. Chuang, W. T. Yang, V. S. Wang, H. C. Chuang, B. Y. Bao, C. S. Liu and T. Y. Chang, Short-term exposure to noise, fine particulate matter and nitrogen oxides on ambulatory blood pressure: A repeated-measure study, *Environ. Res.*, 2015, **140**, 634–640.
34. E. van Kempen and W. Babisch, The quantitative relationship between road traffic noise and hypertension: a meta-analysis, *J. Hypertens.*, 2012, **30**, 1075–1086.
35. W. Babisch, K. Wolf, M. Petz, J. Heinrich, J. Cyrys and A. Peters, Associations between traffic noise, particulate air pollution, hypertension, and isolated systolic hypertension in adults: the KORA study, *Environ. Health Perspect.*, 2014, **122**, 492–498.
36. W. Babisch, G. Wolke, J. Heinrich and W. Straff, Road traffic noise and hypertension–accounting for the location of rooms, *Environ. Res.*, 2014, **133**, 380–387.
37. Y. Cai, B. Zhang, W. Ke, B. Feng, H. Lin, J. Xiao, W. Zeng, X. Li, J. Tao, Z. Yang, W. Ma and T. Liu, Associations of Short-Term and Long-Term Exposure to Ambient Air Pollutants With Hypertension: A Systematic Review and Meta-Analysis, *Hypertension*, 2016, **68**, 62–70.
38. S. A. Stansfeld and M. Shipley, Noise sensitivity and future risk of illness and mortality, *Sci. Total Environ.*, 2015, **520**, 114–119.
39. K. Fuks, S. Moebus, S. Hertel, A. Viehmann, M. Nonnemacher, N. Dragano, S. Mohlenkamp, H. Jakobs, C. Kessler, R. Erbel and B. Hoffmann, Long-term urban particulate air pollution, traffic noise, and arterial blood pressure, *Environ. Health Perspect.*, 2011, **119**, 1706–1711.
40. S. Floud, F. Vigna-Taglianti, A. Hansell, M. Blangiardo, D. Houthuijs, O. Breugelmans, E. Cadum, W. Babisch, J. Selander, G. Pershagen, M. C. Antoniotti, S. Pisani, K. Dimakopoulou, A. S. Haralabidis, V. Velonakis and L. Jarup, Medication use in relation to noise from aircraft and road traffic in six European countries: results of the HYENA study, *Occup. Environ. Med.*, 2011, **68**, 518–524.
41. T. Bodin, M. Albin, J. Ardo, E. Stroh, P. O. Ostergren and J. Bjork, Road traffic noise and hypertension: results from a cross-sectional public health survey in southern Sweden, *Environ. Health*, 2009, **8**, 38.

42. Y. de Kluizenaar, R. T. Gansevoort, H. M. Miedema and P. E. de Jong, Hypertension and road traffic noise exposure, *J. Occup. Environ. Med.*, 2007, **49**, 484–492.
43. M. Sorensen, M. Hvidberg, B. Hoffmann, Z. J. Andersen, R. B. Nordsborg, K. G. Lillelund, J. Jakobsen, A. Tjonneland, K. Overvad and O. Raaschou-Nielsen, Exposure to road traffic and railway noise and associations with blood pressure and self-reported hypertension: a cohort study, *Environ. Health*, 2011, **10**, 92.
44. S. L. Kingsley, M. N. Eliot, E. A. Whitsel, Y. Wang, B. A. Coull, L. Hou, H. G. Margolis, K. L. Margolis, L. Mu, W. C. Wu, K. C. Johnson, M. A. Allison, J. E. Manson, C. B. Eaton and G. A. Wellenius, Residential proximity to major roadways and incident hypertension in postmenopausal women, *Environ. Res.*, 2015, **142**, 522–528.
45. A. Tobias, A. Recio, J. Diaz and C. Linares, Noise levels and cardiovascular mortality: a case-crossover analysis, *Eur. J. Prev. Cardiol.*, 2015, **22**, 496–502.
46. A. Recio, C. Linares, J. R. Banegas and J. Diaz, The short-term association of road traffic noise with cardiovascular, respiratory, and diabetes-related mortality, *Environ. Res.*, 2016, **150**, 383–390.
47. R. Beelen, G. Hoek, D. Houthuijs, P. A. van den Brandt, R. A. Goldbohm, P. Fischer, L. J. Schouten, B. Armstrong and B. Brunekreef, The joint association of air pollution and noise from road traffic with cardiovascular mortality in a cohort study, *Occup. Environ. Med.*, 2009, **66**, 243–250.
48. M. Sorensen, M. Hvidberg, Z. J. Andersen, R. B. Nordsborg, K. G. Lillelund, J. Jakobsen, A. Tjonneland, K. Overvad and O. Raaschou-Nielsen, Road traffic noise and stroke: a prospective cohort study, *Eur. Heart J.*, 2011, **32**, 737–744.
49. C. Tonne, J. I. Halonen, S. D. Beevers, D. Dajnak, J. Gulliver, F. J. Kelly, P. Wilkinson and H. R. Anderson, Long-term traffic air and noise pollution in relation to mortality and hospital readmission among myocardial infarction survivors, *Int. J. Hyg. Environ. Health*, 2016, **219**, 72–78.
50. W. Babisch, H. Ising, B. Kruppa and D. Wiens, The incidence of myocardial infarction and its relation to road traffic noise—the Berlin case-control studies, *Environ. Int.*, 1994, **20**, 469–474.
51. W. Babisch, H. Ising, J. E. Gallacher, D. S. Sharp and I. A. Baker, Traffic noise and cardiovascular risk: the Speedwell study, first phase. Outdoor noise levels and risk factors, *Arch. Environ. Health*, 1993, **48**, 401–405.
52. W. Babisch, H. Ising, J. E. Gallacher, P. M. Sweetnam and P. C. Elwood, Traffic noise and cardiovascular risk: the Caerphilly and Speedwell studies, third phase–10-year follow up, *Arch. Environ. Health*, 1999, **54**, 210–216.
53. W. Babisch, B. Beule, M. Schust, N. Kersten and H. Ising, Traffic noise and risk of myocardial infarction, *Epidemiology*, 2005, **16**, 33–40.
54. J. Selander, M. E. Nilsson, G. Bluhm, M. Rosenlund, M. Lindqvist, G. Nise and G. Pershagen, Long-term exposure to road traffic noise and myocardial infarction, *Epidemiology*, 2009, **20**, 272–279.

55. T. Bodin, J. Bjork, K. Mattisson, M. Bottai, R. Rittner, P. Gustavsson, K. Jakobsson, P. O. Ostergren and M. Albin, Road traffic noise, air pollution and myocardial infarction: a prospective cohort study, *Int. Arch. Occup. Environ. Health*, 2016, **89**, 793–802.
56. M. Sorensen, Z. J. Andersen, R. B. Nordsborg, S. S. Jensen, K. G. Lillelund, R. Beelen, E. B. Schmidt, A. Tjonneland, K. Overvad and O. Raaschou-Nielsen, Road traffic noise and incident myocardial infarction: a prospective cohort study, *PLoS One*, 2012, **7**, e39283.
57. W. Babisch, Updated exposure-response relationship between road traffic noise and coronary heart diseases: a meta-analysis, *Noise Health*, 2014, **16**, 1–9.
58. D. Vienneau, C. Schindler, L. Perez, N. Probst-Hensch and M. Roosli, The relationship between transportation noise exposure and ischemic heart disease: a meta-analysis, *Environ. Res.*, 2015, **138**, 372–380.
59. A. Seidler, M. Wagner, M. Schubert, P. Droge, J. Pons-Kuhnemann, E. Swart, H. Zeeb and J. Hegewald, Myocardial Infarction Risk Due to Aircraft, Road, and Rail Traffic Noise, *Dtsch. Arztebl. Int.*, 2016, **113**, 407–414.
60. M. Sorensen, P. Luhdorf, M. Ketzel, Z. J. Andersen, A. Tjonneland, K. Overvad and O. Raaschou-Nielsen, Combined effects of road traffic noise and ambient air pollution in relation to risk for stroke? *Environ. Res.*, 2014, **133**, 49–55.
61. Y. de Kluizenaar, F. J. van Lenthe, A. J. Visschedijk, P. Y. Zandveld, H. M. Miedema and J. P. Mackenbach, Road traffic noise, air pollution components and cardiovascular events, *Noise Health*, 2013, **15**, 388–397.
62. S. Floud, M. Blangiardo, C. Clark, K. de Hoogh, W. Babisch, D. Houthuijs, W. Swart, G. Pershagen, K. Katsouyanni, M. Velonakis, F. Vigna-Taglianti, E. Cadum and A. L. Hansell, Exposure to aircraft and road traffic noise and associations with heart disease and stroke in six European countries: a cross-sectional study, *Environ. Health*, 2013, **12**, 89.
63. A. M. Dzhambov and D. D. Dimitrova, Exposure-response relationship between traffic noise and the risk of stroke: a systematic review with meta-analysis, *Arh. Hig. Rada Toksikol.*, 2016, **67**, 136–151.
64. M. Monrad, A. Sajadieh, J. S. Christensen, M. Ketzel, O. Raaschou-Nielsen, A. Tjonneland, K. Overvad, S. Loft and M. Sorensen, Residential exposure to traffic noise and risk of incident atrial fibrillation: A cohort study, *Environ. Int.*, 2016, **92–93**, 457–463.
65. H. Kalsch, F. Hennig, S. Moebus, S. Mohlenkamp, N. Dragano, H. Jakobs, M. Memmesheimer, R. Erbel, K. H. Jockel and B. Hoffmann, Are air pollution and traffic noise independently associated with atherosclerosis: the Heinz Nixdorf Recall Study, *Eur. Heart J.*, 2014, **35**, 853–860.
66. B. Oftedal, N. H. Krog, A. Pyko, C. Eriksson, S. Graff-Iversen, M. Haugen, P. Schwarze, G. Pershagen and G. M. Aasvang, Road traffic noise and markers of obesity - a population-based study, *Environ. Res.*, 2015, **138**, 144–153.

67. A. Pyko, C. Eriksson, B. Oftedal, A. Hilding, C. G. Ostenson, N. H. Krog, B. Julin, G. M. Aasvang and G. Pershagen, Exposure to traffic noise and markers of obesity, *Occup. Environ. Med.*, 2015, **72**, 594–601.
68. J. S. Christensen, O. Raaschou-Nielsen, A. Tjonneland, K. Overvad, R. B. Nordsborg, M. Ketzel, T. Sorensen and M. Sorensen, Road Traffic and Railway Noise Exposures and Adiposity in Adults: A Cross-Sectional Analysis of the Danish Diet, Cancer, and Health Cohort, *Environ. Health Perspect.*, 2016, **124**, 329–335.
69. J. S. Christensen, O. Raaschou-Nielsen, A. Tjonneland, R. B. Nordsborg, S. S. Jensen, T. I. Sorensen and M. Sorensen, Long-term exposure to residential traffic noise and changes in body weight and waist circumference: A cohort study, *Environ. Res.*, 2015, **143**, 154–161.
70. J. S. Christensen, D. Hjortebjerg, O. Raaschou-Nielsen, M. Ketzel, T. I. Sorensen and M. Sorensen, Pregnancy and childhood exposure to residential traffic noise and overweight at 7 years of age, *Environ. Int.*, 2016, **94**, 170–176.
71. M. Sorensen, Z. J. Andersen, R. B. Nordsborg, T. Becker, A. Tjonneland, K. Overvad and O. Raaschou-Nielsen, Long-term exposure to road traffic noise and incident diabetes: a cohort study, *Environ. Health Perspect.*, 2013, **121**, 217–222.
72. A. M. Dzhambov and D. D. Dimitrova, Exposures to road traffic, noise, and air pollution as risk factors for type 2 diabetes: A feasibility study in Bulgaria, *Noise Health*, 2016, **18**, 133–142.
73. M. Sorensen, D. Hjortebjerg, K. T. Eriksen, M. Ketzel, A. Tjonneland, K. Overvad and O. Raaschou-Nielsen, Exposure to long-term air pollution and road traffic noise in relation to cholesterol: A cross-sectional study, *Environ. Int.*, 2015, **85**, 238–243.
74. D. Fecht, A. L. Hansell, D. Morley, D. Dajnak, D. Vienneau, S. Beevers, M. B. Toledano, F. J. Kelly, H. R. Anderson and J. Gulliver, Spatial and temporal associations of road traffic noise and air pollution in London: Implications for epidemiological studies, *Environ. Int.*, 2016, **88**, 235–242.
75. W. Babisch, G. Pershagen, J. Selander, D. Houthuijs, O. Breugelmans, E. Cadum, F. Vigna-Taglianti, K. Katsouyanni, A. S. Haralabidis, K. Dimakopoulou, P. Sourtzi, S. Floud and A. L. Hansell, Noise annoyance–a modifier of the association between noise level and cardiovascular health? *Sci. Total Environ.*, 2013, **452–453**, 50–57.
76. S. A. Stansfeld, D. S. Sharp, J. Gallacher and W. Babisch, Road traffic noise, noise sensitivity and psychological disorder, *Psychol. Med.*, 1993, **23**, 977–985.
77. J. I. Halonen, J. Vahtera, S. Stansfeld, T. Yli-Tuomi, P. Salo, J. Pentti, M. Kivimaki and T. Lanki, Associations between nighttime traffic noise and sleep: the Finnish public sector study, *Environ. Health Perspect.*, 2012, **120**, 1391–1396.
78. K. Paunovic, B. Jakovljevic and G. Belojevic, Predictors of noise annoyance in noisy and quiet urban streets, *Sci. Total Environ.*, 2009, **407**, 3707–3711.

Environmental Impact of Hybrid and Electric Vehicles

BILLY WU* AND GREGORY J. OFFER

ABSTRACT

Hybrid and electric vehicles play a critical role in reducing global greenhouse gas emissions, with transport estimated to contribute to 14% of the 49 GtCO_2eq produced annually. Analysis of only the conversion efficiency of powertrain technologies can be misleading, with pure battery electric and hybrid vehicles reporting average efficiencies of 92% and 35% in comparison with 21% for internal combustion engine vehicles. A fairer comparison would be to consider the well-to-wheel efficiency, which reduces the numbers to 21–67%, 25% and 12%, respectively. The large variation in well-to-wheel efficiency of pure battery electric vehicles highlights the importance of renewable energy generation in order to achieve true environmental benefits. When calculating the energy return on investment of the various technologies based on the current energy generation mix, hybrid vehicles show the greatest environmental benefits, although this would change if electricity was made with high amounts of renewables. In an extreme scenario with heavy coal generation, the CO_2eq return on investment can actually be negative for pure electric vehicles, highlighting the importance of renewable energy generation further. The energy impact of production is generally small (∼6% of lifetime energy) and, similarly, recycling is of a comparable magnitude, but it is less well studied.

---

*Corresponding author.

Issues in Environmental Science and Technology No. 44
Environmental Impacts of Road Vehicles: Past, Present and Future
Edited by R.E. Hester and R.M. Harrison
© The Royal Society of Chemistry 2017
Published by the Royal Society of Chemistry, www.rsc.org

1 Introduction

In order to achieve reductions in greenhouse gas (GHG) emission from the transport sector, hybrid and electric vehicle (EV) technology will be essential.[1–3] The Intergovernmental Panel on Climate Change estimated that in 2010, 49 GtCO$_2$eq was produced globally, and of this, 14% was attributed to transport.[4] The International Energy Agency (IEA) highlighted the importance of reducing GHG emissions in 2009 by stating that if current trends were to continue, transport-related CO$_2$ emissions would increase by 80% by 2050, making it extremely difficult to maintain atmospheric concentrations below a target of 450 ppm.[5] As of 2015, 1.3 million EVs were in use globally, which represents a compound average growth rate (CAGR) of 67% since 2012.[6] Yet, despite this rapid growth, EV sales are still only a small proportion of the 90 million automobiles that are produced annually,[7] although this is expected to shift in the coming years and will have regional differences. Future EV volumes vary from source to source, and there is no definitive forecast, but indicative targets from the IEA suggest that if 140 million EVs are deployed by 2030, there would be a 50% chance of limiting average global temperature increases to 2 °C.[8]

In 2010, the Europe-wide fleet-averaged emissions from passenger vehicles were approximately 160 gCO$_2$ km^{-1}. Through a combination of engine downsizing and vehicle lightweighting, it is predicted that the approximate lower limit for a fossil fuel diesel internal combustion engine (ICE) is 85 gCO$_2$ km^{-1}. Through hybridisation of the diesel ICE with energy storage technologies, this is envisaged to decrease to approximately 60 gCO$_2$ km^{-1}; however, carbon reductions beyond this towards longer-term targets of 20 gCO$_2$ km^{-1} can only be achieved with pure battery EVs (BEV).[9] Nevertheless, it is important to understand that EV introduction is only part of the solution to reducing transport-based GHG emissions, and other factors, such as renewable energy generation, also need to be considered.

To analyse the potential environmental impact of EVs, it is important to understand the variations in technology types and their characteristics. EVs as a whole can broadly be divided into three main categories: hydrogen fuel cell vehicles (FCVs), pure BEVs and hybrid EVs (HEV), with HEVs having further subdivisions based on the degree of hybridisation and powertrain configuration. FCVs, which are close to commercialisation, convert hydrogen into electrical energy through a proton exchange membrane fuel cell (FC), with the only by-product being water. Full BEVs only use electrical energy provided from a battery and have the advantage of being zero emission at the point of use; however, they still suffer from problems such as limited range, long charging times and higher capital costs compared to ICE-powered vehicles. HEVs, which combine an energy storage element with an ICE, whilst not fully zero emission, do offer improvements in fuel economy. This is achieved through engine downsizing, reduced ICE transient loads and operation at a more efficient point by means of load shifting *via* the energy storage device. The absolute efficiency gains vary depending on the powertrain design and applied drive cycle.[10]

Table 1 Comparison of petrol, hybrid and electrical storage systems in four leading vehicles. Reproduced from B. G. Pollet, I. Staffell and J. L. Shang, Current status of hybrid, battery and fuel cell electric vehicles: From electrochemistry to market prospects, *Electrochim. Acta*, **84**, 235–249. Copyright (2012) with permission from Elsevier.[11]

	Conventional	Hybrid	Hydrogen	Battery
Reference vehicle	Volkswagen Golf VI	Toyota Prius III	Honda FCX Clarity	Nissan Leaf
Fuel weight (kg)	40.8	33.3	4.1	171
Storage capacity (kWh)	500	409	137	24
Specific energy (Wh primary kg^{-1} fuel)	12 264	12 264	33 320	140
Storage system weight (kg)	48	40	93	300
Specific energy (Wh primary kg^{-1} of storage)	10 408	10 261	1469	80
Net power (kW)	90	100	100	80
Power plant and auxiliary weight (kg)	233	253	222	100
Specific energy (Wh primary kg^{-1} total equipment)	1782	1398	315	60
Average conversion efficiency (%)	21	35	60	92
Effective storage capacity (kWh useable)	105.0	143.1	82.0	22.1
Specific energy (Wh usable kg^{-1} total equipment)	374	486	260	55

A high-level comparison by Pollet *et al.*[11] of the various current/close to commercialisation vehicle technologies is presented in Table 1 in order to highlight some of the key metrics that are characteristic of different vehicle powertrains.

From Table 1, the most evident contrast between conventional ICE vehicles and BEVs is the difference in the specific energy. If only considering the specific energy (Wh primary kg^{-1} fuel), then there is a nearly two orders of magnitude difference between the technologies. However, only considering these raw metrics is unfair, as the conversion efficiency of a battery can be up to four-times greater than that of an ICE. The specific energy (Wh usable kg^{-1} total equipment) is therefore a more suitable comparison metric. Whilst BEVs are still lower than ICEs in this regard, the difference is less than one order of magnitude, which could potentially be surmounted with innovations in battery chemistries and pack engineering. Focusing on the specific energy (Wh usable kg^{-1} total equipment), FCVs are already competitive and HEVs are superior; however, the key challenges here include cost, lifetime and refuelling infrastructure.

The analysis shown in Table 1, which includes the average conversion efficiency, whilst useful, can also be misleading from an overall efficiency perspective. Table 2 shows the well-to-wheel (WtW) efficiency of the various powertrain technologies which is more useful to consider from an environmental perspective. The large variation in the WtW efficiency of BEVs

Table 2 Typical well-to-tank, tank-to-wheel and well-to-wheel efficiencies of each technology. Reproduced from B. G. Pollet, I. Staffell and J. L. Shang, Current status of hybrid, battery and fuel cell electric vehicles: From electrochemistry to market prospects, *Electrochim. Acta*, **84**, 235–249. Copyright (2012) with permission from Elsevier.[11]

Vehicle type	Well to tank	Tank to wheel				Well to wheel
BEV	32–100%	Charger 90%	Battery 92%	Inverter 96%	Motor 91%	Mechanical 92%	21.3–66.5%
H2 FCV	75–100%	Fuel cell 51.8%		Inverter 96%	Motor 91%	Mechanical 92%	31.2–41.6%
Hybrid	82.2%	30.2%					24.8%
Diesel	88.6%	17.8%					15.8%
Petrol	82.2%	15.1%					12.4%

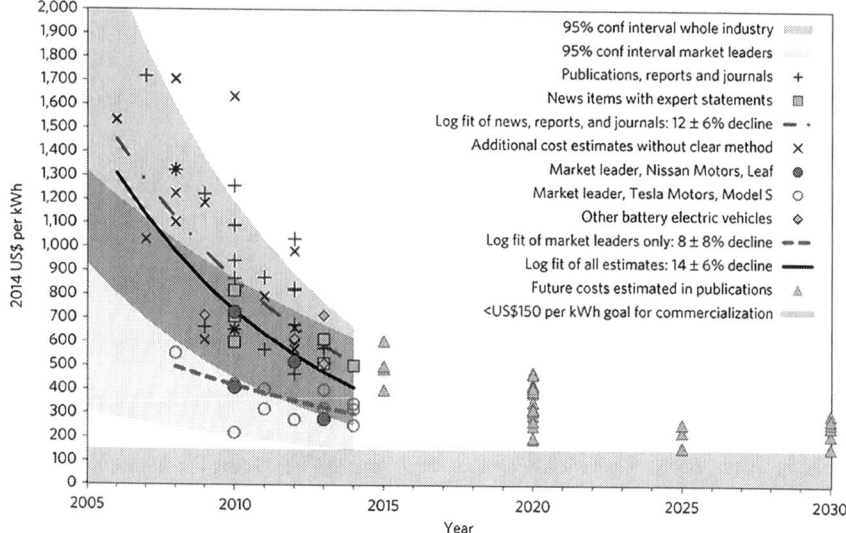

Figure 1 The falling cost of lithium-ion batteries. Reprinted by permission from Macmillan Publishers Ltd: *Nat. Clim. Change*, ref. 12. Copyright (2015).

is mainly due to the range in well-to-tank efficiency and highlights the importance of considering how the electricity for BEVs is generated.

Acknowledging the efficiency benefits of BEVs, one of the key barriers to their mainstream adoption is cost. From a historical perspective, the economies of scale and learning curves play a large part in the price of the batteries, which has shown a significant decrease in recent years, with an estimated industry average cost of approximately $1300 kWh^{-1} in 2006 falling to $400 kWh^{-1} in 2014 as shown in Figure 1. Whilst the cost reductions have been significant, it is generally understood that in order to be cost competitive with ICEs, the cost of the battery has to fall to approximately $150 kWh^{-1}.

Considering the difference between the current costs, the required cost targets and the need for manufacturing scale-up, the Automotive Council UK has outlined a technology roadmap for low-carbon passenger vehicles.[13] Here, it outlines that in order to transition to full BEVs, micro/mild hybrids will be the first introduced technology to improve the fuel economy of ICE vehicles. Full hybrids and plug-in hybrids, whereby the size of the battery increases with the ICE being downsized to become a range extender, is then seen as the next horizon. Finally, it is forecasted that the EV charging infrastructure will eventually be developed, allowing for mass market penetration of BEV technology.

Alongside technology shifts, a common theme will be vehicle weight reduction. For ICE vehicles, there is a strong correlation between vehicle weight and fuel economy,[14] which is well known. However, whilst weight reduction will always be beneficial to improving fuel economy, the correlations differ for BEVs, HEVs and FCVs. For BEVs and HEVs, the ability to recover a large proportion of the kinetic energy through regenerative braking mitigates some of the importance of weight reduction. Pure FCVs, on the other hand, cannot recover this kinetic energy, and thus the importance of weight reduction is much more significant. Nevertheless, research and development in battery weight reduction is an active and important area. This will likely come about through three main routes: innovations in battery chemistries; improvements in battery pack designs; and more intelligent use of the energy storage system.

- Battery chemistry innovations: improvement in battery materials is one of the main drivers behind the progression of BEVs from the early days of milk floats with lead acid batteries to the Tesla Model S with lithium-ion batteries. The majority of current lithium-ion batteries use a graphite anode combined with a cathode commonly composed of a combination of nickel, cobalt and manganese. Near-term improvements in energy density will be achieved through optimisation of these chemistries; however, medium-term energy density gains are envisaged to come about through the use of silicon anodes,[15,16] sulphur-based cathodes[17,18] and nickel-rich layered oxide cathodes.[19] In the long term, the application of metal-air batteries such as lithium-air[20] and zinc-air[21] shows promise; however, their poor lifetimes and efficiencies remain challenges. In terms of the environmental impact of producing these new battery chemistries, very little has to date been reported.
- Improvements in battery pack design: often when discussing battery technologies the energy density of the material is cited; however, once the materials are integrated into a cell, the gravimetric energy density decreases due to the weight associated with inactive phases such as current collectors, binders, separators and packaging. Integrating a cell

into a pack and system then incurs additional weight penalties through cooling systems, physical enclosures and electrical connections, which can be substantial. Optimising these engineering aspects of batteries will inevitably result in weight savings and better utilisation of the energy storage capacity, which will impact the life cycle analysis (LCA) of the system.

- More intelligent use of the energy storage system: starting from the energy density of an individual cell and linearly scaling to a pack and including the additional mass of ancillary components does not necessarily result in the available energy. Often, cell-to-cell variations[22] result in an underutilised capacity to avoid over-charging and over-discharging a battery. In addition to the cell-to-cell variations, the full state-of-charge (SOC) range of a battery is rarely used in automotive applications, as it is often difficult to extract charge at low SOCs due to large voltage variations and increased cell resistance. Currently, extending these self-imposed operational limits is an area of active research.[23,24]

It is generally acknowledged that HEVs and BEVs will be needed in order to achieve proposed emission targets. Yet, there are many challenges that need to be addressed before this is a reality, and the environmental impact of the technologies needs to be fully understood. This chapter will: review the current state of the art with respect to automotive energy storage technologies and their application in hybrid and electric vehicles; discuss the influence of different load cycles; consider the life cycle assessment of the storage technology; and consider any potential global warming potential of the technology.

2 Energy Storage and Conversion Technologies

The progressive uptake of hybrid and electric vehicles has been underpinned by advances in energy storage technologies. These can come in various types, but in the context of automotive technologies, these are mainly electrochemical technologies such as batteries and supercapacitors, though there are notable exceptions, such as mechanical flywheels.[25] Whilst hydrocarbon-based fuels are a form of energy storage, a key distinguishing feature of electrochemical energy storage devices is their ability to recharge on-board from the regenerative braking energy of the electric motors. This ability to recover waste kinetic energy can increase the fuel efficiency of the vehicle by 20–50% depending on the size of the motor and the drive cycle.[26] In addition, the electrochemical nature of the energy conversion means that higher efficiencies can be achieved over combustion-based fuels, which are limited by the Carnot efficiency.

In general, batteries are the energy storage technology most frequently employed in hybrid and electric vehicles. Whilst supercapacitors and flywheels are also suitable in theory, they are not practical as the prime mover for a vehicle due to their low energy density and therefore extremely low range. This is true except for a few unique examples, such as supercapacitors buses,[27] which are likely to belong to a very specialised niche due to the infrastructure costs and requirement for regular timetabled stops for recharging. Therefore, both flywheels and supercapacitors tend to be hybridised with another energy storage technology or conversion device in order to reduce transient loads.[28]

An overview of the different battery technologies considered for automotive applications is shown in Table 3. The earliest EVs employed lead acid-based chemistries; however, their low energy efficiency and energy density limited their more widespread use. In the case of lead acid batteries, whilst the lead is toxic, the recyclability is high. Nickel–cadmium batteries achieved moderate uptake in consumer electronics; however, the toxicity of the cadmium and the memory effect associated with the chemistry ultimately limited their uptake. Nickel–metal hydride (Ni–MH)-based batteries were the first chemistry to see appreciable commercial uptake in electric and hybrid vehicles; however, this area is now transitioning to lithium-ion-based chemistries due to their superior energy density. Beyond lithium-ion batteries, there is extensive research on metal-air-based batteries, although their lifetimes and energy efficiencies remain limiting problems for commercialisation.

As batteries are zero emission at the point of use, the influencing factors that determine their environmental impact are the materials used in their construction, their disposal, the storage/conversion efficiency and how the electricity was generated. Toyota, who were one of the early adopters of hybrid and electric vehicle technology, used Ni–MH-based batteries in their earlier Prius model range before recently transitioning to using more lithium-ion-based chemistries. On a kgCO_2eq basis, Majeau-Bettez et al.[30] showed that Ni–MH-based batteries produced approximately double the emissions (considering production and use) compared with lithium-ion battery chemistries based on nickel–cobalt–manganese and lithium–iron phosphate. This analysis, however, does not consider the environmental impact of the disposal.

Comparing and contrasting the different storage technologies can be problematic and often confusing. For example, there have been many attempts to compare hydrogen FCVs with BEVs on a like-for-like basis, and there has been significant conflict between the two research communities over the years. Some have attempted to demonstrate their synergies[31] rather than use their differences to argue that one technology is superior to the other for a one-size-fits-all solution.[32] However, comparisons are necessary and there are many papers and reports that compare these technologies with each other and the incumbents, often trying to predict a winner. This is fraught with difficulties, and it has been shown that the uncertainties in the assumptions used by most authors allow any technology to be predicted as the winner by carefully selecting the assumptions in order to predetermine the answer.[33]

Table 3 Comparison of different battery-based energy storage technologies for automotive applications. Adapted from Macmillan Publishers Ltd: *Nature*, ref. 29. Copyright (2008).

Battery type and approximate period of use	Features	Environmental impact	Practical energy density (Wh kg^{-1})
Lead acid (1859–1909)	Poor energy density, moderate power rate, low cost	Lead is toxic, but recycling is efficient to 95%	37[29]
Nickel-cadmium (1909–1975)	Low voltage, poor/moderate energy density, relatively high cost, memory effect	Cadmium is a toxic heavy metal.[30,31] Nickel is not green (difficult extraction/unsustainable) and is toxic. Not rare, but limited recyclability	50–75[31]
Nickel-metal hydride (1975–1990)	Low voltage, moderate energy density, high power density	Nickel is not green (difficult extraction/unsustainable) and is toxic. Not rare, but limited recyclability	60–70[32]
Lithium-ion (1990–present)	High energy density, power rate cycle life, costly	Depletable elements (cobalt) in most applications; replacements of manganese and iron are green (abundant and sustainable), lithium chemistry is relatively green (abundant, but the chemistry needs to be improved). Recycling is feasible, but at an extra energy cost	100–150[33]
Zinc-air (future)	Medium energy density, high power density	Mostly primary or mechanically rechargeable. Zinc smelting is not green, especially if primary. Easily recyclable.	350–500 (1086 theoretical)[a,20]
Lithium-air (future)	High energy density, but poor energy efficiency and rate capability	Rechargability to be proven. Excellent carbon footprint. Renewable electrodes. Easy recycling	Unclear (3458 theoretical)[a,20]

[a]Includes the mass of oxygen.

In recent years, it has become clear that vehicles powered by electricity stored in batteries are currently the frontrunners to replace the ICE and liquid hydrocarbon tank, with hydrogen FCVs being the next most likely contender, although there is still uncertainty over whether plug-in HEVs (PHEV) or BEVs will dominate. This is because vehicles with batteries have a head-start due to the presence of a mature lithium-ion battery supply chain that is capable of delivering the capacity needed by the automotive industry in order to scale-up production quickly at a reasonable price. For example, the lithium-ion battery industry produced an estimated 35 GWh of cells in 2015,[34] from which the Statista service estimated the global lithium-ion battery market to be $9.8 billion. Assuming that this is true, this means that Tesla, with an average battery pack size of 80 kWh and sales in 2015 of 50 580 vehicles, bought 12% of the global lithium-ion battery production in 2015. Statista predicts the battery industry to grow to $15.6 billion by 2020 (a CAGR of 60%), although they acknowledge that this is a low EV uptake scenario, and others predict growth to reach anywhere between $30 and $40 billion.[34] Assuming costs reduce to $200 kWh^{-1}, the upper estimate of $40 billion would mean annual production of 200 GWh year^{-1}, which represents a CAGR of 41%, but it is not known if the investment plans of the major battery producers are commensurate with these estimates. However, the gigafactory being built by Tesla and Panasonic alone will have the capacity to produce 35 GWh year^{-1} by 2020 (equivalent to the production of the entire world in 2015). This factory alone would be enough to sustain the production of 437 500 Tesla vehicles a year.

In contrast, although FC system costs are following a downward trajectory according to the US Department of Energy (DOE), and look likely to reach a cost of $90–160 kW^{-1} by 2020, the actual cost for FC systems remains high. In 2016, FCs were estimated to be $24 000 for an 85 kW system, or $280 kW^{-1}, assuming a manufacturing volume of 20 000 systems year^{-1}.[35] However, these values are misleading since cost reductions, due to volume manufacturing, have yet to be implemented. In addition, this is still well short of the US DOE target of $30–40 kW^{-1} that they estimate is needed to be competitive with ICEs. FCs also lag behind batteries in terms of volumes. Global total cumulative FC installations reached 1 GW by 2014, with the global FC industry projected to install the next 1 GW by 2016/17.[36] For comparison, 1 GW of FCs would be roughly 12 500 vehicles (assuming 80 kW stacks). This puts them roughly 10 years behind batteries in terms of the number of vehicles the FC industry is likely to be capable of supplying.

3 Hybrid Vehicles

There are several different forms of hybrid vehicles; however, the fundamental concept is the combination of two or more power sources to provide tractive power to the vehicle powertrain, with the aim of providing combined benefits that neither system would be able to achieve in isolation.

The often-cited advantages of hybrid systems over their pure ICE counterparts include:

- Increasing fuel efficiency of the ICE by reducing engine transients and allowing it to operate at its most efficient point;
- The ability to recover regenerative braking energy;
- Reducing engine idling losses;
- Allowing for engine downsizing whilst maintaining total vehicle equivalent performance and thus reducing the frictional losses in the ICE.

Whilst there are many specific variations of hybrid vehicle configurations, there are four main classifications: series, parallel, series–parallel and complex hybrids.[37] These vary based on their control and configuration. The environmental impacts of the different configurations will have some variation due to increased efficiencies of one over another; however, the most profound differences in emissions are due to the degree of hybridisation, which defines how much electrical storage is installed. Offer et al.[38] showed, for a FCV, that as the degree of hybridisation increased, the average lifetime emissions from small to large vehicles decreased. However, as highlighted in Figure 2, there is a law of diminishing returns above a battery size of 5 kWh for small vehicles and 25 kWh for large vehicles. As the degree of hybridisation increased, it was also highlighted that to realise emission reductions, the decarbonisation of electricity was an increasingly important parameter, as shown in Figure 2. Thus, electrification of road transport must also be accompanied by decarbonisation of electricity generation in order to reduce WtW emissions.

In terms of the fuel consumption benefits of an ICE-hybrid vehicle, these vary depending on the vehicle, degree of hybridisation and load cycle. Fontaras et al.[39] tested a Toyota Prius II and Honda Civic IMA, which are classified as full and mild hybrids, respectively, under different load cycles. Results showed that the higher the degree of hybridisation, the larger the fuel economy benefits under urban driving conditions. Above 60 KPH, the mild and full hybrids exhibited similar fuel consumptions, and above 90 KPH, the fuel consumption was similar to that of the equivalent ICE vehicle. Under urban driving conditions, fuel consumption was found to be 40–60% lower than the average equivalent ICE vehicle. These benefits are enhanced for drive cycles with very low average speeds and frequent stop-and-go events.

4 Impact of Different Usage Cases

One of the major challenges for automotive applications is the wide usage range that a vehicle needs to be designed for. Often this is captured and accounted for in the form of vehicle drive cycles, which are time–velocity traces that powertrains are validated against. The use of drive cycles in assessing the true fuel economy of a vehicle powertrain is often highlighted and shows significant differences compared to the application of 'nominal

Environmental Impact of Hybrid and Electric Vehicles 143

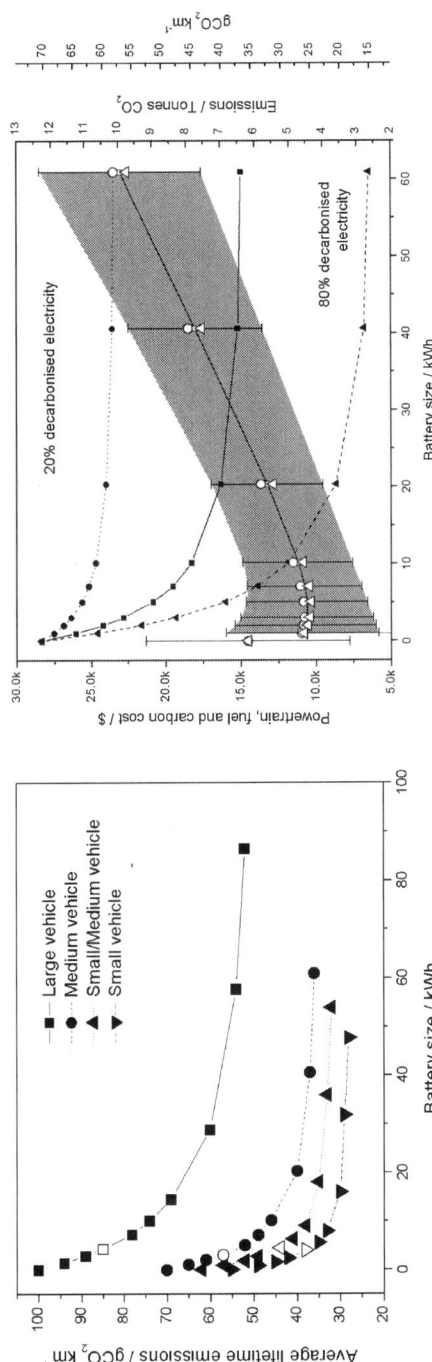

Figure 2 (left) The CO_2 emissions for different vehicle types: large (■), medium (●), small/medium (▲) and small (▼). (right) Sensitivity to the extent of decarbonisation of electricity generation with electricity decarbonisation assumptions set at low (○) and high (△). The CO_2 emissions are overlaid with the axis on the right and assumptions set at low (●), average (■) and high (▲). Reprinted from G. J. Offer, M. Contestabile, D. A. Howey, R. Clague and N. P. Brandon, Techno-economic and behavioural analysis of battery electric hydrogen FC and hybrid vehicles in a future sustainable road transport system in the UK, *Energy Policy*, **39**(4), 1939–1950. Copyright (2011) with permission from Elsevier.[38]

loads'.[40] André[41] presented a comprehensive overview of different vehicle drive cycles with the aim of deriving a common set of reference real-world driving cycles. Here, he showed 12 types of European driving patterns ranging from congested urban to motorway, steady speed. Typical metrics used to characterise drive cycles include average/peak velocity, duration, average acceleration and peak acceleration.[41,42]

Whilst the drive cycles presented by André[41] are indicative of real-world driving, current automotive benchmarking is often performed with the New European Drive Cycle as an industrial standard. However, there are criticisms of this, as it is not very indicative of real-world driving conditions, and thus there have been efforts to introduce other standard drive cycles, such as the New York City Cycle (NYCC) and Highway Fuel Efficiency Test (HWFET) (shown in Figure 3), which represent urban city driving and highway driving, respectively.

This can have dramatic effects on the fuel efficiency of vehicles and can change the optimum powertrain selection. For instance, Karabasoglu and Michalek[10] analysed the influence of driving patterns on life cycle cost and emissions of HEVs. They showed that under urban drive cycles such as the NYCC, the life cycle emissions of a HEV can be 60% lower than those of a conventional ICE vehicle. In contrast, the same HEV was shown to have marginal emission reductions under highway drive cycles such as the HWFET.

By converting the time–velocity profiles into time–power profiles *via* a vehicle model and analysing the results, the differences in the energy/power requirements become even more apparent. Figure 4 shows histogram plots of the normalised cumulative energy requirements of the NYCC and HWFET drive cycles against the power. In the NYCC drive cycle, it is apparent that a significant amount of regenerative braking energy is available. Here it should be noted that it is not always possible to recapture all the regenerative braking energy, especially at high power and low motor speeds due to charging limitations of the battery and inefficiencies in the motor/power converters under certain operating regions, respectively. In contrast, the HWFET drive cycle shows an insignificant amount of regenerative braking energy to be available due to increased air resistance dissipating the kinetic energy, as well as fewer deceleration events. Thus, for highway

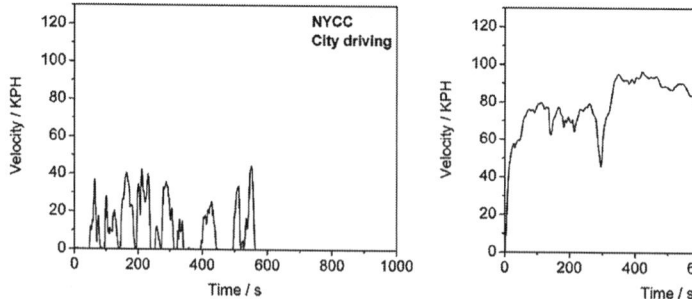

Figure 3 Time–velocity traces representative of (left) urban driving (NYCC) and (right) highway driving (HWFET).

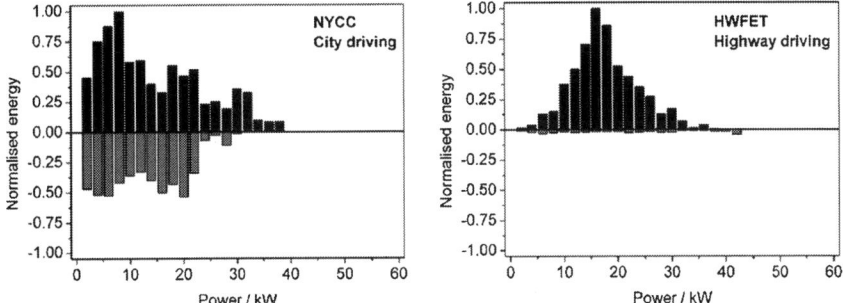

Figure 4 Histograms showing the normalised energy *vs.* power for (left) the NYCC and (right) the HWFET drive cycles.[43]

driving, the fuel economy benefits of HEVs is negligible; however, for urban city driving, these can be significant, in part due to the braking energy that can be recovered.

5 Life Cycle Assessment

See also Chapter 9 for details of life cycle assessment methods. With respect to the environmental impact of HEVs and BEVs, much of the academic literature has focused on WtW studies comparing fossil fuel and electricity use, as it is often viewed that the in-use phase of vehicle usage dominates the environmental impact.[3] For instance, Campanari *et al.*[40] conducted a detailed study of the WtW emissions of the various powertrains. Intuitively, the study showed that BEVs powered by electricity generated renewably offered the lowest-emission option. However, when considering the BEV WtW performance with an average energy mix or 100% coal/natural gas, the performance was much lower. Here, FCVs became much more favourable from the perspective of efficiency and CO_2 emissions, especially for vehicles with longer ranges due to the increased vehicle mass.

However, it would be unfair to disregard the energy consumption requirements of producing and disposing of the storage devices in the analysis. A recent review that looked at 79 LCA studies on lithium-ion batteries[44] found that, on average, producing 1 kWh of storage capacity was associated with a cumulative energy demand of 328 kWh and caused GHG emissions of 110 kgCO_2eq. However, these authors also concluded that although the majority of existing studies focus on GHG emissions or energy demand, impacts in other categories such as toxicity may be even more important.

The energy return on investment (EROI) is often used to assess energy production technologies, such as solar panels. However, batteries are fundamentally different, as they do not produce energy, they merely store it. Therefore, in order to calculate the EROI for a battery, it is necessary to calculate the energy saved during operation of the vehicle, as shown in eqn (1). For example, using the model developed by Contestabile *et al.*,[33] if a HEV contains 2 kWh of batteries, and assuming the vehicle efficiency increased by

12% from 58 to 66 mpg, with a lifetime mileage of 109 000, it would save over 10.5 MWh of energy over its lifetime, giving it an EROI of over 16.

$$\text{EROI} = \frac{\text{kWh}_{\text{saved}}}{\text{kWh}_{\text{production}}} \quad (1)$$

The same equation can be used for emissions by replacing kWh with tonnesCO_2eq, and assuming 3.0 kgCO_2 per litre for petrol. Thus, on a CO_2eq emission basis, HEVs would produce 3.2 tonnesCO_2eq less, saving almost 15-times the emissions produced in making the batteries.

The analysis for a BEV is slightly different. For the Nissan LEAF with a 24 kWh capacity, the battery manufacturing cost alone would amount to 7.9 MWh of energy and 2.64 tonnesCO_2eq. Assuming the same lifetime mileage of 109 000 and an energy consumption of 0.27 kWh mile^{-1}, it would save almost 54 MWh of energy over its lifetime, giving it an EROI of 6.8. Taking the 2.64 tonnesCO_2eq emissions due to battery production alone gives an equivalent vehicle emission of 15 gCO_2 km^{-1}, which, although low compared to most other vehicles, is not completely zero and does vary depending on the electricity generation method. Thus, the claim that BEVs are zero emission is only true if zero-emission electricity is used both to recharge the vehicle and throughout every stage of the manufacture of the batteries and raw materials. In addition, this is not the full story, as there are other additional components such as electrical machines and power electronics that must also be taken into account.

A study by Notter et al.[45] showed that in a BEV, the lithium-ion battery production is only responsible for around 6% of the cumulative energy demand and 8% of the global warming potential over the lifetime of its use. The rest of the car was shown to account for approximately 3.5-times more, and the remainder was attributed to operations that were highly subjective to the local electricity production and the drive cycle. A more recent study by Ellingsen et al.[46] explored the effect of the type of electricity production in more detail, as well as the effects of vehicle size and mileage. They found that smaller vehicles and electricity produced by non-renewable sources had the greatest environmental impact and that the higher manufacturing impacts of EVs were compensated for by the lower environmental impact when using the vehicle, unless the electricity was produced by coal. If the electricity was produced by natural gas, the total vehicle mileage needed to be greater than 100 000 to be of benefit, but a total vehicle mileage of just 30 000 was required if the electricity was produced by wind. The review by Nordelöf et al.[47] concluded that electricity production is the main cause of environmental impact for PHEVs. Using the assumptions from Ellingsen et al.[46]—the energy efficiencies of coal- and gas-fired electricity generation of 33% and 45%, respectively—and the model of Contestabile et al.,[33] Table 4 has been produced.

It should be noted that the results in Table 4 are highly dependent upon the input assumptions, and therefore these results are only to be used as a

Table 4 Key energy metrics for different vehicle types.

	Lifetime fuel energy (MWh)	Battery energy (MWh)	Battery EROI	Fuel emissions (tonnesCO$_2$eq)	Battery emissions (tonnesCO$_2$eq)	Battery CO$_2$ROI
Petrol	83.0	N/A	N/A	25.7	N/A	N/A
Hybrid	72.5	0.66	16.0	22.4	0.22	14.7
Electric coal	88.1	7.9	−0.7	29.9	2.64	−1.6
Electric gas	64.6	7.9	2.3	17.3	2.64	3.2
Electric renewable	29.1	7.9	6.8	0.6	2.64	9.5
PHEV gas	73.0	2.5	4.1	21.16	0.83	5.4
PHEV renewable	54.3	2.5	11.6	12.4	0.83	16.1

qualitative comparative guide. Therefore, eqn (1) can be used for other vehicles, but the kWh$_{saved}$ must take into account the energy displaced compared to a conventional powertrain according to eqn (2).

$$kWh_{saved} = kWh_{conventional} - kWh_{alternative} \qquad (2)$$

The effect of different electricity production and degrees of electrification can be seen quite clearly in Figure 5, taken from a recent review of the environmental impacts of HEVs, PHEVs and BEVs.[47] The sensitivity of the BEV environmental impact to the electricity generation source is perhaps the most important factor to consider. Hawkins et al.,[3] for instance, showed that in regions that are dependent on coal electricity generation, an increasing trend in SO$_x$ emissions was observed.

5.1 Battery Utilisation

Different vehicles use their batteries to varying extents, with hybrids often having a far greater utilisation than BEVs. Battery utilisation was shown by Contestabile et al. to be strongly affected by the size of the battery pack and the way that the vehicles are used.[33] With current behavioural patterns, an average daily mileage of 20–30 miles can be met with a modest battery size of between 5 and 15 kWh depending on the size of the vehicle.[38] In contrast, battery utilisation for larger battery packs in BEVs drops significantly, as shown in Figure 6. If the same total lifetime mileage is assumed, this affects the results considerably, as the displaced energy kWh$_{saved}$ will only increase slightly, but kWh$_{production}$ increases considerably, as shown in Figure 7. This is why PHEVs have a far higher EROI and CO$_2$ROI than EVs, as shown in Table 4.

However, changes in business models and behavioural patterns, and/or the introduction of autonomous vehicles, could change this again, and by increasing battery utilisation, this makes BEVs more favourable again.[48]

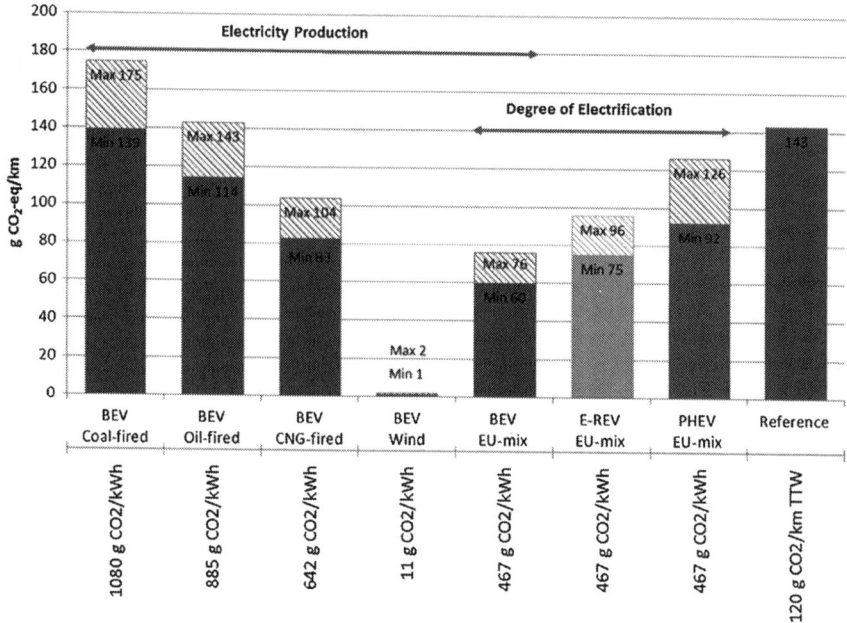

Figure 5 WtW GHG emissions for different electricity production methods and degrees of electrification.
Reproduced from A. Nordelöf, M. Messagie, A.-M. Tillman, M. L. Söderman and J. Van Mierlo,[47] Environmental impacts of hybrid, plug-in hybrid, and battery electric vehicles – what can be learn from life cycle assessment? *Int. J. Life Cycle Assess.*, **19**(11), 2014, 1866–1890, with permission from Springer © 2014.[47]

5.2 Vehicle-to-grid

Vehicle-to-grid (VtG) balancing involves, at its simplest, using smart meters to charge BEVs or PHEVs at a time that suits the grid operators. More advanced systems feed energy back from the vehicles into the grid at times of high demand, helping the grid operators balance the system. First proposed by Amori Lovins in 1995,[49] this approach has been discussed for many years,[50,51] and it has been theorised as one of the ways to enable high penetration of intermittent renewable electricity generation on the grid.[50] VtG technology is therefore one of the potential benefits of large-scale electrification of vehicles, but this has yet to be demonstrated or implemented at a practical scale. There are also other challenges to overcome, such as battery degradation, communication and changes to infrastructure.[52] However, if VtG technology could be implemented at scale, this would improve the environmental impact of BEVs powered by non-renewable energy generation, since there would be expected grid-scale benefits to having balancing capability with respect to removing inefficient peaker plant generation (a peaker plant is a power generation facility which is often only run in periods of high electricity demand). For instance, Sioshansi and Denholm[53] suggested that the introduction of VtG could potentially eliminate more than 80% of the CO_2 generated due to the additional generation required to support PHEVs.

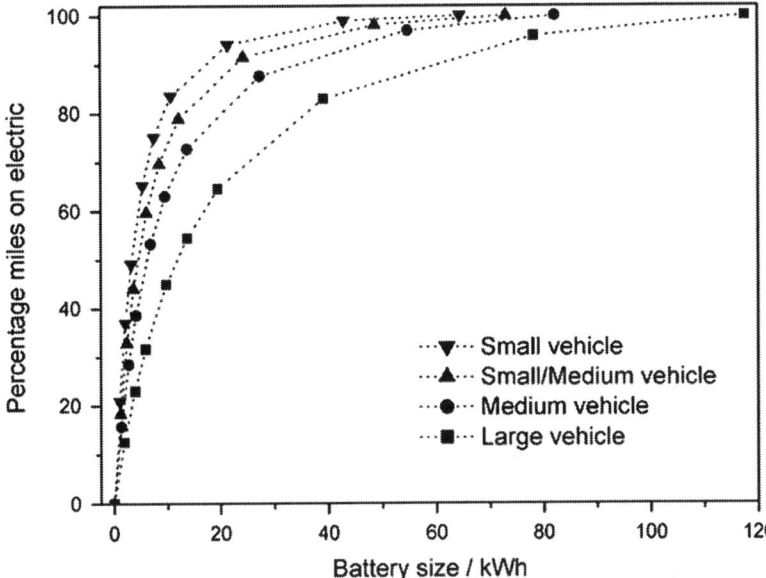

Figure 6 The percentage of miles that can be driven using electricity as a function of battery size according to current behavioural patterns.
Reprinted from G. J. Offer, M. Contestabile, D. A. Howey, R. Clague and N. P. Brandon, Techno-economic and behavioural analysis of battery electric hydrogen FC and hybrid vehicles in a future sustainable road transport system in the UK, *Energy Policy*, **39**(4), 1939–1950. Copyright (2011) with permission from Elsevier.[38]

5.3 Battery Lifetime and Degradation

In addition to the factors already considered, the lifetime of the battery pack (*i.e.* the rate of degradation) is important. All of the studies above either explicitly or implicitly assume that the battery pack will not need to be changed over the lifetime of the vehicle. In reality, some degree of degradation will likely occur, and not accounting for this effect can have significant bearing on the environmental impact. For example if the battery needs replacing once in a vehicle's lifetime, this will halve the EROI and CO_2ROI.

Another factor is the cycle life of the battery, which can be taken into account with a total lifetime energy throughput according to eqn (3).

$$kWh_{lifetime \cdot throughput} = kWh_{capacity} \times cycles \qquad (3)$$

This assumes that each cycle uses the full capacity of the battery (*i.e.* 100% depth of discharge [DOD]). If this is not true, then eqn (4) can be used instead.

$$kWh_{lifetime \cdot throughput} = kWh_{capacity} \times \%_{DOD} \times cycles \qquad (4)$$

This is important because the rate of degradation can often be significantly slowed down by reducing the DOD of a battery,[54] and hence the

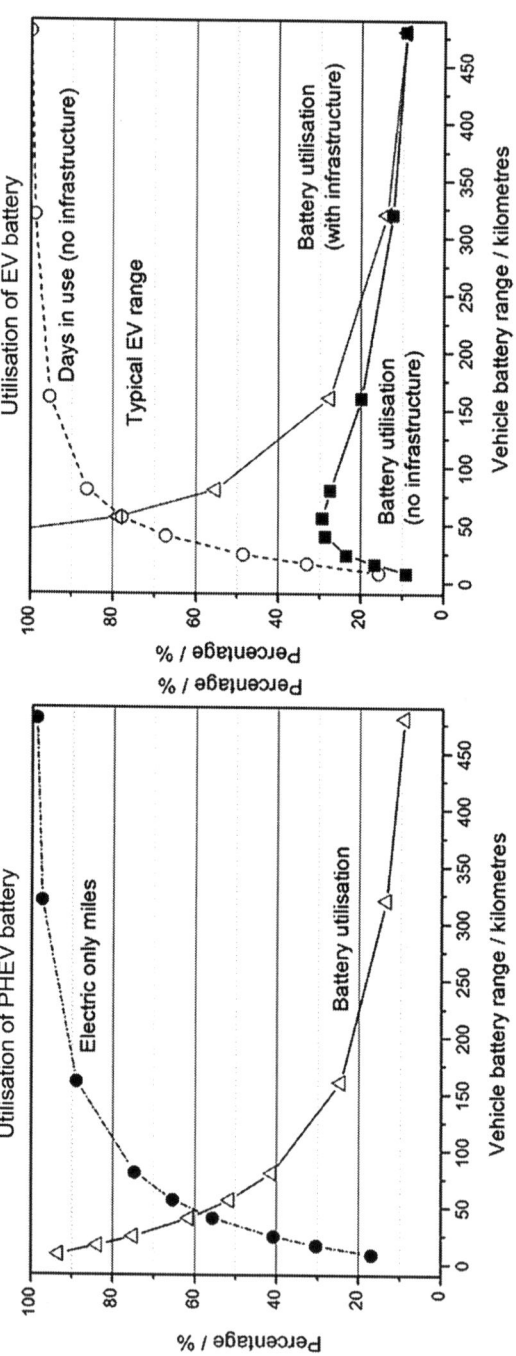

Figure 7 (left) Battery utilisation as a function of vehicle range for a PHEV, also showing the electric-only range, and (right) for an EV with and without infrastructure (*i.e.* fast charging), also showing days in use when without infrastructure. Reproduced from ref. 48 with permission from The Royal Society of Chemistry.

$kWh_{lifetime \cdot throughput}$ can be increased by changing the way the battery is used. This is done in most HEV applications to extend the life of the battery, and to a lesser extent even in BEVs. The $kWh_{lifetime \cdot throughput}$ can also be helpful when trying to take into consideration second life applications.

5.4 Recycling and Second Life

Whilst the materials used in the construction of lithium-ion batteries are finite, it is unlikely that the adoption of BEV and HEV technology in the future will deplete global reserves.[55] The lithium content of a lithium-ion battery only accounts for approximately 0.7% of the mass, and the current extraction processes from brines are relatively simple and have low energy demands.[45] Yet recycling is necessary to reduce the impact on base material mining such as aluminium, copper, cobalt, manganese and nickel. If not considered, this has economic, energy and emissions costs, decreasing the EROI and CO_2ROI. Second life batteries can have both an economic benefit and increase the $kWh_{lifetime \cdot throughput}$, resulting in improvements in the EROI and CO_2ROI.

Therefore, wherever possible, second life should be considered before recycling, although challenges for real-world implementation include screening and matching of cells with consistent characteristics.[56] Manufacturers of HEVs and BEVs often suggest battery replacement when the remaining energy capacity reaches 70–80%,[57] meaning useful capacity still remains at the point of disposal for transport applications. However, it has also been highlighted that the point of replacement is different for BEVs and HEVs. For instance, Wood et al.[58] highlighted the fact that PHEVs can blend the delivered power, meaning that performance can be maintained at a slight cost of efficiency. Thus, battery replacement should only be considered if there is a significant improvement in performance/efficiency. BEVs are not able to blend power delivery and thus the end-of-life point will often be at a higher remaining capacity. Despite some exploratory works into second life batteries, there are few detailed studies of their environmental impacts due to the variability of degraded cells. Thus, technical barriers such as estimation of degradation need to be overcome,[59] and practical diagnostic techniques[60–62] and economic incentives are also required[57] before second life batteries are implemented at scale.

There has been slightly more work with regards to the recycling of lithium-ion batteries, although this has not been extensive regarding environmental impact. A recent study by Oliveira et al.[63] confirmed again that the use phase dominates the impact, but also showed that the recycling stage can be just as important as the manufacturing stage in terms of environmental impact, as shown in Figure 8.

Dunn et al.[64] concluded that avoiding or reducing SO_x emissions and water contamination from metal recovery for cathode materials should be the key motivator for lithium-ion battery recycling, regardless of the energy intensity of assembly.

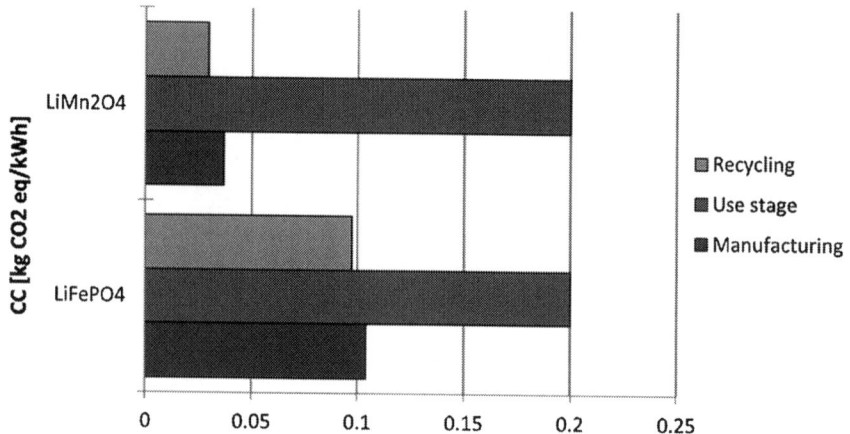

Figure 8 Climate change impacts of two common lithium-ion batteries, including recycling, use stage and manufacturing.
Reproduced from L. Oliveira, M. Messagie and S. Rangaraju, Key issues of lithium-ion batteries – from resource depletion to environmental performance indicators, *J. Cleaner Prod.*, **108**, 354–362. Copyright (2015) with permission from Elsevier.[63]

However, of concern is that a recent review of recycling methods concluded that most of the research achievements are still only at the pilot or laboratory scale and that there is still a need to establish firm collection systems, large-scale treatment plans and legislation covering the life cycle of lithium-ion batteries.[65] For example, in China, only 2% of non-lead acid battery waste is properly disposed of and most is simply dumped in landfills or piled in warehouses. Gaines[66] concludes that recycling automotive batteries is more complicated and not yet established because few end-of-life batteries have needed recycling, and it will be another 5–10 years before large numbers reach their end-of-life. However, despite this, there is a need to act now in order to put in place economic and sustainable options for recycling. Gaines[66] also describes the many problems with lithium-ion batteries entering the current lead acid waste streams, such as them causing fires and explosions at lead smelters.

6 Conclusion

It is undeniable that in order to achieve global GHG emission reduction targets, hybrid and electric vehicles must be implemented. However, transitioning from ICE-based powertrains to electric will require further improvements in technology and cost reductions. In the near term, the environmental benefits of hybrid vehicles have been shown to be greater than those of pure BEVs in the majority of regions. This is due to the large proportion of coal and gas power generation in these regions offsetting the benefits of the local zero-emission characteristics. In the case of a purely coal-powered electrical network, the net CO_2ROI can actually be negative. However, it is also important to consider that the relative benefit is also highly sensitive on the usage cycle, with HEVs showing significant fuel economy gains of 20–50% in urban driving modes, but limited benefit in highway driving due to the significantly reduced amount of regenerative braking energy available.

High utilisation of the battery benefits the environment, assuming a net positive CO_2ROI. It has been suggested that the implementation of VtG technologies could offset the additional CO_2 associated with the increased generation requirements of EVs by more than 80%; however, this is yet to be significantly implemented. In the longer term, BEVs will have a greater environmental benefit, but they rely on the energy mix of a region shifting towards renewables. The results of various works thus suggest that the electrification of transport needs to be accompanied by shifts in electricity generation to more renewable sources in order to avoid being counterproductive.

Research efforts into the environmental impact of lithium-ion battery production suggest that this accounts for only approximately 6% of the lifetime energy consumption of a BEV, with the majority dictated by the in-use phase. Whilst the common perception is to consider the impact of lithium in the battery, this only accounts for approximately 0.7% of the mass, and consideration of elements such as copper, aluminium, cobalt, manganese and nickel are more important. The area of energy consumption during production is reasonably well researched; however, research into lithium-ion battery recycling is much more limited, but the few studies that have been conducted suggest that the environmental impact could be in the same order as the production, although this is not yet conclusive. Second life batteries have been shown to have more of a positive environmental impact than recycling at the end of life for EV batteries. However, trials have been few due to challenges in estimating degradation, manual screening processes and limited economic incentives.

It is therefore clear that the current status of BEVs and HEVs is a long way from truly zero emission, and this should therefore catalyse efforts in generating renewable electricity, producing hybrid and electric vehicles and establishing suitable supply chains and policies for handling end-of-life batteries.

References

1. M. Granovskii, I. Dincer and M. A. Rosen, *J. Power Sources*, 2006, **159**, 1186–1193.
2. T. R. Hawkins, B. Singh, G. Majeau-Bettez and A. H. Strømman, *J. Ind. Ecol.*, 2013, **17**, 53–64.
3. T. R. Hawkins, O. M. Gausen and A. H. Strømman, *Int. J. Life Cycle Assess.*, 2012, **17**, 997–1014.

4. Intergovernmental Panel on Climate Change, *Climate Change 2014 - Mitigation of Climate Change*, 2014.
5. International Energy Agency, *Transport, Energy and CO2 - Moving Towards Sustainability*, 2009.
6. Worldwide number of hybrid and electric vehicles, Statista, 2016, https://www.statista.com/statistics/270603/worldwide-number-of-hybrid-and-electric-vehicles-since-2009/, accessed 30/10/2016.
7. Worldwide automobile production since 2000, Statista, 2016, https://www.statista.com/statistics/262747/worldwide-automobile-production-since-2000/, accessed 30/10/2016.
8. International Energy Agency, *Global EV Outlook 2016 - Beyond One Million Electric Cars*, 2016.
9. D. Howey, R. North and R. Martinez-botas, Grantham Inst. Clim. Chang. - Brief. Pap., 2010.
10. O. Karabasoglu and J. Michalek, *Energy Policy*, 2013, **60**, 445–461.
11. B. G. Pollet, I. Staffell and J. L. Shang, *Electrochim. Acta*, 2012, **84**, 235–249.
12. B. Nykvist and M. Nilsson, *Nat. Clim. Change*, 2015, **5**, 329–332.
13. Automotive Council UK, *Automotive Technology Roadmaps*, 2013.
14. F. An and D. J. Santini, in *SAE Technical Paper*, SAE International, 2004.
15. C. Erk, T. Brezesinski, H. Sommer, R. Schneider and J. Janek, *ACS Appl. Mater. Interfaces*, 2013, **5**, 7299–7307.
16. W.-J. Zhang, *J. Power Sources*, 2011, **196**, 13–24.
17. L. Chen and L. L. Shaw, *J. Power Sources*, 2014, **267**, 770–783.
18. M. Wild, L. O'Neill, T. Zhang, R. Purkayastha, G. Minton, M. Marinescu and G. J. Offer, *Energy Environ. Sci.*, 2015, **8**, 3477–3494.
19. A. Manthiram, B. Song and W. Li, *Energy Storage Mater.*, 2017, **6**, 125–139.
20. D. Aurbach, B. D. McCloskey, L. F. Nazar and P. G. Bruce, *Nat. Energy*, 2016, **1**, 16128.
21. Y. Li and H. Dai, *Chem. Soc. Rev.*, 2014, **43**, 5257–5275.
22. B. Wu, V. Yufit, M. Marinescu, G. J. Offer, R. F. Martinez-Botas and N. P. Brandon, *J. Power Sources*, 2013, **243**, 544–554.
23. S. Nagashima, K. Takahashi, T. Yabumoto, S. Shiga and Y. Watakabe, *J. Power Sources*, 2006, **158**, 1166–1172.
24. W. Waag, C. Fleischer and D. U. Sauer, *J. Power Sources*, 2014, **258**, 321–339.
25. A. Dhand and K. Pullen, *Int. J. Automot. Technol.*, 2013, **14**, 797–804.
26. J. K. Ahn, K. H. Jung, D. H. Kim, H. B. Jin, H. S. Kim and S. H. Hwang, *Int. J. Automot. Technol.*, 2009, **10**, 229–234.
27. D. P. Dubal, Y. P. Wu and R. Holze, *ChemTexts*, 2016, **2**, 13.
28. B. Wu, M. A. Parkes, V. Yufit, L. De Benedetti, S. Veismann, C. Wirsching, F. Vesper, R. F. Martinez-Botas, A. J. Marquis, G. J. Offer and N. P. Brandon, *Int. J. Hydrogen Energy*, 2014, **39**, 7885–7896.
29. M. Armand and J.-M. Tarascon, *Nature*, 2008, **451**, 652–657.
30. G. Majeau-Bettez, T. R. Hawkins and A. H. Strømman, *Environ. Sci. Technol.*, 2011, **45**, 4548–4554.

31. G. J. Offer, D. Howey, M. Contestabile, R. Clague and N. P. Brandon, *Energy Policy*, 2010, **38**, 24–29.
32. U. Bossel, *Proc. IEEE*, 2006, **94**, 1826–1837.
33. M. Contestabile, G. J. Offer, R. Slade, F. Jaeger and M. Thoennes, *Energy Environ. Sci.*, 2011, **4**, 3754.
34. B. Nykvist and M. Nilsson, *Nat. Clim. Change*, 2015, **5**, 329–332.
35. US Department of Energy Fuel Cells Technologies Office Record, *Fuel Cell System Cost*, 2014.
36. K. Ann, *Fuel Cell and Hydrogen Annual Review 2015*, 2015.
37. C. C. Chan, *Proc. IEEE*, 2002, **90**, 247–275.
38. G. J. Offer, M. Contestabile, D. A. Howey, R. Clague and N. P. Brandon, *Energy Policy*, 2011, **39**, 1939–1950.
39. G. Fontaras, P. Pistikopoulos and Z. Samaras, *Atmos. Environ.*, 2008, **42**, 4023–4035.
40. S. Campanari, G. Manzolini and F. Garcia de la Iglesia, *J. Power Sources*, 2009, **186**, 464–477.
41. M. André, *Sci. Total Environ.*, 2004, **334–335**, 73–84.
42. R. J. North, R. B. Noland, W. Y. Ochieng and J. W. Polak, *Transp. Res., Part D*, 2006, **11**, 344–357.
43. B. Wu, PhD thesis, Imperial College London, 2014.
44. J. F. Peters, M. J. Baumann, J. Braun and M. Weil, *Renewable Sustainable Energy Rev.*, 2015, 491–506, DOI: 10.1016/j.rser.2016.08.039.
45. D. A. Notter, M. Gauch, R. Widmer, P. Wäger, A. Stamp, R. Zah and H.-J. Althaus, *Environ. Sci. Technol.*, 2010, **44**, 6550–6556.
46. L. A.-W. Ellingsen, B. Singh and A. H. Strømman, *Environ. Res. Lett.*, 2016, **11**, 54010.
47. A. Nordelöf, M. Messagie, A. M. Tillman, M. Ljunggren Söderman and J. Van Mierlo, *Int. J. Life Cycle Assess.*, 2014, **19**, 1866–1890.
48. G. J. Offer, *Energy Environ. Sci.*, 2015, **8**, 26–30.
49. F. Nemry, G. Leduc and A. Muñoz, *Plug-in Hybrid and Battery-Electric Vehicles: State of the research and development and comparative analysis of energy and cost efficiency*, Joint Research Centre: Institute of Prospective Technologies, 2009, JRC 54699.
50. H. Lund and W. Kempton, *Energy Policy*, 2008, **36**, 3578–3587.
51. B. D. Williams and K. S. Kurani, *J. Power Sources*, 2007, **166**, 549–566.
52. S. Habib, M. Kamran and U. Rashid, *J. Power Sources*, 2015, **277**, 205–214.
53. R. Sioshansi and P. Denholm, *Environ. Sci. Technol.*, 2009, **43**, 1199–1204.
54. S. Saxena, C. Hendricks and M. Pecht, *J. Power Sources*, 2016, **327**, 394–400.
55. M. C. McManus, *Appl. Energy*, 2012, **93**, 288–295.
56. S. J. Tong, A. Same, M. A. Kootstra and J. W. Park, *Appl. Energy*, 2013, **104**, 740–750.
57. V. V. Viswanathan and M. Kintner-Meyer, *IEEE Trans. Veh. Technol.*, 2011, **60**, 2963–2970.
58. E. Wood, M. Alexander and T. H. Bradley, *J. Power Sources*, 2011, **196**, 5147–5154.

59. J. Neubauer and A. Pesaran, *J. Power Sources*, 2011, **196**, 10351–10358.
60. Y. Merla, B. Wu, V. Yufit, N. P. Brandon, R. F. Martinez-Botas and G. J. Offer, *J. Power Sources*, 2016, **331**, 224–231.
61. B. Wu, V. Yufit, Y. Merla, R. F. Martinez-Botas, N. P. Brandon and G. J. Offer, *J. Power Sources*, 2015, **273**, 495–501.
62. Y. Merla, B. Wu, V. Yufit, N. P. Brandon, R. F. Martinez-Botas and G. J. Offer, *J. Power Sources*, 2016, **307**, 308–319.
63. L. Oliveira, M. Messagie, S. Rangaraju, J. Sanfelix, M. Hernandez Rivas and J. Van Mierlo, *J. Cleaner Prod.*, 2015, **108**, 354–362.
64. J. B. Dunn, L. Gaines, J. C. Kelly, C. James and K. G. Gallagher, *Energy Environ. Sci.*, 2015, **8**, 158–168.
65. X. Zeng, J. Li and N. Singh, *Crit. Rev. Environ. Sci. Technol.*, 2014, **44**, 1129–1165.
66. L. Gaines, *Sustainable Mater. Technol.*, 2014, **1**, 2–7.

Development Implications for Malaysia: Hydrogen as a Road Transport Fuel

ANGELINA F. AMBROSE,* RAJAH RASIAH AND ABUL QUASEM AL-AMIN

ABSTRACT

Renewable energy policies targeting the adoption of hydrogen as road transport fuel have been argued to have significant impacts on meeting both energy security needs and climate change goals. Transportation is among the most challenging sectors to decarbonise as it is fraught with uncertainties associated with the high cost of renewable energy. This chapter seeks to conceptualise socio-technical transition in order to analyse the usefulness of hydrogen fuel cell vehicles (FCVs) as a backstop technology in a broad sense so as to support effective transition policies by estimating gross damage and other environmental effects from FCV prioritisation. Projections analyse the potential economic–environment impact of CO_2 emissions by switching fuel utilisation from fossil fuels to hydrogen FCVs in Malaysia over the period 2015–2030. The results reveal that compared to slow or moderate adaptation, aggressive hydrogen FCV adaptation is necessary to strengthen the economy while effectively reducing gross damage and absolute CO_2 emissions. Potential hydrogen demand in Malaysia is expected to be 55 million metric tons by 2030.

---

*Corresponding author.

1 Introduction

We live in exciting times of solution-targeted research for some of the most daunting problems related to global warming. In this sense, the special problem of meeting the growing demand of the transport sector comes into focus in this chapter. There is a movement of foresight research that focuses on the decarbonisation of the transport sector. For example, the International Energy Agency (IEA) Technology Roadmap indicated that the move to low-carbon solutions and changes from the current path will require an energy revolution. Andrews and Shabani[1] stress that there is a need to shift focus from an exclusive hydrogen economy by re-envisioning the specific role that hydrogen will play in sustainability planning. A hydrogen transition study in the USA projects that along with other alternative vehicle solutions, hydrogen will be able to meet both climate change and energy security goals.[2] Similar outcomes are predicted for hydrogen transition for the UK.[3] Growing international concerns over climate change and dependence on oil imports have led to great interest in the viability of alternative sources of energy, with car makers launching hydrogen fuel cell vehicles (FCVs). Apart from technological development, hydrogen fuelling station infrastructure has been a major component that supported the growth of hydrogen fuel cell-powered cars.[4] It was found that there were 224 hydrogen stations in 28 countries in 2013.

It is encouraging indeed to note that we no longer need to view the environment in terms of a trade-off against economic growth, as the latter can be achieved while preserving the earth through the use of sustainable fuels. We have had the choice of transitioning to clean and more sustainable fuels for a long time. After all, renewable energy is not a new concept. The big question is: why has it not happened yet? How can nations commit to making the transition to clean energy through the development of backstop technologies?

We will lay out some of the factors that have pushed recent interest in hydrogen energy to the forefront, especially its application in the transportation sector. In doing so, we seek to build the argument further regarding the relevance and potential of hydrogen as a non-polluting future energy source to meet the challenges of both energy security and global warming by deploying an adapted computable general equilibrium (CGE) model in order to demonstrate the implications of adopting hydrogen FCVs for Malaysia. The projections will also take account of abatement costs that would arise from the transition. However, in the long run, no country can afford not to transition to sustainable energy solutions, such as hydrogen FCVs.

2 Energy Demand, Economic Growth and CO_2 Emissions

Energy demand is expected to face continued growth with fuel economic development.[5] The challenge faced by policy makers globally for several decades has been how to simultaneously promote clean energy transition policies and economic growth. Fortunately, this economic growth and environmental degradation trade-off has recently taken on a new dimension. Shifting the energy used to support human activity from environment-unfriendly fuels to environment-friendly fuels can still support high rates of economic growth.[6,7]

However, the transition from brown to green fuels has moved slowly. One the one hand, both developed and developing nations have pledged to reduce per capita emission, as outlined in the Intended Nationally Determined Contributions (INDC) commitments to pre-industrial level. On the other hand, absolute emissions continue to rise at an alarming rate, especially in developing nations. Figure 1 shows the CO_2 emissions of upper-middle-income countries from 1960 to 2012, where both absolute and per capita emissions have experienced sharp rises.[8]

In the case of Malaysia, the growth in CO_2 emission is rising and is higher as compared to per capita emissions, as shown in Figure 2.[8]

Developing countries have experienced rapid motorisation in the past two decades, which has been the main cause of emissions from the transportation sector increasing at a faster rate than in other sectors. The transport sector provides crucial support for social and economic growth. It not only connects distances, but also drives competitiveness. However, hazardous emissions from the transport sector are growing fastest in developing non-Organisation for Economic Co-operation and Development (OECD) countries. Non-OECD countries' transportation energy demands will nearly double and are expected to account for 61% of global energy consumption by 2040.[9] Therefore, choosing sustainable pathways is crucial for developing countries not just to meet the rapid increase in motorisation demand, but also to avoid a massive rise in emissions.

Figure 3 shows that the number of registered vehicles in Malaysia has risen sharply from 11 million in 2000 to 24 million in 2013. Malaysia's

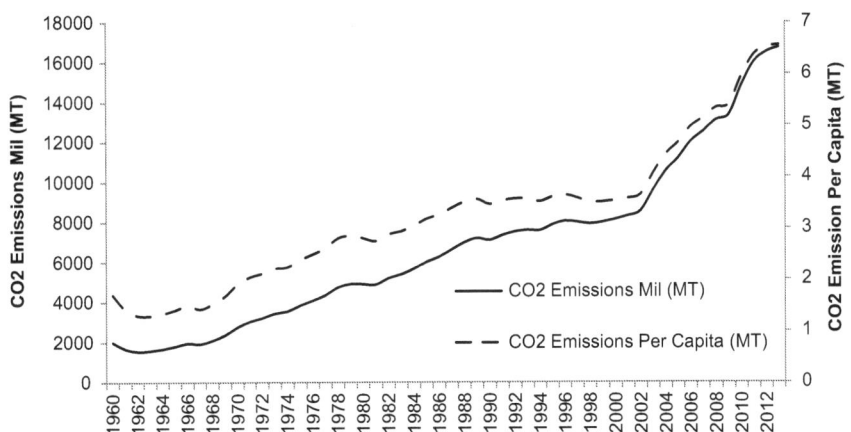

Figure 1 CO_2 emissions of upper-middle-income countries, 1960–2012 (note: Mil = millions).

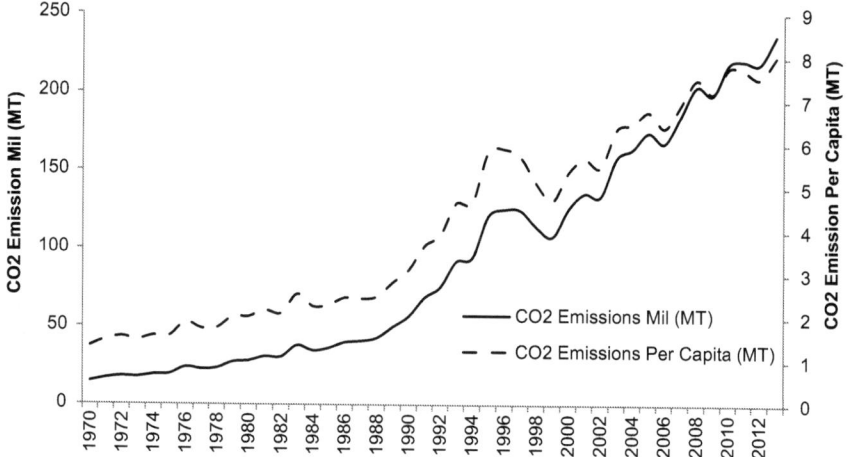

Figure 2 CO_2 emissions of Malaysia, 1970–2012 (note: Mil = millions).

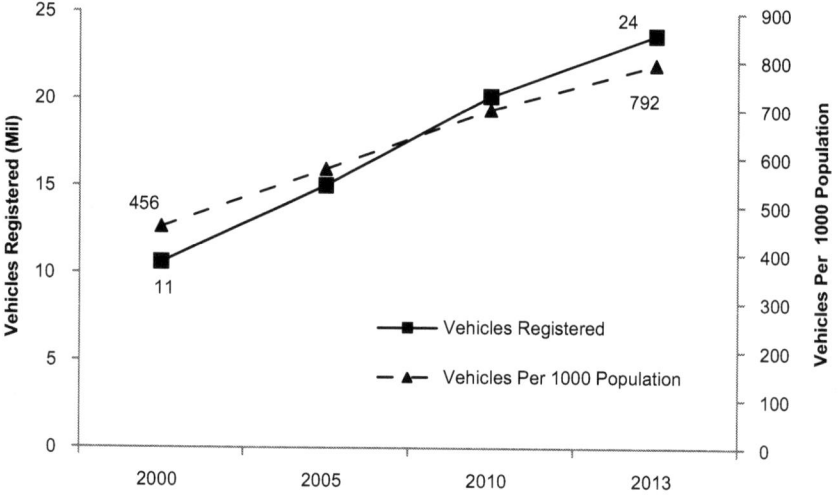

Figure 3 Vehicles registered and vehicles per 1000 members of the population in Malaysia, 2000–2013.

vehicle ownership per 1000 members of the population is among the highest in the region at 792 vehicles per 1000 members of the population in 2013 compared to 456 vehicles per 1000 members of the population in 2010. This increased vehicle ownership intensity inevitably puts pressure on the environment.[10]

In Malaysia, the transport sector is also the largest energy consumption sector at 37%, followed by industry consumption at 30% of total energy consumption. High consumption of fossil-based fuel is expected to continue in Malaysia as vehicle sales are promoted aggressively.

3 Hydrogen Fuel Cell Vehicles and Hydrogen Pathways

There have been debates surrounding the potential role that the hydrogen economy might play in fuelling sustainable economic growth. Despite much scepticism that the hype for the hydrogen economy has come and gone, there is still the expectation that the transport sector may be a good starting point in order to gradually introduce hydrogen into national economies.[11] The re-envisioning of the specific role that hydrogen can play in decarbonising the transport sector, along with other alternatives rather than it being the only solution, is worth considering as part of a sustainable strategy.[1,12]

Compared to other low-carbon technologies, the necessary socio-technical components for hydrogen are less developed. In addition, the adoption of alternative vehicle technologies, particularly using hydrogen fuel cells, has remained a challenge in developing countries. One key dilemma in developing countries relates to designing transport policies that have an impact on fostering hydrogen energy systems in the shorter term (2020), rather than pushing it to the longer term (2050).[13]

Despite these challenges and uncertainties, the IEA projects that demand for hydrogen FCVs is expected to be 150 million by 2050.[14] The added advantage of zero emissions from hydrogen FCVs makes this option very attractive compared to other alternative vehicle solutions in the transport sector. However, the potential for zero emissions depends largely upon the renewable hydrogen pathways chosen by nations and on timely investment in infrastructure.

Using a CGE life cycle analysis approach, for Japan, the potential positive impacts of hydrogen application sectors such as hydrogen FCVs, fuel cells and hydrogen fuelling stations were found to be greater compared to hydrogen generation sectors such as biohydrogen, steam reforming and electrolysis.[15]

Hydrogen is an abundant element and can be produced from various primary and secondary sources. The hydrogen FCV pathway depends on readily available sources in given regions. The pathways include: steam reforming of natural gas and corn ethanol; water electrolysis using grid generation and solar electricity; and coal gasification with and without carbon sequestration. Currently, 50 million tonnes of hydrogen are produced globally, and 96% of hydrogen production comes from natural gas reforming. In order to reduce the environmental impact, CO_2 needs to be captured and sequestered. However, a predominantly non-renewable source is ultimately unsustainable. The potential benefits of zero emissions can be realised from well-to-wheel depending upon the utilisation of renewable resources for the production of hydrogen.

In Asia, biohydrogen is emerging in the bioeconomy and is expected to play an important role in meeting future renewable energy demands. Malaysia is seen as being most sensitive in terms of GDP growth when investing in bioindustries. The fastest-growing biohydrogen industries in Asia are expected to be in India, China and Malaysia by 2050.[16]

Table 1 Hydrogen production methods.

Electrolysis
Plasma arc decomposition
Thermolysis
Thermochemical processes
Photovoltaic electrolysis
Photocatalysis
Photoelectrochemical
Dark fermentation
High-temperature electrolysis
Hybrid thermochemical cycles
Coal gasification
Fossil fuel reforming
Biophotolysis
Photofermentation
Artificial photosynthesis
Photoelectrolysis

Although hydrogen itself is non-polluting, the source and method of production must be taken into account. Hydrogen generation pathways and choice of local feedstock can significantly impact emissions, and this has the added advantage of reducing dependence on imported fossil fuels. Table 1 lists the various hydrogen production methods.[17]

4 Concepts in Fostering Hydrogen in Transportation

The area of sustainability transition of socio-technical systems has gained much attention lately. However, it is in opposition with existing systems that have been locked in due to path dependency.[18] Given compelling arguments for a more sustainable socio-technical transport system, the unwillingness or even inability of governments to take the lead on such critical new technology like hydrogen serves as a reminder of the key impediment in the value system placed on current *versus* future energy systems.

Socio-technical systems are interdependent social and technical systems at a societal level. For transportation, it involves the linking of clusters of several elements, such as regulation, infrastructure, users, markets, production systems and culture, which make up the socio-technical system for land transport as shown by Figure 4.[19]

Energy theorist Cesare Marchetti explained that a new technology needs a very special condition to root.[11] The importance of technology and renewable energy in solving pressing environmental problems is undeniable. This is particularly true for developing Asian economies and the role of innovation and innovative capabilities in critical socio-technical systems.[20] However, the emphasis of this chapter revolves around the role of innovation and the adoption of technology in socio-technical systems that are already available, which is referred to as backstop technology.

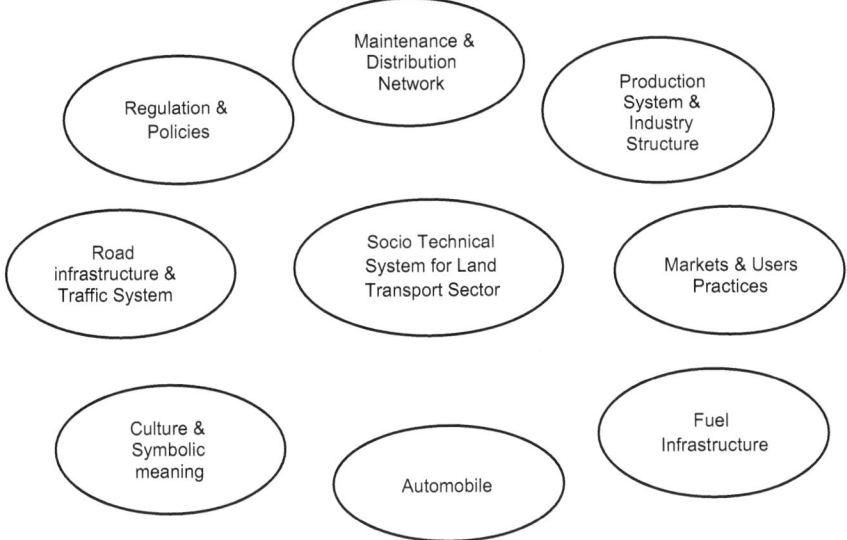

Figure 4 Socio-technical systems for land transport.

Technology as a concept in neoclassical economics is concerned with production function, outputs from various inputs of labour and capital. However, backstop technologies are environmentally friendly technologies that are already available but have yet to be used widely because of their high cost.[21] Economic modelling based on technological change is relatively new, having been developed in the last two decades.[22] In addition, to formulate a successful transition to a sustainable energy pathway, a profound understanding of technical change regarding the scaling of backstop technologies is necessary.[23] Backstop technologies are critical to decarbonising the transport sector as they help governments to pursue climate mitigation and adaptation strategies effectively. Also, backstop technologies offer energy security against supply side shocks because they are free from factor market volatility.[24]

Energy security is an important consideration in energy policy formulation. The IEA defines energy security as "the uninterrupted availability of energy sources at an affordable price." Broader conceptualisation of energy security linking both mitigation and specifically technology adaptation measures in the energy policy literature is relatively new. There is still much confusion in the interpretation of mitigation *versus* adaptation in the transport sector. According to the Intergovernmental Panel on Climate Change, adaptation is adjustment in human systems to actual expected climatic stimuli in order to minimise negative impacts and maximise any benefits. On the other hand, mitigation refers to interventions to reduce causes of climate change. In the current context, for example, mitigation involves fuel subsidy removal for fossil fuels and imposing a carbon tax to discourage the burning of fossil fuels that adversely affect the environment. Meanwhile, adaptation involves building a resilient and sustainable

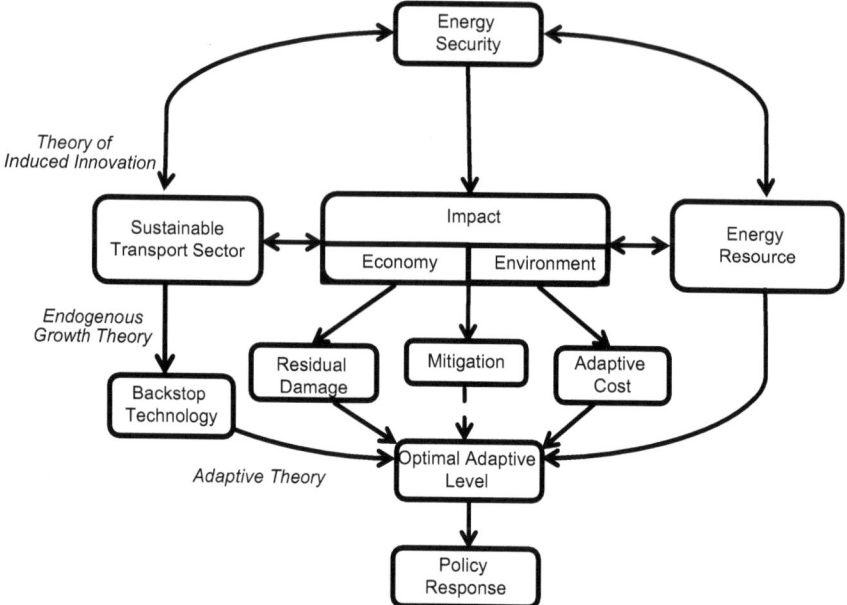

Figure 5 Conceptual framework.

transport system that minimises CO_2 and promotes zero emissions while at the same time meeting energy security and economic growth goals.

Adaptive capacity depends on natural resources, particularly climate variability, and on economic resources. Impacts and damages depend on capability as well as having a two-way relationship between climate variability and economic resources. This damage in monetary terms can be expressed as the summation of residual damage costs and costs of adaptation. By considering the climate variability and economic resources with associated damages, the optimum adaptation level can be obtained. The cost of damage with this optimum level can be used to select the appropriate policy response. Figure 5 shows this conceptual framework.

5 Simulation Experiments

5.1 Dynamic Computable General Equilibrium Model

A dynamic CGE (DCGE) model has been developed for this study. The national sectors are aggregated into non-energy and energy sectors to reflect the fact that hydrogen is used mainly as fuel for fuel cells in transportation. The energy sectors are aggregated into fossil fuel as conventional and hydrogen FCVs as renewable energy sectors. Hydrogen is used as an imperfect substitution with fossil fuels in the production structure followed by an applied general equilibrium framework.[25,26]

The DCGE model investigates the policy responses of the economy by applying selective levels of adoption of hydrogen FCVs and measures the

Development Implications for Malaysia: Hydrogen as a Road Transport Fuel 165

responses of the economy with the economy-wide climatic and macro-economic impacts, together with potentially important solutions, in order to check carbon emissions from the transport sector while at the same time addressing energy security concerns and cost effects.

5.2 Malaysian Social Accounting Matrix

The DCGE model utilises economic inquiry to convert all economic activities and impacts into a common unit based on the total amount as monetary value and then calibrating in common units based on present economic data generated from the Malaysian Social Accounting Matrix.

The DCGE model consists of nine sectors: three institutional agents, two primary factor production methods and the rest of the world. The nine sectors were aggregated from the 2010 Malaysian Input–Output Table that initially comprised 124 sectors. For the purposes of this study, the transport sector is disaggregated into land transport, which represents land motor vehicles, and other transport, which represents the air and water modes of transport. The sectors are:

i. Agriculture
ii. Mining
iii. Manufacturing
iv. Electricity, gas and water
v. Construction
vi. Wholesale and retail trade
vii. Land transport
viii. Other transport
ix. Services

5.3 Model Specifications

The introduction of hydrogen FCV adoption and decarbonised transportation along with the concept of impacts of climate change are entered into our study model in terms of monetary values. The aggregate monetised gross damage (GD) using conventional fossil fuel in transportation is modelled as a function of the climate and carbon variables as:

$$GD_t = \alpha_i \Delta T_t^2 \qquad (1)$$

where the change of global mean temperature ΔT_t^2 compared to a base year is used and α_i is the parameter of the damage function.

Carbon emissions (*e.g.* from transportation) that affect the economy are considered to be quadratic (or at least the power is greater than 1). This allows for increasing impacts on costs when temperatures rise. The climate impact function is:

$$T_t = \alpha_j T_{t-1} + \alpha_k EM_t \qquad (2)$$

Here, EM_t is carbon emissions, α represent a quadratic coefficient over time with temperature rise and the climate impact function is considered in the current year by T_t and from previous years by T_{t-1}.

This study assumed that exogenous shocks, such as an increase in carbon emissions (EM_t) by a certain amount, lead to an increase in the climate impact function (T_t) compared to the level of the period before. The damages grow linearly with transportation output Q_x, a constant fraction of national commodities (denoted Ω as the output coefficient, μ_t as substitution/mitigation efforts and c in the standard applied general equilibrium modeling equations). This linear trend can be influenced by further factors shifting the amount of damages up and resulting in a change of affecting valuation as:

$$EM_t = \Omega \cdot Qx_t(1 - \mu_t - AL_t) \quad (3)$$

The adaptation cost of hydrogen FCV adoption depends on the output as well as the adoption level (AL). The GD depends on output and emission values (ω) as:

$$\frac{AC_t}{Qx_t} = \gamma_{1-t} \cdot AL_t^{\gamma_2} \quad \text{where} \quad \gamma_1 > 0 \quad \text{and} \quad \gamma_2 > 1 \quad (4)$$

$$\frac{GD_t}{Qx_t} = \omega \cdot M_t \quad (5)$$

We assume that the GD function takes the form given in:

$$\frac{GD_t}{Qx_t} = \alpha_1 \Delta T_t + \alpha_2 \Delta T_t^{\alpha_3} \quad \text{where} \quad \alpha_2 > 0 \quad \text{and} \quad \alpha_3 > 1 \quad (6)$$

This is the most commonly used form for damage costs of climate change, where α_3 generally takes a value between 1 and 3.

The monetary value of GDs (GD_t) as a percentage of output (Q_x) is considered to be a summation of RD_t (residual damage) and AC_t (adoption costs) as:

$$\frac{GD_t}{Qx_t} = \frac{RD_t(GD_t, AL_t, AB_t)}{Qx_t} + \frac{AC_t(AL_t, AB_t)}{Qx_t} \quad (7)$$

GD as a percentage of output depends on the residual damage and AL of hydrogen FCV costs for a certain level of action. Thus, the value of net residual damage depends on GD_t and the AL of hydrogen FCV, AL_t.

Finally, decarbonising the transportation sector is influenced by shifting the AL of hydrogen FCVs in the overall transport sector over time with technological change (δB) and the possibility of knowledge (δh) with energy efficiency ($H_{e,t}$), the cost of hydrogen FCVs ($H_{B,t}$) and innovation possibilities or the accumulation of research and development $hR_{i,t}$ as:

$$H_{i,t} = hR_{i,t} + (1 - \delta h) \cdot H_{i,t-1} \quad (8)$$

$$i = E, B$$

6 Scenarios and Results

Table 2 shows a summary of simulations for business as usual (BAU) and six scenarios divided into slow (S1 and S2), moderate (S3 and S4) and aggressive (S5 and S6) of different hydrogen FCV adaptation levels in the transportation sector from 2015 to 2030. The simulation period follows the Malaysian Hydrogen Roadmap and commercialisation by 2030. Hydrogen production in Malaysia is set at conservative, already-available production methods using steam reforming. However, the potential of biohydrogen production is expected to play a greater role in the future. Hydrogen FCVs (HFCVs) began to be introduced in 2015m with the adaptive costs of HFCV infrastructure for all six scenarios beginning almost the same, at between RM6.9 and RM7.7 billion. However, the rate of adaptation has a big impact on the economy and emissions, as shown below.

Figure 6 shows that the estimated GD in a BAU scenario will increase substantially from RM3.43 billion in 2015 to RM7.14 billion in 2030. In

Table 2 Hydrogen FCV adaptation simulation scenarios.

Scenario	BAU	S1 Slow	S2	S3 Moderate	S4	S5 Aggressive	S6
Adaptive level	0	5%	15%	20%	30%	40%	50%
Initial HFCV cost (RM billion)[a]	—	6.9	7.0	7.1	7.2	7.3	7.7
Real GDP growth rate (% change)[b]	—	−0.012	−0.013	−0.014	−0.015	−0.017	−0.021
GD	7.14	1.96	1.87	1.79	1.70	1.53	1.18
CO_2 (MT) from transport	214	203	194	185	176	158	122
CO_2 per capita	15.5	14.7	14.1	13.4	12.8	11.5	8.9

[a]Initial cost for the introduction of HFCV in the transport sector in 2015.
[b]Real GDP growth rate percentage change relative to baseline.

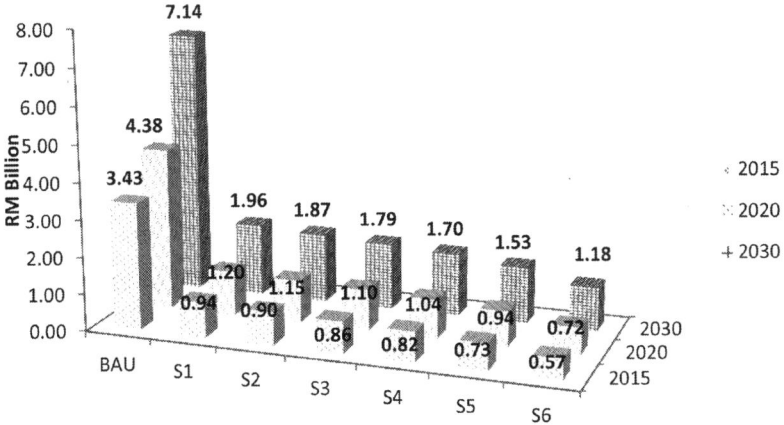

Figure 6 Estimated GDs, 2015–2030.

comparison to BAU scenario, all others scenarios (slow, moderate and aggressive) show sharp reductions of GDs from 2015 to 2030. Slow adaptation S1 and S2 show reductions of between RM0.90 and RM0.94 billion in 2015, which slightly increase to between RM1.87 and RM1.96 billion in 2030. Moderate adaptation S3 and S4 show reduced GDs of between RM0.82 and 0.86 in 2015, increasing to between RM1.70 and 1.79 billion in 2030. However, the potential to reduce GD is best for aggressive scenario S5 and S6, with reductions of between RM0.57 and RM0.73 billion in 2015, only increasing to between RM1.18 and RM1.53 billion in 2030. The GDs can be reduced by between 78% and 83% in 2030 compared to BAU. This is consistent with the literature in which simulated GDs were found to be very high, especially in developing countries. This implies the need for higher levels of adaptation measures and policies that require government subsidies targeted at ramping up hydrogen fuel infrastructure and hydrogen FCVs in Malaysia and other developing countries.

The Malaysian GDP is expected to increase by 5% annually to 2030. It was observed that the real GDP would have a slight loss in all scenarios except at baseline. The range of real GDP percentage losses is negligible, ranging from −0.012% (S1) to −0.021% in (S6), as shown in Figure 7. This is mainly because the costs of hydrogen FCVs are high and cannot be offset completely.[27]

However, as shown in Figure 8, the HFCV benefit is highest in S6, which is attributed to the expected overall cost reduction in an aggressive adaptation scenario. This is also due to the fact that an aggressive scenario strengthens the sectoral impact as compared to a slow or moderate scenario. On a national level, the total hydrogen energy demand is expected to reach 55 million metric tons under aggressive adaptation S6.

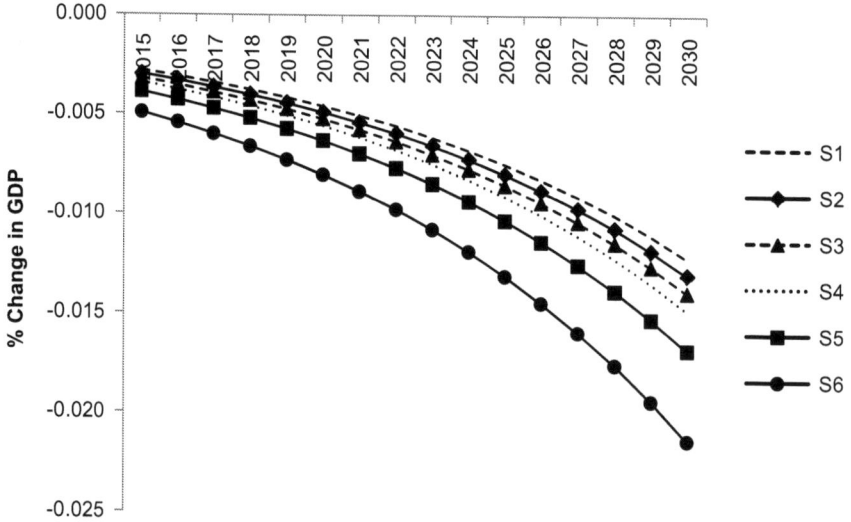

Figure 7 Real GDP, 2015–2030.

Development Implications for Malaysia: Hydrogen as a Road Transport Fuel

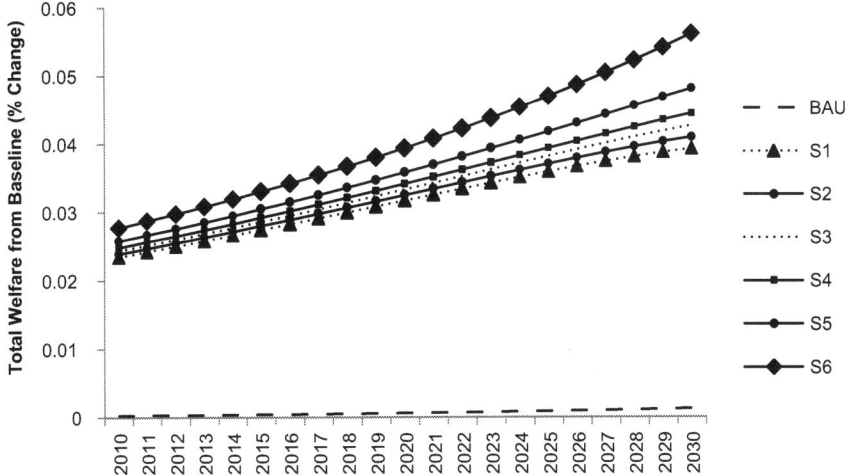

Figure 8 Total welfare, 2015–2030.

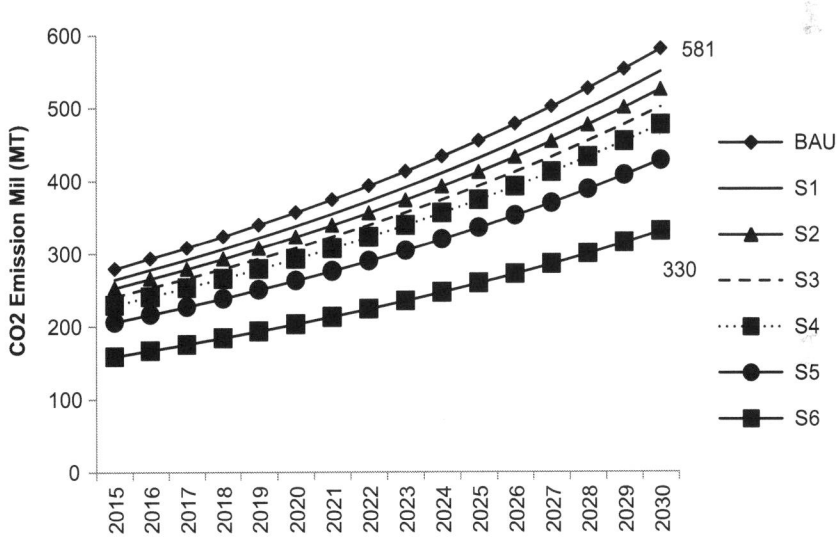

Figure 9 Projected CO_2 emissions of the total Malaysian economy, 2015–2030 (note: Mil = millions).

Figure 9 shows the projected CO_2 emissions of the total Malaysian economy. BAU emissions in 2030 are estimated at 581 $MtCO_2$eq as compared S6 emissions of 330 $MtCO_2$eq. In 2030 BAU, CO_2 emissions will increase by 233% relative to 2005 emissions levels of 174 $MtCO_2$eq. S6 reduced total CO_2 emission by 144% compared to BAU in 2030. This result illustrates the important role that targeted adaptation measures in the transport sector can play in reducing the overall absolute CO_2 emissions of the Malaysian economy.

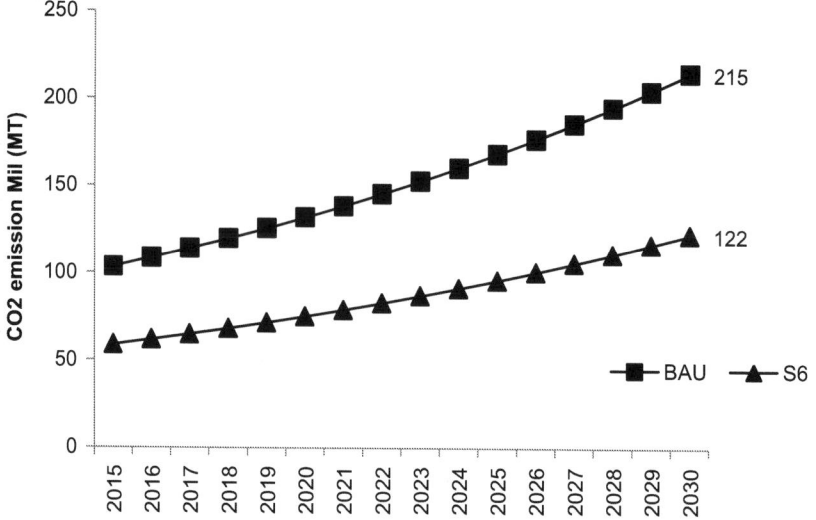

Figure 10 Projected CO_2 emission from the Malaysian transport sector, 2015–2030 (note: Mil = millions).

Given that the total welfare of the economy is shown to be best for S6, Figures 9–11 illustrate only BAU and aggressive S6 results for comparison. Figure 10 shows that the projected CO_2 emissions from the Malaysian transport sector is almost reduced by half. BAU emissions are estimated at 215 $MtCO_2eq$ compared to S6 emissions of 122 $MtCO_2eq$. These emission reductions are only as compelling as the hydrogen production feedstock that can be produced from renewable sources. The aggressive scenario also has an impact on fostering central hydrogen production facilities.

Figure 11 shows the projected CO_2 per capita for Malaysia. In 2005, CO_2 per capita for Malaysia was 6.7 $MtCO_2eq$. BAU projects that CO_2 per capita will increase to 11 $MtCO_2eq$ (64%) in 2020 and 15.5 6.7 $MtCO_2eq$ (131%) in 2030. However, in S6 hydrogen FCV adaptation, CO_2 per capita is expected to be reduced to 6.3 $MtCO_2eq$ (−5.9%) and 8.9 $MtCO_2eq$ (32%) in 2020 and 2030, respectively. Malaysia aims to become an advanced nation by 2020, and under an aggressive adaptation scenario, an important milestone can be achieved by reducing per capita emissions to below 2005 levels. However, due to higher rates of consumption, emissions per capita are expected to rise more for Malaysia by 2030, even under the aggressive adaptation approach of S6.

Figure 12 shows the projected absolute *versus* per capita emissions for Malaysia with and without adaptation of hydrogen FCVs. The BAU scenario shows that CO_2 per capita will increase to 15.5 $MtCO_2eq$ as compared to 8.9 $MtCO_2eq$ for S6. However, the gap in absolute emissions over time increases considerably at 581 $MtCO_2eq$ as compared with S6 emissions of 330 $MtCO_2eq$.

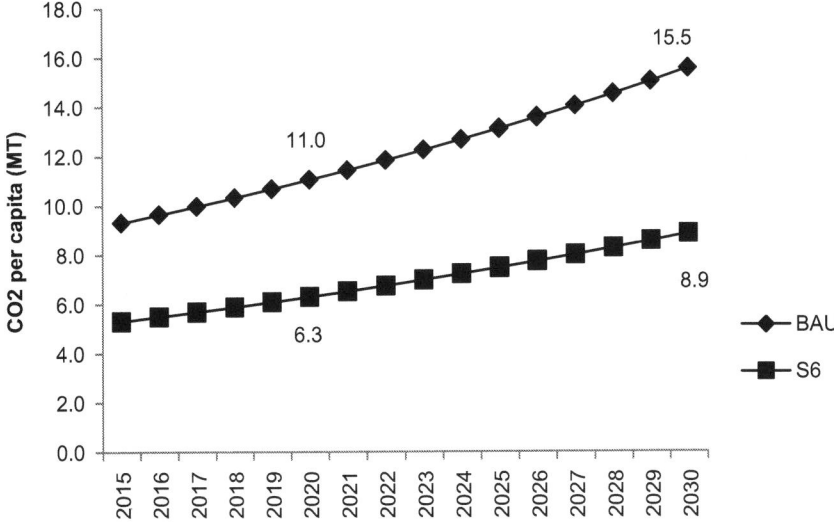

Figure 11 Projected CO_2 per capita in Malaysia, 2015–2030.

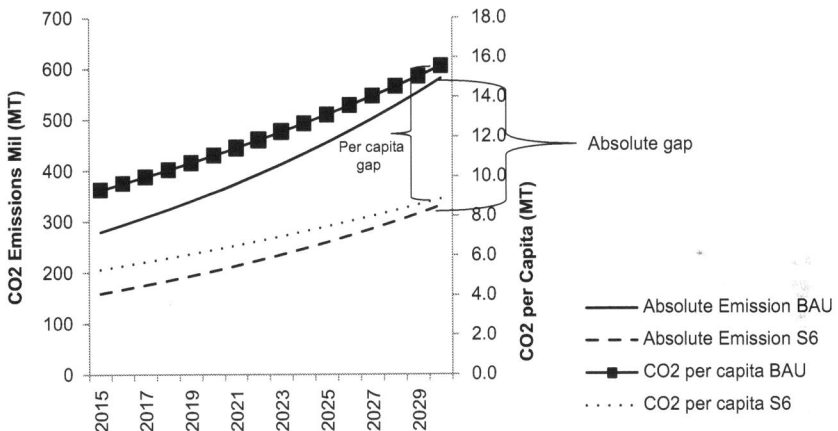

Figure 12 Projected absolute *versus* per capita emissions in Malaysia, 2015–2030 (note: Mil = millions).

7 The Way Forward

There is much uncertainty in the adaptation levels of hydrogen FCVs in Malaysia. This chapter has sought to evaluate the potential impact of backstop technologies, such as hydrogen FCVs, in reducing the uncertainty associated with the diffusion of environment-friendly technologies. We examine the GDs, GDP and emission trajectories of hydrogen FCVs from 2015 to 2030. The simulations done were a BAU scenario, which is dependent on fossil fuels, as well as six adaptation scenarios: slow (S1 and

S2), moderate (S3 and S4) and aggressive (S5 and S6). Our findings show that while all scenarios except BAU show improvements, the aggressive adaptation approach of S6 is best for reducing high levels of GDs and emissions while maintaining overall economic welfare in Malaysia. This is unlike the mitigation efforts in the Malaysian transport sector that tend to prefer a slow or moderate approach, such as reducing fuel subsidies. Slow and moderate adaptation scenarios in the transport sector show negative impacts. Aggressive adaptation is recommended for adopting hydrogen FCVs, which are necessary to offset sectoral impacts. The development of infrastructure for hydrogen FCVs will help Malaysia to meet the hydrogen energy demand of 55 million metric tons by 2030.

Although BAU, slow and moderate scenarios indicate that there will be increases in per capita emissions, absolute emissions will increase at an alarming rate, except for in the aggressive adaptation scenarios. It must also be noted that emissions in the BAU scenario may be higher without infrastructure investment for carbon capture and sequestration. In a US study, which predominantly relied on fossil fuel, even an aggressive adaptation of hydrogen would lead to higher emissions in absolute terms in 2050 compared to 2000 levels. The importance of renewable feedstock in hydrogen production is stressed.[2]

Our simulation results give important insights beyond the COP 21 targets. Given that the atmospheric build-up of emissions and their climatic impacts are attributed to absolute emissions, this illustrates the importance of energy policy context for the adaptation of hydrogen FCVs. Reducing absolute emissions requires adaptive capacity to adopt hydrogen FCVs in order to meet sustainable development goals.

Malaysia needs the development of a hydrogen energy system and fuel cells in the transportation sector in order to be able to go beyond the COP 21 commitment of 45% emissions reductions by 2030 relative to emissions intensities relative to GDP in 2005. A total of 35% of this reduction is on an unconditional basis, and a 10% reduction is conditional upon receiving climate finance, technology transfer and capacity building from developed countries. The long-term prospects of Malaysia and other developing countries' energy system transitions hinge upon their ability to receive and implement these technology transfers in order to decarbonise the fast-growing transport sectors, which will simultaneously reduce absolute emissions and ensure energy security.

While the outcome of this study is illuminating, it has some limitations. Although many variables were analysed in order to meet the objective of the study, some other variables were excluded (*e.g.* emissions at the hydrogen production stage and how specific sectors in the economy might be impacted). Future studies in these areas should be considered in order to obtain a better understanding of the influences and ultimate consequences of adopting hydrogen FCVs in Malaysia.

References

1. J. Andrews and B. Shabani, *Int. J. Hydrogen Energy*, 2012, **37**, 1184–1203.
2. W. Dougherty, S. Kartha, C. Rajan, M. Lazarus, A. Bailie, B. Runkle and A. Fencl, *Energy Policy*, 2009, **37**, 56–67.
3. N. Balta-Ozkan and E. Baldwin, *Int. J. Hydrogen Energy*, 2013, **38**, 1209–1224.
4. J. Alazemi and J. Andrews, *Renewable Sustainable Energy Rev.*, 2015, **48**, 483–499.
5. B. D. Solomon and A. Banerjee, *Energy Policy*, 2006, **34**, 781–792.
6. N. H. Stern, *The Economics of Climate Change: The Stern Review*, Cambridge University Press, 2007.
7. W. D. Nordhaus, *J. Econ. Lit.*, 2007, **45**, 686–702.
8. The World Bank, The World Bank Indicators, CO_2 Emissions, 2016.
9. U.S. Energy Information Administration, International Energy Outlook, 2016.
10. Economic Planning Unit. Malaysia, *Eleventh Malaysia plan (11th MP)*, Putrajaya, 2011.
11. P. Hoffmann and B. Dorgan, *Tomorrow's Energy: Hydrogen, Fuel Cells, and the Prospects for a Cleaner Planet*, MIT Press, 2012.
12. M. Balat, *Int. J. Hydrogen Energy*, 2008, **33**, 4013–4029.
13. L. K. Mytelka and G. Boyle, *Making Choices About Hydrogen: Transport Issues for Developing Countries*, IDRC, 2008.
14. International Energy Agency, *Technology Roadmap Hydrogen Fuel Cells*, 2015.
15. D.-H. Lee, *Int. J. Hydrogen Energy*, 2014, **39**, 19294–19310.
16. D.-H. Lee, *Int. J. Hydrogen Energy*, 2016, **41**, 4333–4346.
17. I. Dincer and C. Acar, *Int. J. Hydrogen Energy*, 2015, **40**, 11094–11111.
18. J. Markard, R. Raven and B. Truffer, *Res. Policy*, 2012, **41**, 955–967.
19. F. W. Geels, *Technological Transitions and System Innovations: A Co-evolutionary and Socio-technical Analysis*, Edward Elgar Publishing, 2005.
20. F. Berkhout, D. Angel and A. J. Wieczorek, *Technol. Forecast. Soc. Change*, 2009, **76**, 218–228.
21. W. D. Nordhaus, H. Houthakker and R. Solow, *Brookings Pap. Econ. Act.*, 1973, **1973**, 529–576.
22. G. M. Grossman and E. Helpman, *Am. Econ. Rev.*, 1990, **80**, 86–91.
23. N. Rosenberg, *Inside the Black Box: Technology and Economics*, Cambridge University Press, 1982.
24. M. Liski and P. Murto, *HECER WP-Paper*, 2006.
25. S. Robinson, A. Yùnez-Naude, R. Hinojosa-Ojeda, J. D. Lewis and S. Devarajan, *North Am. J. Econ. Finance*, 1999, **10**, 5–38.
26. K. A Reinert and D. W Roland-Holst, *Applied Methods for Trade Policy Analysis: A Handbook*, 1997, pp. 94–121.
27. G. Wang, *Int. J. Hydrogen Energy*, 2011, **36**, 1766–1774.

Latest Trends and New Challenges in End-of-life Vehicle Recycling

JEONGSOO YU,* SHUOYAO WANG,* KOSUKE TOSHIKI,* KEVIN ROY B. SERRONA,* GENGYAO FAN* AND BAATAR ERDENEDALAI*

ABSTRACT

Rapid urbanization combined with increases in per capita income have led to the world's motorization. Car manufacturers have been increasing their outputs in response to demands for fuel-efficient cars. New-generation vehicles (NGVs) like hybrid and electronic vehicles have emerged with significantly reduced weights and robust computerized systems. Likewise, the use of efficient batteries like nickel–hydrogen or lithium-ion batteries has been integrated into NGVs. Expectedly, a large volume of these cars will be classified as end-of-life vehicles (ELVs) in the near future. The European Union, Japan and Korea have developed ELV recycling systems based on the extended producer responsibility principle, designed to recover used car parts, scrap metal, batteries, *etc*. However, countries like Mongolia and the Philippines are inundated with imported used cars, including used NGVs, which damage the environment due to their poor emissions and hazardous waste components. Air pollution and lead soil contamination are some of the environmental problems associated with the proliferation of ELVs. This paper tackles the opportunities and challenges of ELV recycling as espoused by the European Union, Japan and Korea and its potential application in developing countries. It also discusses emerging trends in the effective utilization of waste batteries from NGVs.

*Corresponding authors.

1 Introduction

Worldwide car ownership was about 124 million units in 2014. This is 1.4-times as many as 10 years ago. The increase is largely due to rapid motorization in Asian countries. Next-generation vehicles (NGVs) are the backbone of this market today. Meanwhile, a large quantity of end-of-life vehicles (ELVs) will be dumped in developing countries in the near future.

Developed countries/regions like the European Union (EU) and Japan have had ELV recycling systems since 2000 to ensure proper treatment of harmful materials and to recycle valuable resources. This system is based on the extended producer responsibility (EPR) principle; however, it is difficult to closely monitor and control the export of used cars, which is nearly equal to the number of ELVs in developing countries. In fact, it is difficult to pin down the responsibility and cost or burden of ELV recycling.

On the other hand, most automobile manufacturers have focused energy on developing eco-friendly vehicles that have features like low fuel consumption, weight savings and computerized systems. It is necessary to develop new recycling technologies for the NGVs of the distant future because they will be composed of different substances and parts.

It is worth noting that Asian developing countries remain heavily dependent on used car imports from Japan for their motorization. Nobody cares about potential cross-border environmental problems, since such imports are legal transactions under international trading rules and law. Remarkably, old hybrid cars continue to increase at a rapid pace in Mongolia. Therefore, there is a need to enquire as to whether such a scheme will be able to reduce air pollution and save energy, or whether they will all end up in disposal facilities.

2 Legislation on End-of-life Vehicle Recycling and Its Implications

2.1 Background on the Evolution of Legal Systems

As a result of the motorization around the world, ELVs started to emerge dramatically in the 1980s and the proper treatment of ELVs has become a serious challenge.

Since 2000, developed countries like Germany, France, Japan and Korea have passed legislation on ELV recycling to ensure that valuable parts are recycled and hazardous wastes are treated and disposed of properly. Generally, about 75–80% of an ELV's weight will be recycled in developed countries and the remaining 20–25% will be landfilled in the form of automobile shredder residue (ASR).[1] However, the price of scrap iron has plummeted since the 1990s and ELVs are no longer treated as valuable resources. Therefore, there were concerns that ASR, chlorofluorocarbons (CFCs), airbags and other waste from ELVs might be dumped illegally or treated inappropriately.[2]

In the 1990s, Professor Thomas Lindhgvist from Lund University proposed the concept of EPR for the very first time.[3] Afterwards, the Organisation for Economic Co-operation and Development (OECD) provided further insights into EPR and defined it as "a policy approach under which producers accept significant responsibility (financial and/or physical) for the treatment or disposal of post-consumer products."[4]

Based on the principles of EPR, the EU executed an ELV Directive on the 21 October 2000 stipulating that vehicle manufacturers should undertake financial and/or physical responsibility in ELV recycling. Pursuant to the said Directive, EU member nations also made their respective municipal laws in order to ensure the proper management of ELVs.[5]

Influenced by the EU's ELV Directive and the 'Teshima Affair' (which refers to the largest illegal industrial waste dumping case in the postwar period),[6] Japan implemented the 'End-of-Life Vehicle Recycling Law' in January 2005 to recycle ELVs with focus on ASR, CFCs and airbags.

Korea enforced the 'Act on the Resource Circulation and Electronic Equipment and Vehicles' in 2008 and focused on strictly monitoring the process of ELV recycling.

China also promulgated the 'Technical Policies for the Recycling of Automotive Products' in 2006. Although it is not a law yet, the country is evidently planning to introduce their own EPR-based ELV recycling law in the next few years.

In this section, the actual operational status and the problems associated with EPR-based ELV recycling of each country mentioned will be explained.

2.2 Comparison of EPR-based ELV Recycling Laws

Ever since the concept of EPR was put forward in the 1990s, multiple countries have published their own EPR-based ELV recycling laws. In this section, discussions will be centered on countries or regions that are leading in the production, consumption and registration of vehicles, such as the EU and East Asia (mainly Japan, China and Korea). The USA is exclude, because its vehicle recycling law is not based on EPR.

2.2.1 Implementation of EPR in the EU.

The EU's ELV Directive was adopted by the European Parliament in September 2000 and came into force from October 2000. The Directive mandated vehicle manufacturers must share the responsibility of taking back ELVs without commission and to recycle them while bearing all or a substantial part of the recycling costs. It required that the recycling rate of ELVs should surpass 85% (the energy recovery ratio should be less than 5%) by 2006, and surpass 95% (the energy recovery ratio should be less than 10%) by 2015. Meanwhile, about 9 million vehicles were deregistered throughout the EU in the year 2000, and about 25% of these vehicles were landfilled as ASR without proper treatment in advance.[7]

Figure 1[8] shows the recycling rates of 24 EU member countries and the GDP per person in 2016.[9] Since multiple municipal laws were enforced among its members, the EU's goal for the ELV recycling rate for 2006 was considered to be 85% (energy recovery rate excluded). It is clear that among the 24 EU countries, only 12 reached the goal or an achievement rate of only 50%. Additionally, according to the results of the regression analysis, there is no connection between the ELV recycling rate of each country and their nominal GDP per person; that is to say, the increment in nominal GDP per person does not guarantee the increment of the ELV recycling rate. Moreover, the major vehicle manufacturers of the EU are mainly located in Germany, France, England, Italy and Sweden. Supposing that vehicle manufacturers' efforts lead to an increment in the ELV recycling rate, the ELV recycling rate of these five countries should be higher than other countries. However, according to Figure 1, three out of these five countries failed to accomplish their goal (Italy stands at 72.7%, France at 81% and England at 82.3%). Besides, only recycled ELVs with deleted registrations were covered in the calculation of the ELV recycling rate, and so the actual ELV recycling rate may be even lower.

Based on the EU's ELV Directive and the actual conditions in each member country, vehicle manufacturers gradually built cooperative relationships with recycling companies (including dismantling, shredding and sorting operators) to take back ELV without compensation and recycle them properly. However, with free trading and the policy shift (even if the value of ELVs is comparatively high, vehicle manufacturers will not offer any economic compensation to ELV owners), a large amount of ELVs were being exported to developing countries in Eastern Europe, like Poland, from developed countries, like Germany. In fact, about 80% of ELVs in Germany were exported, and even though some of these importing countries may have enforced their national laws based on the Directive, the actual implementation is insufficient and many ELVs were actually recycled by

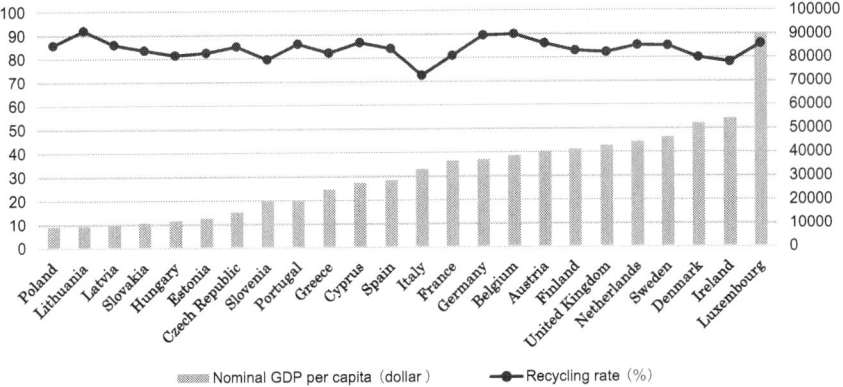

Figure 1 Recycling rate and nominal GDP per person in the EU member states (2006).

non-authorized recycling companies, causing severe pollution. For example, about 90% of ELVs were being recycled by non-authorized recycling facilities in Poland (which is the major recipient country of second-hand vehicles in Eastern Europe).[10]

2.2.2 Implementation of EPR in Japan. Japan amended the 'Waste Management and Public Cleansing Law' in 1994, in which ASR was mandated to be landfilled from 1995. However, the reduction in the remaining capacity of final disposal sites led to a sharp rise in ASR disposal fees. Furthermore, the price of scrap iron slumped greatly from the 1990s to 2001, and as a result, ELVs became a burden because they have a very low resource value, and there was a significant risk that ELVs or ASR may be dumped illegally.[11] In response, vehicle manufacturers and related traders were obliged to take the responsibility for ELV recycling, and so there was an urgent need to establish an EPR-based ELV recycling system.

As a result, Japan enacted the 'End-of-life Vehicle Recycling Law' in 2005 and demanded that automobile manufacturers share the responsibility to take back and properly treat ASR, CFCs and airbags from ELVs. It was also stipulated that the recycling fee should be paid by the last owner of the vehicle.

There are two teams that share the responsibility for recycling ELVs in Japan: the TH team, which includes Toyota and Honda, and the ART team, which includes Nissan, Mitsubishi, Mazda and so on. According to the trial balance sheets of these two teams, they had already started to achieve profits of hundreds of millions of yen every year by recycling ASR, CFCs and airbags. Table 1 shows the income and expenditure balances of Toyota and Nissan.

However, according to the cost burden of recycling and the subjects that were required to be recycled, the law is not based on the principle of EPR in a strict sense. According to Table 1, the reason why the recycling rate in Japan is high is mainly because of the recycling fee, which was levied on vehicle owners. In fact, the surplus of the recycling fund has reached 13.6 billion yen and it was not being utilized adequately.[13]

Table 1 ELV recycling rates and income/expenditure balances of Toyota and Nissan.[12]

Vehicle manufacturer	Item	2013	2014	2015
Toyota	Profit (hundred million yen)	9.42	4.25	3.08
	Recycling rate for ASR	96%	97%	97%
	Recycling rate for airbags	94%	94%	93%
	Recycling rate	99%	99%	99%
Nissan	Profit (hundred million yen)	8.86	7.84	5.25
	Recycling rate for ASR	97.2%	97.6%	98%
	Recycling rate for airbags	94.1%	94.2%	93.3%
	Recycling rate	99%	99%	99%

2.2.3 Implementation of EPR in Korea. Korea enacted the 'Act on the Resource Circulation of Electronic Equipment and Vehicles' in April 2007, and enforced this law from January 2008. However, the existing circumstances of Korea's recycling industry were not fully considered and a 1 year grace period was set. Despite this, the necessary infrastructure to ensure proper recycling of ELVs was not established and the monitoring system was too strict, while the allocation of cost and responsibility has not been decided clearly. In effect, the law failed to be fully operationalized despite it being implemented over the last 7 years.[14]

To reverse this situation, the Ministry of Environment in Korea tried to strengthen EPR by showing the achievements of the monitoring system and by constructing a large-scale recycling company in 2012. However, dismantling companies struggled with the heavy responsibility of recycling ELVs. Based on the ELV recycling law in Korea, as long as ELVs are not chargeable,[15] automobile manufacturers do not have the responsibility to recycle ELVs for free, so discussion on the proper treatment method to recycle ASR, CFCs, and airbags and the cost burden for each stakeholder is continuing. It is important to prescribe the responsibility and associated costs in ELV recycling without necessarily knowing the exact cost of recycling a vehicle.[14]

2.2.4 EPR System in China. China enacted the 'Technical Policies for the Recycling of Automotive Products' in 2006, which aims to establish ELV recycling regulations based on EPR. China has not only focused on the recovery of recyclable materials and the reduction of environmental load during the recycling process, but also on the medium- and long-term industrial development plan with recycling companies.

The provisions for punishment and economic incentives were set out clearly in these guidelines, but these policies have no actual legal force. As a result, the responsibility of automobile manufacturers and the plans that were mentioned in the guidelines became invalid. The responsibility for recycling of ELVs lies entirely with dismantling companies, and automobile manufacturers provide limited information on ELV recycling. In addition, only 30% of deregistered vehicles were recycled by authorized operators in China, while the remaining 70% were dismantled by non-authorized operators or were transported to remote areas.[16]

Furthermore, according to China's 'Regulation on the Disposal of End-of-life Vehicles', the resale of used engines, steering apparatus, transmission gears, axles and frames from ELVs is prohibited for safety reasons. With these limitations, dismantling and used parts sales are not competitive businesses in China. Therefore, it is difficult for China's dismantling companies to introduce the necessary equipment to ensure environmental conservation or to enhance recycling efficiency. Manual dismantling is typical in China and the efficiency of this process is low.

Meanwhile, China is planning to enact the 'New Regulation on the Disposal of End-of-life Vehicles', which will permit the sale,

3 Popularization of Next-generation Vehicles and Their Impact on Vehicle Recycling

3.1 Significant Developments in the Popularization of Next-generation Vehicles

The sale of next-generation vehicles has been increasing rapidly in the last couple of years. About 2.34 million NGVs, including hybrid vehicles (HVs), plug-in hybrid vehicles (PHVs) and electronic vehicles (EVs), were sold around the world in 2015. This number is about 2.5-times larger than 5 years prior.[17] As well as developed countries and regions like Japan, the USA and the EU, developing countries in Asia, such as China and India, are developing NGVs, too. The USA and China are aiming at developing EVs because they are being left behind in HVs compared to Japan, and hence EVs will come into wider use than HVs in the near future. The quantity of end-of-life NGVs is small at present, but is expected to increase drastically along with the popularization of NGVs.

Two significant features of NGVs are weight reduction and computerization. In fact, to improve the fuel efficiency of a vehicle and avoid accidents, the use of plastic and alloy metal and electronic equipment in vehicles has already increased. To promote NGVs' performance, HVs, PHVs and EVs are also equipped with high-efficiency batteries, mainly the nickel–hydrogen battery (NiMH) battery or the lithium-ion battery (LIB). As a result, the constitution of NGVs differs from those of pre-existing vehicles. For example, when comparing vehicles that were manufactured in 2011 with vehicles that were manufactured in the 1980s, the proportion of steel drops by 14%, while the proportion of plastic rises by 5%.[18] Logically, this will lead to a change in the composition of ASR. Thereupon, it is necessary to probe new technologies and approaches to recycling end-of-life NGVs properly.

One of the most important components of a vehicle—the battery—is also changing. For many years, vehicles were only equipped with a lead battery' however, since the Prius was launched onto the market in 1997, the NiMH battery started to be integrated in HVs, and then with the sale of advanced HVs, PHVs and EVs, the LIB also came into wide use. As for the lead battery, despite its high content of harmful substances and its enormous volume, it can be recycled properly without causing any pollution under careful recycling procedures. Considering that the utilization of used NiMH batteries will emerge in the near future and that the LIB is also being widely used of late, it is urgent to explore methods to utilize and recycle NiMH batteries and LIBs.

In this section, the sale conditions and characteristics of NGVs will be discussed, as well as the changes in vehicle content and components (especially the vehicles' batteries) and their impacts on the vehicle recycling industry.

3.2 Trends in NGV Popularization

The sale of next-generation vehicles started with the Prius. In 1997, the PHV, EV and fuel cell vehicle (FCV) came onto the market. However, the most popular next-generation vehicle is still the HV.

Increments in the number of HVs accelerated in 2011 and paved the way for the actual popularization of NGVs. From 2010 to 2015, the sales of HVs increased twofold and dominated about 80% of the NGV market share.[19] In Japan, the number of NGVs, particularly HVs, has boomed since 2009 as a result of Japan's tax reduction and subsidies policy for environmentally friendly vehicles. In 2015, Japan held about 67% of the HV global market share.[20] About 20% of new car sales in Japan were for NGVs in 2014, and this number will increase to 50–70% by 2030 according to the Government's goal.[21] Apart from that, Japanese automobile manufacturer Toyota is also aiming at developing HVs and EVs and is planning to reduce obsolete gasoline and diesel engines by 2050.[22]

However, the growth of HVs is slowing, while sales of EVs are beginning to increase. The American automobile manufacturer Tesla Motors is planning to put its Model 3 on sale in 2017, and by reducing the price by 50%, the Model 3 may lead to brisk sales of EVs.[23]

The largest vehicle market—China—is also developing 'new energy vehicles', which refer to PHVs, EVs and FCVs, and is planning to manufacture and sell 5 million EVs and PHVs by 2020.[24] Thus, as shown in Figures 2[25] and 3,[26] the growth rate of EVs has already surpassed the growth rate of HVs and is predicted to take 10% of the global market for NGVs by 2020. To conclude, NGVs such as HVs and EVs are increasing sharply in number, and in particular, EVs are increasing even faster than HVs. In spite of the fact that the main type of end-of-life NGV is the HV at present, from a long-term point of view, end-of-life EVs will also emerge enormously in the near future.

3.2.1 Changing Composition of Vehicles.
Technologies to improve vehicle fuel efficiency and to avoid accidents automatically have been widely used in NGVs. Moreover, the weight reduction and computerization of NGVs are believed to be more significant than in pre-existing vehicles. However,

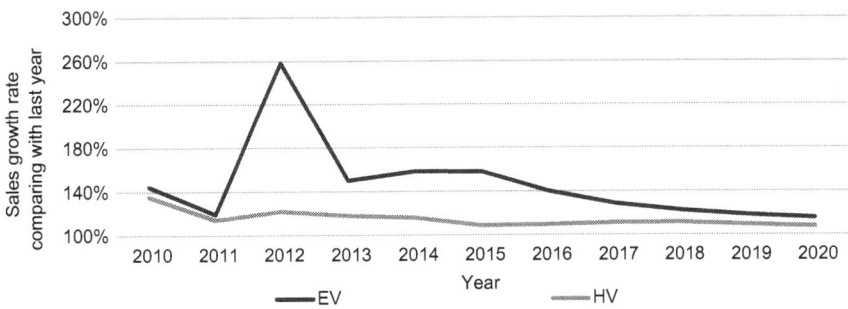

Figure 2 Time series of global sales growth rates for HVs/EVs.

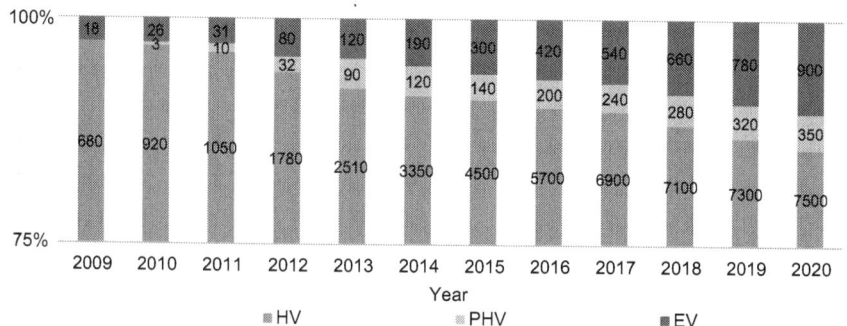

Figure 3 Time series and predictions of market shares for HVs/PHVs/EVs (unit: thousands).

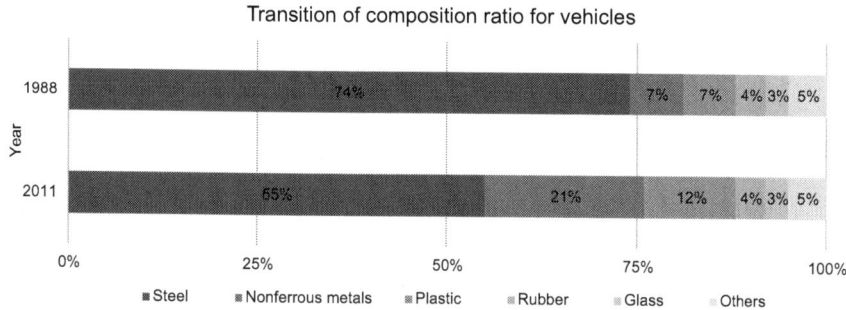

Figure 4 Transition of composition ratios for vehicles.

the composition of pre-existing vehicles has already changed greatly due to the massive use of plastic and precious metals in such vehicles.

As shown in Figure 4,[28] when comparing vehicles manufactured in 2011 with vehicles manufactured in 1988, the steel content has dropped by 19%, whereas the content of nonferrous metals has increased by 14%, and the plastic content has risen by 5%.[27]

Considering that NGVs are also equipped with a NiMH battery or a LIB, especially for EVs that have no engine at all, the steel content could be less than 50%. As a result of the decrease in the content of steel in ASR and the increase in the use of plastics, it is difficult to recycle NGVs efficiently by traditional measures.

3.2.2 Trends in the NGV Battery Market. There are mainly two kinds of battery for NGVs: the NiMH battery and the LIB. As shown in Table 2, the former is used for HVs and the latter is for PHVs and EVs. However, some of the latest model HVs are also equipped with LIBs; therefore, it is believed that in the future, the NGV battery will mainly be LIBs. Furthermore, although PHVs and a subset of HVs comprise 20% of the global LIB market, the EV share of the global LIB market is and will remain at

Table 2 Comparison of the NiMH battery and the LIB.[30]

	NiMH battery	LIB
Power storage capacity	Little	Large
Usage state	HV	EV/PHV/HV

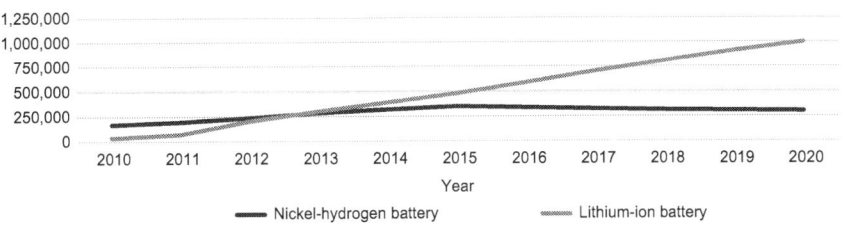

Figure 5 Transitions and predictions of global sales for NiMH batteries/LIBs (unit: million yen).[31]

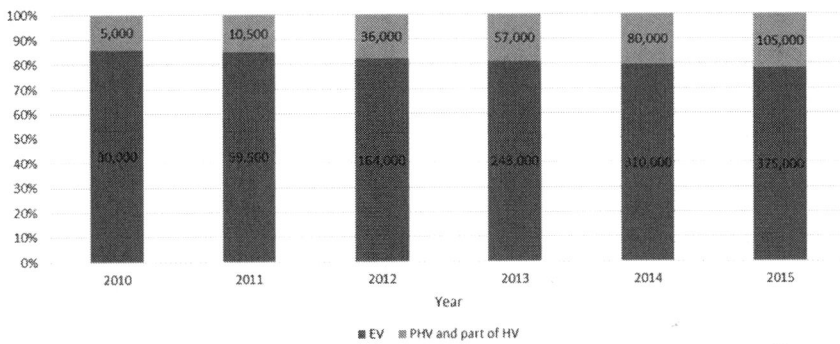

Figure 6 Time series and predictions of global market shares of LIBs for EVs/PHVs/HVs (unit: million yen).[32]

about 80%.[29] That is to say that, in view of the popularization of EVs, the number of LIBs will also increase accordingly Figures 5 and 6.

3.2.3 Future of the End-of-life NGV Battery. Although the NGV battery is changing from the NiMH battery to the LIB, early versions of HVs, which have already been in widespread use, were equipped with NiMH batteries. Therefore, the most common end-of-life battery from NGVs should be the NiMH battery at present. Also, since the NiMH battery in HVs only works as an auxiliary power system, battery degradation is slow, and so the lifespans of NiMH batteries and HVs are generally believed to be the same.

On the other hand, developed countries (especially Japan) export a huge number of second-hand or even virtually end-of-life HVs every year. As a result, the amount of used NiMH batteries is small at present, but it may be larger than 50 000 sets a year by around 2020–2025 due to the massive increase in end-of-life HVs.

In the EV case, in which LIBs serve as the primary power source, the residual content of the battery needs to be maintained at least at 70% to ensure that the vehicle is functional. However, vehicles need high output power at startup and during acceleration, and when the battery is degraded down to 30%, it is difficult to meet this need.[33] Although it may take 10 years or 100 000 km of travel for the battery to degrade to 30%,[34] the lifespan of LIBs can be easily affected by the actual usage (*e.g.* in China, travelling distances are comparatively long and can easily reach 100 000 km). If the battery is in an alternating succession of discharging and recharging, degradation may occur much faster; that is to say, the battery may degrade much faster than expected, and may only take 5–7 years before it needs to be changed. If so, the battery may need to be replaced two or even three times before the vehicle is recycled. In other words, end-of-life LIBs may arise much faster, in greater quantities and more frequently than expected. Hence, massive numbers of end-of-life LIBs may emerge at no later than 2022.

As mentioned above, massive ELV battery numbers may emerge soon, and an effective measures to reuse and recycle these batteries urgently need to be explored. Therefore, Japan's vehicle manufacturers—Toyota, Nissan and Mitsubishi—have already prepared a manual for battery recycling and have started to reuse and/or recycle NiMH batteries and LIBs. Here, we will take the two biggest next-generation vehicle makers in Japan—Toyota and Nissan—as examples in order to introduce the automobile manufacturers' schemes for battery recycling.

3.3 Effective Utilization of Waste Batteries from Next-generation Vehicles

3.3.1 Reuse and Recycling of Nickel–Hydrogen Batteries. Toyota, which leads the popularization of HVs, has already organized a scheme to recycle NiMH batteries[35] and has started a 'battery-to-battery' recycling project.[36] As shown in Figure 7,[37] Toyota classified the recycled NiMH batteries into three ranks. Rank A batteries, which are of the best quality, will be resold as replacement batteries for vehicles, Rank B batteries will be reused as stationary accumulators in vehicle stores, and Rank C batteries, which are poor in quality, will be recycled in order to recover nickel for the battery's positive electrode material.

3.3.2 Reuse and Recycling of LIBs. Lithium and manganese, which are the main materials of the LIB positive electrode active material, are abundant and are cheap in price, and so it is not urgent to recycle LIBs at present.[38] In fact, LIBs are mainly being reused. Nissan and Sumitomo Corporation established the 4R Energy company and utilize LIBs *via* the '4R' method, which refers to 'refabricate', 'resell', 'reuse' and 'recycle'. As shown in Figure 8,[40] batteries with remaining capacities of 70–80% will be reused; 'refabricate' involves making a new battery by changing the

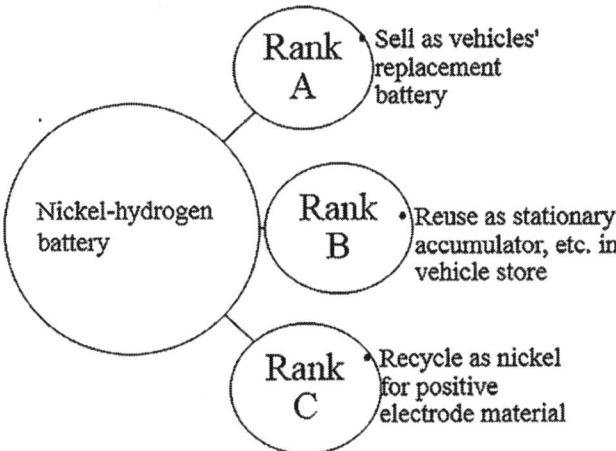

Figure 7 Toyota's scheme for battery reuse/recycling.

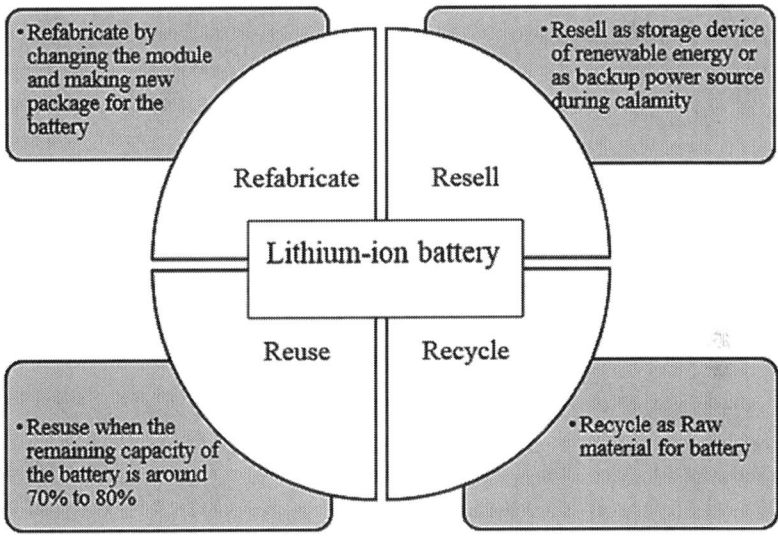

Figure 8 Nissan's scheme for battery reuse/recycling.

module and the packaging of the battery, while 'resell' means that the battery will be resold as a storage device of renewable energy or as a backup power resource in times of disaster. 'Recycle' refers to recycling the battery as raw materials for making batteries.[39]

3.4 Limitations on the Reuse and Recycling of Batteries

As mentioned above, it is common for NiMH batteries and LIBs to be exported along with second-hand vehicles. In fact, the number of end-of-life

Table 3 Actual achievement of automobile manufacturers' recycling schemes for batteries.[42]

	NiMH battery	LIB
The main vehicle makers that organized the battery recycle scheme	Toyota, Nissan, Honda, Mazda, Mitsubishi, Fuji	Toyota, Nissan, Honda, Mazda, Mitsubishi, Fuji, Suzuki
Results of battery recycling in 2013	3083 sets	35 sets

NiMH batteries that were recycled under the automobile manufacturers' schemes was only 3083, and the number of LIBs was only 35 in 2013 (Table 3).[41]

Furthermore, although automobile manufacturers are trying to reuse batteries as accumulators or backup power resources, there will be a limitation to this market. Considering the increment of next-generation vehicles, it is more appropriate to reuse these batteries as batteries again; in other words, 'vehicle-to-vehicle' reuse may be the best approach to utilizing these end-of-life batteries properly.

To summarize, massive numbers of end-of-life NiMH batteries and LIBs are soon to emerge. Therefore, an efficient system to reuse and/or recycle NiMH batteries and LIBs should be enacted in vehicle recycling law, and it is urgent that such an approach is established.

4 Effects of Second-hand Vehicle Exportation on International Resource Circulation and Emerging Cross-border Environmental Problems

4.1 The Two Sides of Second-hand Vehicle Exportation

Vehicle numbers in rapidly developing Asian countries such as China and India are increasing. Similarly, motorization in low-income countries such as Mongolia, Myanmar and Sri Lanka is also accelerating with the increased importation of second-hand vehicles. In Mongolia, over 80% of registered vehicles are second-hand vehicles from Japan and Korea. However, these vehicles are usually of low grade and are being overused, and so it is highly possible that these vehicles will cause environmental and safety problems.

In developed countries, the popularization of NGVs (mostly HVs) began in recent years. In the case of Japan, HVs entered the market in 1997, and the number of HVs reached 4.7 million as of March 2015.[43] Although Japanese automobile manufacturers are making efforts to take back and recycle end-of-life HVs, over 50% of end-of-life HVs are being exported to countries such as Mongolia (vehicles that have been used for more than 6 years) and Russia (vehicles that have been used for less than 5 years). Unfortunately, the treatment of these vehicles after they are discarded is not transparent.[44]

The exportation of second-hand vehicles or second-hand components will certainly improve the recycling rate for developed countries with enforced

ELV recycling laws. However, if importing countries fail to treat or recycle these vehicles properly, large amounts of resources will be wasted, causing cross-border pollution problems. For example, although the beneficial use of second-hand vehicle components prevails in Mongolia, the country does not have the ability to recycle iron from vehicles properly, nor to efficiently recycle nonferrous and precious metals. Furthermore, inappropriate treatment and/or illegal dumping of harmful and high-risk components (*e.g.* airbags, refrigerant gas, waste liquid, waste oil, plastic, electronic components and batteries) are also quite common in Mongolia due to technical and financial limitations. Although the importation of second-hand vehicles can encourage economic development in developing countries, the management of vehicles, ELV recycling regulations and the introduction of recycling technologies and human resources usually do not match the increment of vehicles. Therefore, problems such as the degradation of the metropolitan environment and ineffective utilization of resources need urgent resolution.

ELV recycling regulations in the EU, Japan and Korea are mainly focused on solving the illegal dumping of ELVs and the inappropriate treatment of ASR, but as NGVs came into wide use, now the law is also aiming at the appropriate treatment of ELVs and ensuring useful resources not only from domestic ELVs but also from exported vehicles (which may become ELVs) that are hoped to be recycled efficiently. In other words, we should consider not only the EPR principle or the ELVs that are recycled within these developed countries, but also exported second-hand vehicles and transnational environment problems.

In this section, we will take Mongolia as an example in order to analyze the conditions and problems concerning the international circulation of second-hand vehicles, as well as international resource circulation and cross-border environmental problems (scrap iron and waste lead).

4.2 Conditions and Characteristics of Second-hand Vehicle Exportation in Japan

Second-hand vehicle exports from EU countries (*e.g.* Germany and France), Japan and Korea are increasing as a result of the economic development and increasing need for vehicles in developing countries.

Second-hand vehicle exports in Japan have expanded since the enforcement of the ELV recycling law in 2005. Because of the high quality of Japan's second-hand vehicles, they became extremely popular around the world. As shown in Figure 9, although the export volume decreased temporarily due to the Lehman Brothers shock of 2009, it has gradually bounced back.

Statistics of foreign trade for exported second-hand vehicles became available in Japan from April 2001.[45] According to these statistics, Asian countries occupied 45% of the Japanese second-hand vehicle export market, while Africa stood at 22%, Europe stood at 12%, Oceania stood at 10% and South Africa stood at 8% of the market.

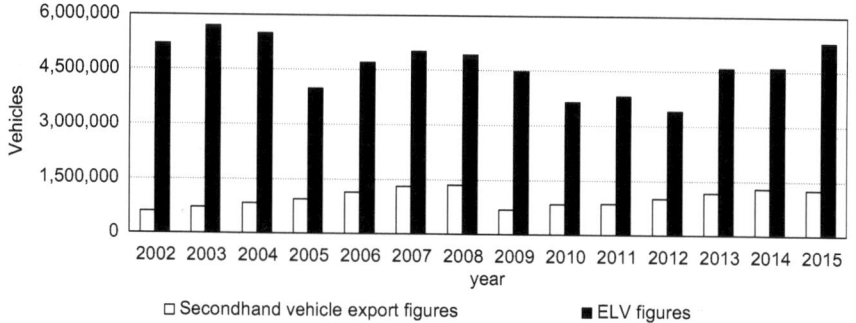

Figure 9 Change in the number of second-hand vehicle exports and number of ELVs in Japan.[63]

Table 4 Major export destinations of Japanese second-hand vehicles.[60]

Country	Ranking	Number of second-hand Japanese vehicle owned by every 1000 people	Number of vehicles owned by every 1000 people
Myanmar	1	2.99	6
Sri Lanka	6	0.20	15
Pakistan	7	1.70	52
Mongolia	12	12.30	133

Second-hand vehicle exports decreased sharply in 2009 as a result of the Lehman Brothers shock and the high exchange rate of the Japanese yen at that time. Even under this situation, exportation increased once again because of Japanese second-hand vehicles' high quality, the development of new exporting methods/markets and the increasing need for second-hand vehicles in Asian and African countries, along with their economic development.

Second-hand vehicle imports in countries such as Russia, Myanmar, the United Arab Emirates (UAE), New Zealand and Chile were massive from 2011 to 2015. It should be noted that despite the small populations in Mongolia and Sri Lanka, their second-hand vehicle import rates are quite large (Table 4).

As shown in Table 5, Mongolia has the highest import rate of second-hand vehicles among Asian countries. On average, every 1000 people in Mongolia own 12.3 second-hand Japanese vehicles. This number is about four-times larger than that of Myanmar, which has the largest rate of second-hand vehicle imports in Asia. On the other hand, the vehicle ownership ratio in Mongolia is 133 for every 1000 people, and this number is about 22-times larger than that of Myanmar and about 8.9-times larger than that of Sri Lanka. Hence, motorization in Mongolia is considered to be led by second-hand Japanese vehicles.

Table 5 Imported used cars from Japan (2015).[61]

Ranking	2011	Number	2012	Number	2013	Number	2014	Number	2015	Number
1	Russia	110 791	Russia	142 412	Russia	167 822	Myanmar	160 437	Myanmar	141 066
2	UAE	80 712	Myanmar	120 836	Myanmar	134 681	Russia	128 312	UAE	136 212
3	Chile	69 473	UAE	87 793	UAE	98 831	UAE	112 827	New Zealand	118 427
4	New Zealand	68 091	Pakistan	64 644	New Zealand	91 322	New Zealand	110 333	Kenya	77 473
5	South Africa	67 458	Chile	61 701	Chile	78 000	Chile	73 364	Chile	64 658
6	Kenya	39 248	New Zealand	61 465	South Africa	62 275	Kenya	67 059	Sri Lanka	59 322
7	Sri Lanka	38 496	South Africa	59 789	Kenya	61 396	South Africa	53 540	Pakistan	49 485
8	Pakistan	37 880	Kenya	44 659	Kyrgyzstan	36 026	Kyrgyzstan	48 351	Russia	49 144
9	Mongolia	35 983	Mongolia	30 172	Mongolia	34 919	Georgia	38 759	South Africa	46 475
10	Uganda	23 791	Uganda	24 837	Tanzania	30 912	Pakistan	38 228	Tanzania	42 738
11	Kyrgyzstan	23 542	Philippines	23 666	Pakistan	28 785	Tanzania	37 343	Philippines	33 964
12	Malaysia	21 791	Malaysia	23 370	Malaysia	27 837	Mongolia	35 367	Mongolia	31 684

4.3 Analysis of the Condition of Second-hand Vehicle Imports in Mongolia

4.3.1 Variations and Characteristics of Second-hand Vehicle Imports.
In this section, the characteristics and problems concerning second-hand vehicle imports in Mongolia will be analyzed.

The number of vehicles in Mongolia was 40 000 in 1990 and reached 450 000 by 2015. Moreover, second-hand vehicle imports have increased rapidly since 2007 and are increasing by 19% every year. Meanwhile, per capita GDP in Mongolia was 1511 US dollars in 1990 and reached 3951 US dollars by 2015.[46] Therefore, it is plain to see that the ownership ratio of automobiles increases along with increments in income.

As shown in Figure 10, before democratization, approximately 95% of the vehicles in Mongolia were from the Soviet Union, and the remaining 5% were from other countries, such as India, Japan and Korea. After the collapse of communism, second-hand vehicles from Japan and Korea increased, leading to motorization in Mongolia (especially in Ulaanbaatar, the capital of Mongolia). Hence, registered vehicles reached 450 000 units in 2015. This number is 10-times larger than the 40 000 units of the 1990s. Second-hand vehicles from Korea were popular during the first half of the 2000s due to their low price and stable supply of components. They even occupied 48.5% of Ulaanbaatar's registered vehicles. Japanese vehicles became popular due to their high quality, despite their comparatively high price. As a result, the proportion of Korean vehicles dropped to 8.5%, with about 80% of the vehicles now being from Japan. Hence, international exchange between Japan and Mongolia became active and trade volumes increased rapidly.

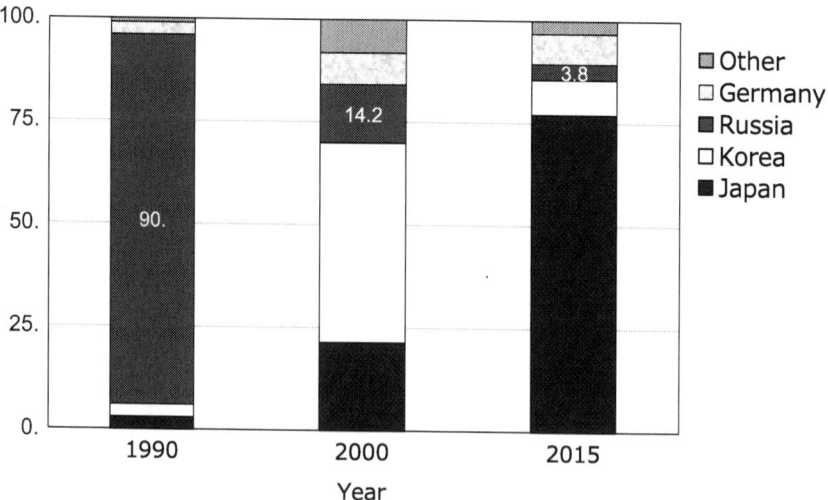

Figure 10 Change in second-hand vehicle imports in Mongolia.[64]

4.3.2 Rapid Increase in Second-hand Vehicle Imports and Related Policies.

One of the characteristics of the transportation sector in Mongolia is the large volume of registered second-hand vehicles. As mentioned above, about 90% of Mongolia's registered vehicles are second-hand vehicles from Japan and Korea. Among these vehicles, over 70% have been used for over 10 years and less than 10% are within 3 years of their registration dates. Although this situation is changing gradually, the proportion of second-hand vehicles remains extremely high (Table 6).

The reason why second-hand vehicles from Japan and Korea boomed in Mongolia is because of their customs duties. Most countries have strict restrictions (*e.g.* vehicle registration year, handle position, *etc.*) when importing second-hand vehicles. However, in Mongolia, although higher taxes are levied for vehicles that have been used for over 10 years, strict restrictions on the importation of these second-hand vehicles have not been enforced. Furthermore, although second-hand vehicle-oriented motorization is developing in Mongolia, proper inspection of vehicles, related regulations and technologies for recycling, and infrastructure (roads, bridges and transportation system) have not been fully prepared and are causing various problems, such as terrible traffic jams and accidents, air and soil pollution and illegal dumping of ELVs.

On the other hand, importation of second-hand HVs has also been increasing in Mongolia since 2010. Behind this transition was the popularization of NGVs in developed countries and regions such as Japan, Europe and the USA. In fact, the number of NGVs had already reached 4.13 million by 2013, of which 95.5% were HVs (about 3.9 million). As for end-of-Life NGVs, the first version of the HV was the Toyota Prius, which was launched in 1997. Therefore, it has already exceeded its expected lifetime, but most end-of-life Priuses have been exported as second-hand vehicles to Mongolia and Russia recently.

Since tax exemption is applicable to eco-friendly vehicles in Mongolia no matter how long the vehicle has been used, importation of HVs has increased rapidly and occupies 30% of registered vehicles in Mongolia. Most of these HVs were from Japan. Thus, it is expected that numerous end-of-life HVs will emerge in the next few years. According to the authors' traffic

Table 6 Vehicle ages of service in Mongolia.[62]

Year	Data	Up to 3 years	4–9 years	10 years and over	Total amount
2011	Number	8585	54 283	191 618	254 486
	%	3.4%	21.3%	75.3%	100%
2012	Number	10 770	46 114	255 658	312 542
	%	3.4%	14.8%	81.8%	100%
2013	Number	20 325	79 022	246 126	345 473
	%	5.9%	22.9%	71.2%	100%
2014	Number	26 492	79 470	278 902	384 864
	%	6.9%	20.6%	72.5%	100%

analysis of Ulaanbaatar in 2015 and August 2016, about 30% of these vehicles are Japanese HVs. HVs that have been used for over 6 years are mainly being exported to Mongolia due to their tax exemption as eco-friendly vehicles (Figure 11).

4.4 Effect of Second-hand Vehicle Imports on Resource Recycling and the Environment

4.4.1 Inadequate Treatment and Inefficient Recycling of ELVs. Most of the imported vehicles in Mongolia are over 15 years of age, and so a large quantity of ELVs will emerge in the next few years. However, it is difficult to manage such vehicles properly because they are being transported to remote areas or sold to nomadic people. In fact, among all of these vehicles, tax evasion vehicles and vehicles without inspection stand at 30%, and it is common for these vehicles to be maintained inappropriately or dumped illegally.[47]

Even if these ELVs were being treated properly, the recycling process of scrap iron is uncertain and the treatment is usually inefficient. Basically, used parts will be put into the reuse market regardless of whether it will be sold or not. However, without knowing the best-selling product and consideration of the balance between supply and demand, the used parts market is usually nonfunctional and inventory management is inefficient. If this situation continues, it is possible that these used parts will be discarded as waste materials eventually (Figure 12).

Figure 11 Japanese hybrid cars in Ulaanbaatar (photographs by Prof. Yu, 26th August 2016).

Latest Trends and New Challenges in End-of-life Vehicle Recycling 193

Figure 12 Used parts market in Mongolia (photographs by Prof. Yu, 26th August 2016).

Looking at the recycling route for scrap iron, exclusively government-managed steel mills are willing to take scrap iron as a raw material (Figure 13), but many problems still exist in this process.

Figure 14 shows the conceptual scheme of the recycling process for scrap iron in Mongolia. Associated problems are: the price of scrap iron was not decided by the market; trustworthy weighing implements are antiquated; the efficiency of the logistical network is inadequate (trucks and railways); intermediate treatment is lagging behind (cutting, sorting, pressure and shredding); and business conditions (for payments and payment dates) are still unreasonable. Therefore, the creation and development of a new and/or related industry by building a sound and efficient recycling system is an urgent priority (including the preparation of legal systems, the introduction of technology, human resources and proper trading modes, *etc.*).

4.4.2 Cross-border Environmental Problems from Untreated Lead Batteries.
Lead batteries are usually classified as hazardous waste and are under the strict management of the 'Basel Convention on the Control of Transboundary Movements of Hazardous Wastes and their Disposal'. For example, about 30% of waste batteries in Japan are transported to Korea, but during this process, strict monitoring on the number of batteries is enforced and only temporary authorization will be offered.[48] Moreover, related stakeholders such as the import trader, exporter, shipping company,

Figure 13 Circulation (treatment) of scrap iron in Mongolia (photographs by Prof. Yu, 26th August 2016).

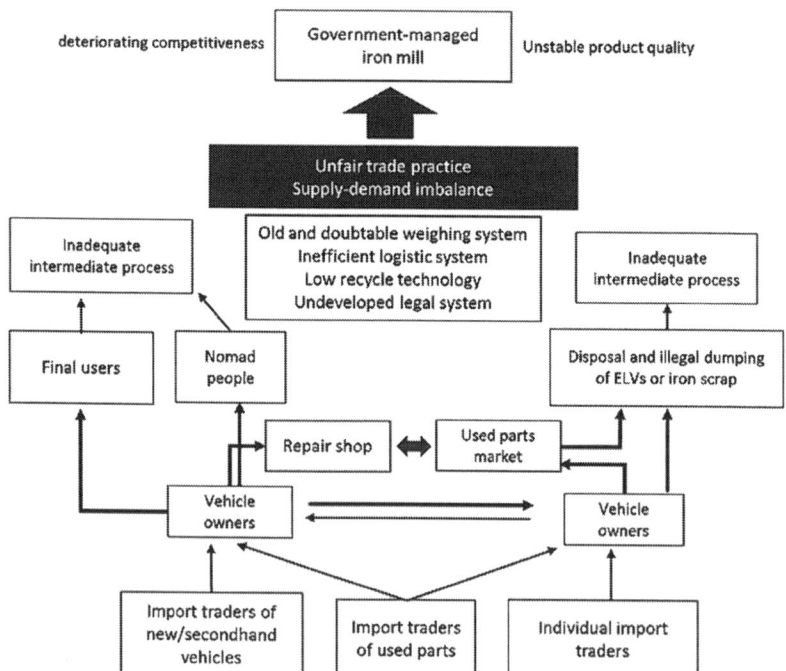

Figure 14 Recycling route for scrap iron in Mongolia.

lead refining operator and hazardous wastes processor will be required to be registered. Therefore, there will be no pollution during the recycling process.

However, when batteries are exported along with second-hand vehicles, they will not be monitored or controlled as waste batteries. In the case of Mongolia, imported vehicles that have been used for over 10 years will need to change their batteries within a year. This is compounded by the fact that in winter, the temperature is always below -20 °C, so these batteries' lifespans are even shorter in Mongolia. Based on our survey, a vehicle's battery needs to be changed every 2 years in Mongolia (in Japan, this figure is over 4 years). Furthermore, nomadic people use automobile batteries as storage batteries for solar power systems and, on average, battery disposal is at a rate of about one battery per year, and most of these batteries will be left in the grassland.[49]

Considering this situation, as vehicle batteries in Mongolia need to be changed frequently, the number of batteries that have been used for a single vehicle may be much higher than in other countries. However, these batteries are not being treated properly in Mongolia and are being refined inappropriately by illegal businesses or are being exported to China through unauthorized routes. Figure 15 shows the recycling route of waste batteries based on an interview investigation with battery stores, vehicle owners, nomadic people and waste battery collectors.

Actually, it is difficult for lead batteries to pass through customs because they may cause pollution; therefore, it is common for these batteries to be refined in advance and then exported as lead to China. However, due to protests and monitoring by citizens' groups in Mongolia, these batteries are now being transported through unauthorized routes after the illegal dumping of dilute sulfuric acid liquid from the batteries. These batteries are then transported to the national border, and this is a new type of cross-border pollution problem that needs to be solved with urgency.

4.4.3 New Challenges in International Resource Recycling of Nickel–Hydrogen Batteries. As mentioned above, NGVs are increasing rapidly in Mongolia, but the scenario is different from those of developed countries. Most of these NGVs have been used for over 15 years, and their travel distances can exceed 100 000 km. Most importantly, the batteries in these vehicles are often seriously degraded.

Take the Prius, which is leading in terms of NGV sales, as an example. Although these vehicles should be dismantled in accordance with Toyota's manual, it is common practice for dismantling operators only to change the broken battery cell in order to extend the vehicle's lifespan in Mongolia. Furthermore, although Japanese automobile manufacturers are trying to reuse and/or recycle NiMH batteries properly, most NiMH batteries are being

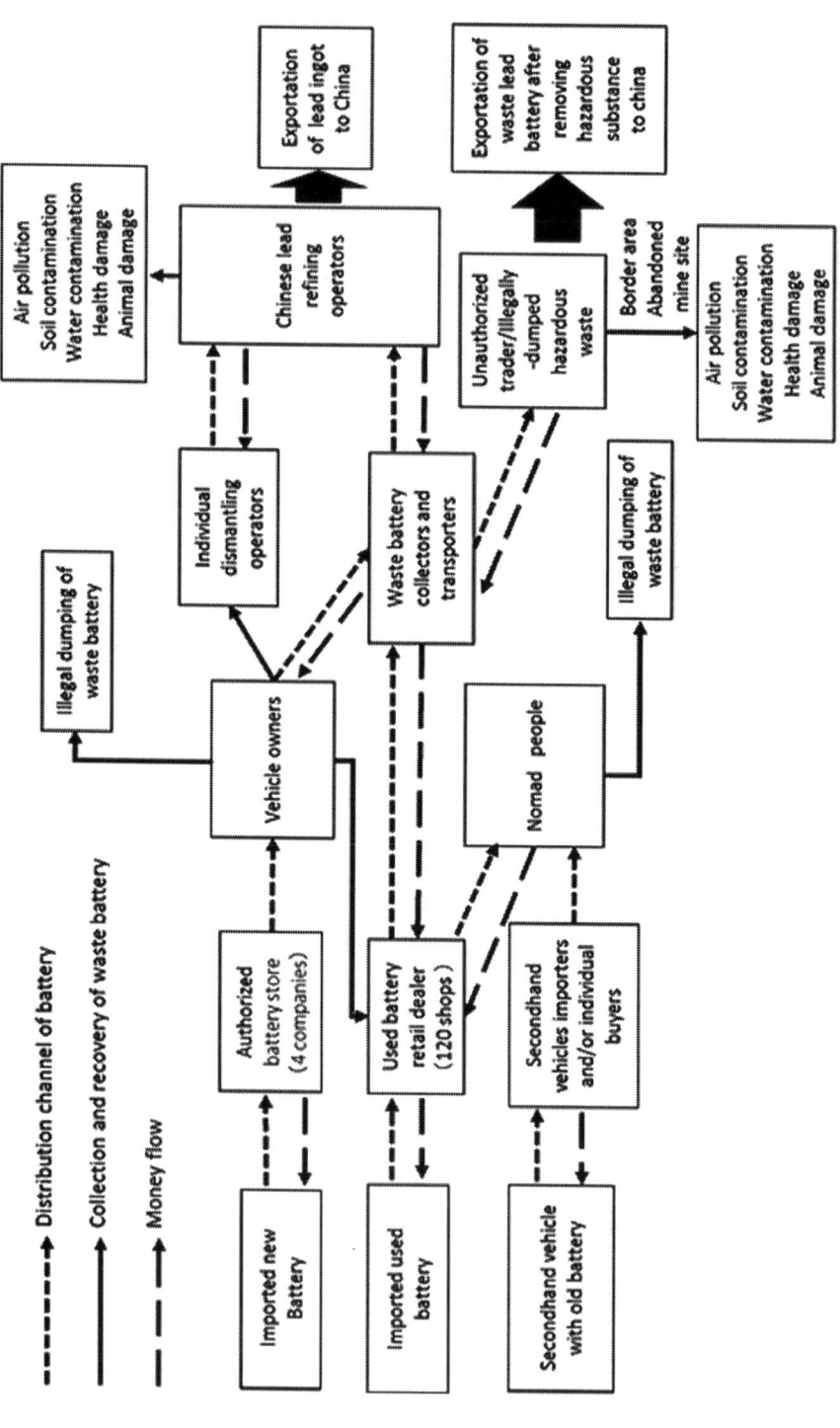

Figure 15 Recycling route of waste batteries in Mongolia.

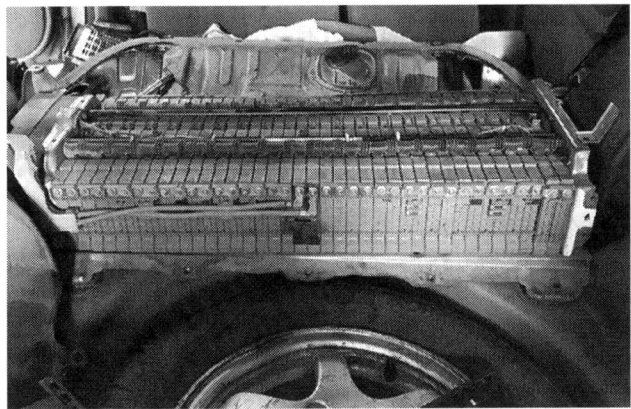

Figure 16 Repair of a HV's NiMH battery in Mongolia (photographs by Prof. Yu, 26th August 2016).

exported along with HVs. However, there are no technologies in Mongolia capable of recycling these batteries properly, and there are high risks that these batteries will be dumped illegally and that repairs will be conducted under risky conditions. Although Toyota is planning to collect and recycle NiMH batteries in Mongolia, the definition of these batteries is also not clear (whether they are hazardous waste or recyclable resources). Considering that the presence of NGVs is increasing around the world, it is important to consider the possibility of international resource recycling for NGV batteries and motors. Moreover, considering the massive volume of HVs in Mongolia and the commercialization of utilizing used HV NiMH batteries as storage batteries by automobile dealers in Japan, we firmly believe that one of the most valuable ways to utilize used NiMH batteries is as storage batteries for nomadic people in Mongolia (Figure 16).

5 Environmental Pollution Caused by Improper End-of-life Vehicle Processing in Developing Countries: A Case Study on Lead Battery Recycling in Mongolia

5.1 Potential of Serious Environmental Damage

Like other developing countries, Mongolia has experienced rapid motorization since the beginning of the 21st century, especially in its capital city, Ulaanbaatar. A large number of cars that are driven in Mongolia are used cars imported from Japan. Previously, the most popular vehicles were off-road vehicles, which could cope with the poorly conditioned urban roads that prevailed. With recent improvements in urban and suburban road conditions, these off-road vehicles have been increasingly replaced by sedans, luxury cars and hybrid cars.

However, the increased import volume of used cars has increased the amount of automotive waste. In particular, the lead batteries of automobiles last for approximately 3–5 years. Because the life cycle of these batteries is considerably shorter than the automobile lifespan, waste battery accumulation has become a major environmental problem.

In the traditional Mongolian yurt or ger, nomads purchase used automotive lead batteries and reuse them as household batteries. The batteries are charged by the electric power generated by solar panels during the daytime and provide power during the night-time. However, the number of waste automotive lead batteries is now exceeding demand. Meanwhile, the international price of lead has soared since the late 2000s and remains high to this day. This trend has driven an increase in companies that collect waste automotive lead batteries in Ulaanbaatar and refine the contained lead for export.

The toxicity of lead is well known. Acute toxicity from large quantities of ingested lead is highly improbable because inorganic lead is only slightly soluble in water. A certain quantity of lead is also excreted in the urine. However, if the long-term ingestion of lead exceeds the amount that can be excreted, the excess lead accumulates in bones. The symptoms of chronic lead toxicity include anemia, headaches, vomiting, loss of appetite and kidney disease.

The useful features of lead are also well documented. For example, lead is soft, heavy and easy to process and is utilized in various products, such as water supply pipes, bullets and fishing weights. Although advanced countries have recently regulated the use of lead because of its toxicity, lead remains the most suitable material for automotive batteries. Specifically, lead batteries allow transient flows of large currents, exhibit no memory effect and are low cost, enabling regular replacement.

The collection and recycling of waste automotive lead batteries was triggered by the increasing motorization of developing countries. However, careless recycling pollutes the surrounding environment, posing a serious risk of lead toxicity.

5.2 *Overview of Field Investigations and Their Results*

In the summer of 2013, we surveyed the current status of waste management in Mongolia. After interviews with landfill officials, we found that cattle frequently enter and graze in places such as landfill sites and illegal dumping sites. Such sites are often contaminated with hazardous waste in developing countries. Fearing the high likelihood that livestock are exposed to hazardous materials, we surveyed previous studies and interviewed nomads and researchers in Mongolia. In the summer of 2014, we learned of a lead battery refinery built on a grassy field in the Nalaikh district of Ulaanbaatar. At that time, the number of sick and dead livestock that had grazed around the area was increasing. Therefore, to clarify the current status of lead pollution, an ongoing investigation is being conducted in collaboration with the researchers who provided the initial information. The investigation is detailed below.

5.2.1 Lead Concentrations of Soils in and around Ulaanbaatar in 2014.

We first conducted a literature review and conducted an interview survey of the nomads.

Batjargal et al.[50] reported the lead and arsenic concentrations in topsoil and subsoil (30 cm sub-surface) samples taken from 11 locations along the major roads in urban Ulaanbaatar in 2007. Although the mean lead concentrations were generally below the maximum permissible level in Mongolia (100 mg kg^{-1}; MSN 5850: 2008), two of the topsoil samples and one subsoil sample contained higher than permissible levels of lead. In addition, the lead contents were higher in the topsoil samples than in the subsoil samples. The main causes were identified as increased volumes of cars and leaded gasoline. However, the soil environment and lead battery refinery were not mentioned in this report.

To address this knowledge gap, we collected 26 topsoil samples from the central area and Nalaikh district of Ulaanbaatar. The sampling sites are indicated in Figure 17.

Collected soils were air-dried for 24 hours and sieved through a 2 mm plastic sieve to remove large debris and plant material. Each soil sample (1 g, <2 mm) was then shaken in 10 mL nitric acid (1 M HNO_3) for 200 minutes on a reciprocating shaker (300 cycles minute^{-1}). The mixture was filtered through quantitative filter paper (grade 2; Whatman). The lead concentrations in all solutions were measured using an atomic absorption spectrometer (ICE 3000; Thermo Fisher Scientific) in the State Central Veterinary Laboratory, Mongolia. The analytical results are presented in Figure 18, for which the horizontal axis is a logarithmic scale.

Lead concentrations were much higher in samples 21–26 than in the other samples. Samples 21–24 were collected from the bottom of the wall surrounding the site of refinery (A). Similarly, samples 25 and 26 were collected from the bottom of the wall surrounding the site of refinery (B), an additional lead refinery uncovered in our interview survey. Sample 23, with the highest lead concentration among our samples, was collected near a 5 cm-diameter hole at the bottom of the wall, suggesting drainage leakage from refinery (A). These refineries clearly did not measure their flue gases or wastewater discharge.

Samples 16–20 were collected from the grassy field close to refinery (A). The east side of this refinery is heavily grazed by livestock such as cattle, horses, sheep and goats, which are attracted to the pond located 330 m northeast of the refinery. In addition, the grassy field slopes down from refinery (A). The lead concentration in the field is nonhomogeneous, and may be related to the distance from refinery (A). However, the source of the soil lead (flue gas or drainage) could not be ascertained.

Samples 10–13 were collected around the factories in Nalaikh district, whose business activities were unknown. The lead concentration was

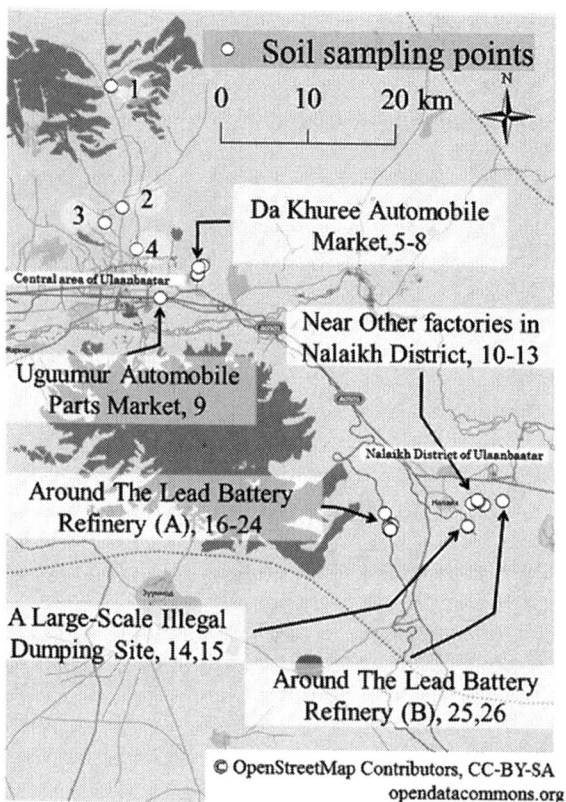

Figure 17 Sampling locations in and around Ulaanbaatar, Mongolia, in September, 2014.

higher in sample 13 than in the other samples and was possibly related to lead battery recycling, including refining. Samples 14 and 15 were collected from a large-scale illegal dumping site in Nalaikh district. This site was too wide for an even sampling over its range, but no waste lead batteries were found.

Samples 5–8 were collected from locations in and around the Da Khuree Automobile Market, and sample 9 was collected from the Uguumur Automobile Parts Market. The lead concentrations of these samples exceeded the permissible level. We infer that lead concentrations are elevated at sites with several cars. Samples 1–4 were collected along the road in the Ger area (central Ulaanbaatar). Among these samples, the lead concentration in sample 3 (collected from the front of an automotive repair shop) was above the permissible level. Lead from the shop had leached into this sample. Our values were similar to those that were reported by Batjargal et al.[50] In summary, lead battery refineries around Ulaanbaatar are concentrated sources of lead pollution. Moreover, in a second interview survey, the nomads living around the grassy field reported irregular operations of these

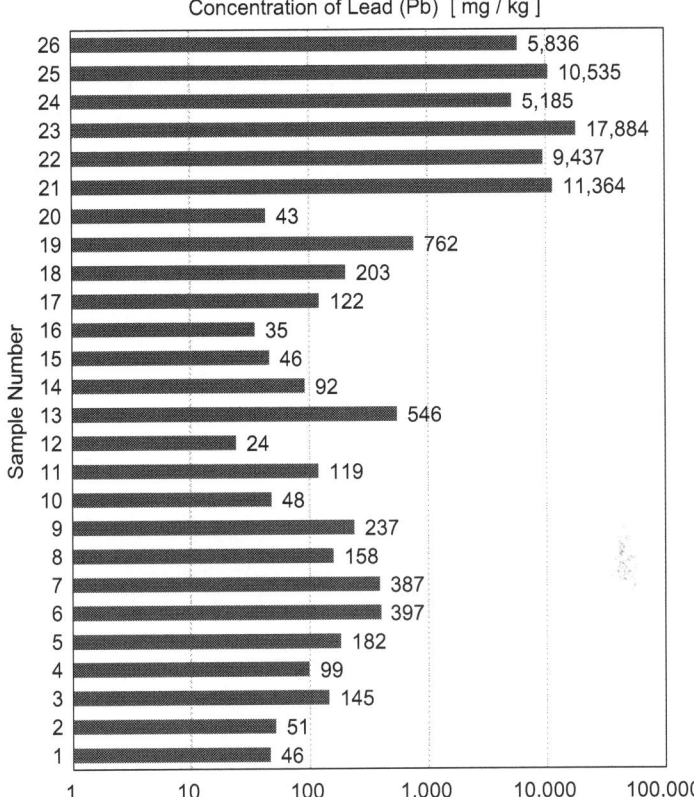

Figure 18 Lead concentrations of topsoil samples taken in and around Ulaanbaatar, Mongolia.

refineries, rather than daily long-term operation. Thus, the high lead concentration may indicate current or recent operation of the refineries.

5.2.2 Lead Concentrations of Soils around Lead Battery Recycling Factories in 2015. To clarify the lead distribution in the grassy field at the east side of refinery (A), we repeated our investigation in the summer of 2015. The sample locations are indicated in Figure 19. Based on the lead concentrations determined in 2014, we collected topsoil and subsoil samples at A1 (just below the chimney), A2 (25 m east of A1), A3 (close to the hole), A4 (1 m east of A3), A5 (the northeast corner of the refinery site), A6 (25 m east of A5) and A7 (25 m east of A6). At site A3, the 20 cm subsoil could not be sampled because of rock.

In addition, an interview with a nomadic family revealed the emission of flue gas from a hut located on the east side of the site. Site A5 is immediately below the outlet of this flue gas.

The collected soils were air-dried for 24 hours and sieved through a 2 mm plastic sieve to remove large debris and plant material. The lead

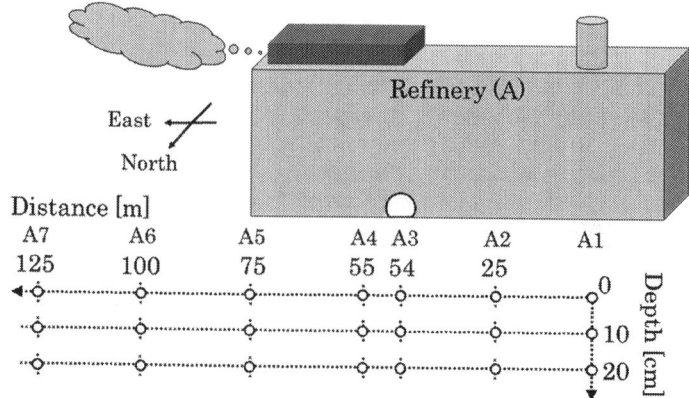

Figure 19 Sampling points around lead battery refinery (A) in Nalaikh District, Ulaanbaatar, Mongolia, in September, 2015.

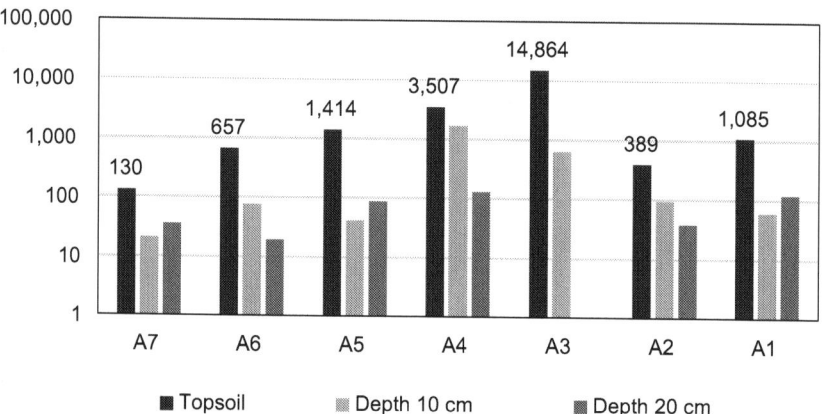

Figure 20 Lead concentrations of topsoil and subsoil samples taken from around lead battery refinery (A).

concentrations were then analyzed by portable X-ray fluorescence (Genius XRF; Skyray Instruments, China). The analysis results are presented in Figure 20, for which the vertical axis is a logarithmic scale.

Here, we are interested in the changing lead concentration of the topsoil. Similar to the analysis results of 2014, the highest lead concentration was found in the topsoil at A3, which is close to the hole. The second highest lead concentration appeared in the topsoil at A4, located 1 m east of A3. At this site, the lead concentration was a third that of site A3. The lead concentration decreased with increasing distance toward the east, reducing to permissible levels ($\sim$100 mg kg^{-1}) by around site A7. Therefore, the lead discharged from refinery (A) extended to sites between A6 and A7 (at least).

On the other hand, almost all of the subsoil samples contained lower than permissible levels of lead (the exceptions were the 10 cm depths at A3 and A4). Therefore, lead is unlikely to penetrate underground and contaminate the groundwater, pond and other surrounding water environments. However, the surface drainage flows depend on the amount of drainage, posing a danger to livestock encountering the drainage from the hole.

5.2.3 Lead Concentrations in the Blood of Livestock in 2015.

In the summer of 2015, we also collected blood samples from livestock and analyzed their lead concentrations. This study clarified whether the livestock in the grassy field around refinery (A) accumulated more lead than livestock in other areas. For example, Pareja-Carrera et al.[51] compared the lead concentrations in the blood of sheep inhabiting a mining area and a reference area in Spain. They obtained lead concentrations of 4.6–13.7 $\mu g\, dL^{-1}$ (44 sheep) in the mining area and <3.0 $\mu g\, dL^{-1}$ (the lower measurement limit) in the reference area. These differences were statistically significant.

The types of livestock assessed are summarized in Table 7. A farmer family resides next to refinery (A), a nomadic family inhabits the grassy field around refinery (A) and another family resides more than 50 km away to the east of refinery (A).

To collect blood samples, we punctured the jugular vein using a blood collection tube containing heparin. Blood samples were dissolved within 24 hours and measured within 48 hours (if refrigerated, they will keep for

Table 7 Information on blood samples taken.

No.	Type of animal	Age (years)	Explanation
Farmer's livestock in the site next to refinery (A), Nalaikh District, Ulaanbaatar			
F1	Calf	<1	
F2	Calf	<1	
F3	Goat	1	Unnatural death of its mother (twins with F6)
F4	Goat	3	Became blind in last winter
F5	Goat	<1	
F6	Goat	<1	Unnatural death of its mother (twins with F3)
F7	Goat	<1	
Nomad's livestock in the grassy field around refinery (A), Nalaikh District, Ulaanbaatar			
N1	Goat	1	Unnatural death of its mother
N2	Goat	2	Unnatural death of its kid
N3	Goat	3	Unnatural death of its kid
N4	Goat	3	Unnatural death of its kid
N5	Sheep	3	Was diseased
Nomad's livestock 50 km or more away from refinery (A), Tov prefecture			
E1	Goat	Mature	
E2	Sheep	Mature	
E3	Sheep	Mature	
E4	Sheep	Mature	

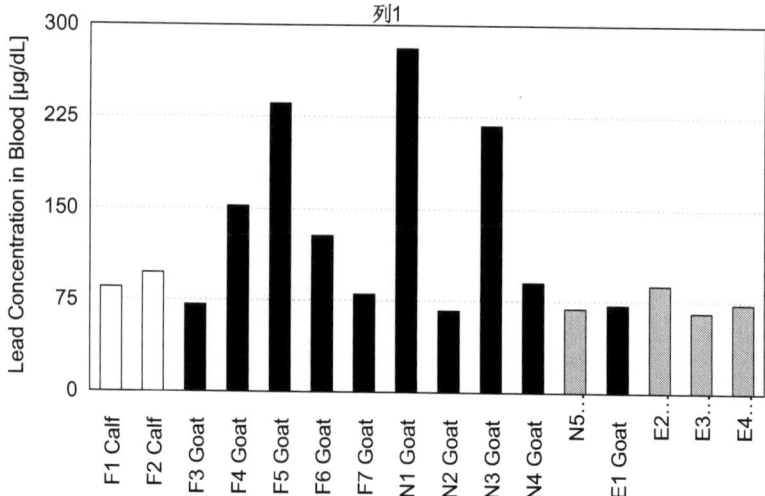

Figure 21 Lead concentrations of blood samples taken from the stock farm next to refinery (A) (F1–F7), a nomad's livestock around refinery (A) in Nalaikh (N1–N4) and a nomad's livestock 50 km or more away from refinery (A) (E1–E4).

1 week). Lead concentrations were measured by a LeadCare® II System (Magellan Diagnostic, MA, USA). Figure 21 shows the lead concentrations in the blood samples of the livestock.

Surprisingly, the lead concentrations of all samples exceeded 50 µg dL^{-1}. Even at a 50 km distance from refinery (A), the lead concentrations in sheep blood were much higher than those reported in Pareja-Carrera et al.'s Spanish study. We confirmed that the distant nomadic family had not inhabited the Nalaikh district and was removed from any other refineries. Thus, other sources of discharge lead cannot be ruled out. Moreover, lead concentrations in the blood samples of three goats near refinery (A) were 200 µg dL^{-1} or higher. Such high concentrations might have been ingested by grazing on grass close to refinery (A). We consider that it is necessary to collect blood samples of livestock and to analyze them.

5.3 Challenges from a Case Study

Two main challenges emerged from our investigation. First, lead battery refineries are a new source of lead discharge and pose a major risk to livestock. In other words, motorization has added an artificial risk in environments previously exposed only to natural risks.

One feature of nomadic stock farming is the absence of any barriers. Therefore, high-risk areas are not physically separated from the grazing fields of livestock. Nomadic people cannot constantly attend to their livestock or prevent them from entering high-risk places. Moreover, we must consider that the risk is directly ingested into the living body of the livestock, thus revealing the relationship between the lead distribution in the grassy fields and the current status of lead exposure in livestock. The permissible levels, emission standards and effluent standards of facilities also need reassessing in grassy fields.

The second challenge is the difficulty of cracking down on illegal operations in grassy fields. For example, refinery (A) is operated only sporadically, when a sufficient number of waste lead batteries have accumulated. If the operating license has been revoked, constantly monitoring the refinery at all times in the grassy field would be a difficult task. Even if a new facility discharges much environmental pollution into a grassy field, the extent of the pollution cannot be immediately known, nor can the emissions be compared against even a strict emission standard. To overcome this problem, we must establish a collection route for waste lead batteries and develop a proper recycling industry.

6 Environmental Problems Associated with the Proliferation of Used Vehicles in Metro Manila, Philippines

6.1 Current State of Used Vehicles in the Philippines

Metro Manila is a megacity with almost 13 million inhabitants as of 2015.[52] The jeepney is the most ubiquitous passenger vehicle in Metro Manila, the national capital region of the Philippines. They are dubbed as the 'king of the road' for their ability to ferry passengers and negotiate either wide or narrow streets. Running costs are also cheap and maintenance is easy. They are also adorned with colorful decorations at the front and sides of the vehicles. Historically, they hit the roads in the 1950s, and they were was rehabilitated from US military vehicles dating from World War 2.[53] Due to their popularity, the number of jeepneys reached its peak at around half a million nationwide in the 1980s. However, this number dwindled to around 160 000 due to decreasing demand and the government stopping issuing new franchises due to pollution concerns.[54] As of 2011, roughly 60 000 units operate in the metropolis. According to the Philippine Department of Transportation (DOTr), 160 000 jeepney units that are operating nationwide will reach their age limit by 2016. This means that they have been operating for 15 years and soon will become ELVs. Figure 22 shows a typical jeepney in Metro Manila.

The problems associated with old jeepneys are myriad. Most of them are operating with very old engines and their emissions are very bad. In fact, the transport sector in the Philippines contributes about 37% of total greenhouse gas (GHG) emissions, 80% of which comes from jeepneys.[55] Aside from jeepneys, the use of vehicles such as buses, trucks, motorcycles, tricycles and trailers also contribute to air pollution in Metro Manila. As of 2011, there were close to 1 million public utility vehicles all over the Philippines. Most of the buses and trucks are second-hand vehicles that are usually imported from Japan and Korea, just like the engines of jeepneys. The reason for this is that it is cheaper to import used cars than to buy new ones. Figure 23 shows an old engine that was recovered from a jeepney.

Figure 22 Typical passenger jeepney in Metro Manila (photographs by Dr Serrona, 20th April 2016).

Figure 23 Old engine from a disassembled jeepney (photographs by Dr Serrona, 20th April 2016).

6.2 Existing Legislation

The Philippines has no specific legislation on ELV recycling yet. The call for the phasing out of old vehicles in order to decongest roads and reduce air pollution in Metro Manila points toward having a framework for used vehicles to be recycled, just as Japan and Korea are doing. But the Philippines has a centerpiece legislation called 'The Philippine Ecological Waste Management Act of 2000' or 'Republic Act 9003', which provides a comprehensive approach to solid waste management with an emphasis on reducing, reusing and recycling. It treats motor vehicle wastes such as batteries, oil and tires as 'special wastes', which are to be handled separately from residential and commercial wastes.

The country is a signatory of the Basel Convention, which is a global treaty aiming at prohibiting the export of hazardous wastes from developed to developing countries for final disposal, reuse, recycling and recovery. The economic condition of the country, however, prevents the full enforcement of the treaty, such that the import of used vehicles with toxic components continues. Domestically, 'Republic Act 6969' or the 'Toxic Substance and Hazardous and Nuclear Waste and Control Act of 1990' was enacted in order to regulate, restrict or prohibit the import, manufacture, processing and disposal of chemical substances, including their entry, transit and storage. As such, the country has the necessary framework for hazardous waste management. Planned phasing out of old vehicles is also consistent with the 'Philippine Clean Air Act of 1999' or 'Republic Act 8749', which enforces reduction of GHG emissions across sectors.

6.3 Current Proposals to Undertake ELV Recycling

The DOTr, which is responsible for the management of the transportation sector in the Philippines, came up with informal proposals on how to go about the phasing out of old jeepneys. In reality, removing them from Philippine roads is not merely a step toward get rid of dilapidated commuter vehicles, but also to ensuring that the livelihood of thousands of jeepney drivers are equally addressed. A proposal that was put forward was the phasing out of 15 year-old jeepneys with a 1 year allowance of extended use in order to allow drivers to transition to a new mode of transportation. The DOTr will then initiate discussions with local banks for flexible terms so that operators can obtain loans for purchasing new vehicles. It will then buy old jeepneys, disassemble them and sell recovered scrap materials to junkshops and manufacturers.[56] This proposal bodes well for the government since it will ensure that old jeepneys are not discarded as waste, but rather as another resource. In an informal interview with a local junkshop dealer in Metro Manila, a retired jeepney can generate about US$100 worth of scrap metal.[57] This amount is small probably because of the crude and manual method of recovering scrap metals and other parts. Data suggest that a single ELV in a developing country can generate about 500 kg of scrap. Other parts like head and tail lights, doors, wheels, engines, transmissions and brakes can be recovered, reused and recycled.[58]

6.4 Future of ELV Recycling in Metro Manila

ELV recycling has huge potential in Metro Manila and all over the country because of the high quantity of used vehicles. Coupled with that is the rising number of new car purchases, which attests to the increasing per capita income of Filipinos, as manifested in the steady economic growth of the country for the past couple of years. The government has been mulling over incorporating recycling in the phasing out of used vehicles, as it sees the potential economic and social benefits of undertaking such a process. Aside from modernizing the transport system and reducing GHG emissions from the transport sector, ELV recycling will promote urban mining and consequently generate jobs in the recycling sector, which currently recovers about 30–40% of the municipal solid waste. The Philippines has no specific policy yet on ELV recycling, and the planned phasing out of jeepneys will contribute to its development. In addition, since this will be a pioneering effort, the phasing out will occur on a gradual basis, starting with for-hire vehicles, which will also include buses, trucks and taxi cabs.

There are key ingredients in order for ELVs in developing countries like the Philippines to prosper. First is the constant volume of ELVs, which will ensure a steady supply of scrap metals and parts. Second is the market scale of used parts and recycled resources. This requires having a good network of recyclers and manufacturers that will absorb products from ELVs. Third is the manner by which harmful materials like ASR are treated and disposed.[59] Again, the country has 'Republic Act 6969', which will take care of hazardous waste, but the question is the availability of infrastructure that will absorb the waste.

6.5 Challenges in Undertaking ELV Recycling in the Philippines

In the Philippine context, it is important to focus on the transition phase in ELV recycling because, as mentioned, there is a social component to it. Operators and drivers of jeepneys will be displaced if there are no alternatives presented. As such, incentives for them should be established in order to encourage them to sell or surrender their used vehicles. There is also a need to investigate existing local technologies, and manual-based recovery provides jobs and increases the value of recovered parts, while machine-based systems promote efficiency. There has to be a balance between these options in order to maintain local jobs and ensure the quality of recovered parts. Profitability is another factor to take into account, since ELV recycling has to run like a business. The demand for raw metals is increasing with the rapid urbanization of cities in Asia, and so this is an opportunity for the Philippines to step up its supply of recyclable resources to markets in Asia.

7 Recommendations and Challenges for the Future

Since the 2000s, the EU, Japan, Korea and China have introduced ELV recycling regulation. However, the responsibility of each stakeholder for their share of the cost has not been decided clearly, and this process is not going smoothly. In addition, the export of used cars from developed countries has taken on unstoppable momentum.

Although automobile manufacturers were not enthusiastic about taking responsibility or bearing the recycling fee until this point, the recent rise in demand for precious and rare metals has piqued their interest in ELV recycling. Moreover, as a result of the ever-growing attention on international resource circulation and global environmental problems, attitudes towards EPR may vary greatly.

Furthermore, due to the popularization of next-generation vehicles (NGVs), vehicle compositions may change greatly. The popularization of NGVs not only affects the automobile manufacturing industry, but also the recycling industry. Therefore, it is necessary that the research and development of NGV recycling, the preparation in law and the discussion about the export of second-hand vehicles are put into practice soon.

In Japan, a huge number of old used cars, including HVs and along with precious resources and batteries, are being exported every year. Since it is difficult to stop vehicle exports, it is important to build up an international reuse/recycle scheme both domestically and abroad as a solution to ensuring the stable provision of useful materials like lead batteries in the future and the appropriate disposal of end-of-life NGV waste.

Unfortunately, developed countries and international automobile manufacturers are not responsible for their exported used cars. Under the EPR principle, large automobile companies are required to reduce environmental burdens when disposing of and recycling cars. When an exported car is disposed of and recycled in another county, especially in a developing country, the government and automobile companies should be responsible for pollution control and effective recycling. This requires a global ELV recycling system based on the results of long and careful study of international resource recycling and cross-border environmental problems.

Notes and References

1. Japan Automobile Recycling Promotion Center, http://www.jarc.or.jp/automobile/law/, (accessed October 2016).
2. Ministry of the Environment, Japan, http://www.env.go.jp/recycle/car/outline1.html, (accessed October 2016).
3. Japan Productivity Center for Socio-Economic Development, *Domestic and overseas regulations of recycling economy and its impact on economic development*, Ministry of Economy, Trade and Industry, 2003, pp. 49–82.
4. Organization for Economic Co-operation and Development, *Extended Producer responsibility: A Guidance Manual for Governments*, Organization for Economic Co-operation and Development, 2001, DOI: 10.1787/9789264189867-en.
5. Holland, *The Management of End-of-Vehicles Decree (Bba)*, 2002, and Germany, *Ordinance on the Transfer, Collection and Environmentally Sound Disposal of End-of-life Vehicles (End-of-life Vehicle Ordinance)*, 2002.
6. Some of Japan's dismantling operators delivered, incinerated and landfilled industrial waste (mainly ASR from ELV) at Teshima, Kagawa Prefecture, since 1978. This affected the health of the people in Teshima. The illegal action was carried out under the auspices of the Kagawa Prefecture government, and it was not stopped until 1990, when the Hyogo police filed cases against them. Teshima residents applied for

pollution arbitration in 1990. Subsequently, the governor of Kagawa Prefecture apologized and promised to restore the damaged environment to its original state in July 2000. Togawa Kenichi, *Current Status of Japan's Automobile Recycling System*, Asian Automotive Environmental Forum, Japan, October 2015, pp. 23–25.

7. A. Terazono, ELV recycling in German, *Mater. Cycles Waste Manage. Res.*, 2002, **13**(No. 4), 210–220.
8. Figure 1 was created by reference to Eurostat, http://ec.europa.eu/eurostat (recycling rate), and GLOBAL NOTE, http://www.globalnote.jp/ (GDP).
9. Liechtenstein is excluded due to a lack of data and the small level of ELV emergence.
10. Abe Shin, *ELV recycling in Europe*, Monthly SEIBIKAI, 2006, vol. 37, pp. 42–45, http://homepage3.nifty.com/ariken/seibikai029.html, (accessed October 2016).
11. In Teshima's case, the industrial waste mainly consisted of ASR form ELVs, and this case has a stimulating effect on the enforcement of ELV recycling law. Shown before in reference 6.
12. Table 1 was created by reference to Toyota, http://www.toyota.co.jp/, and Nissan, http://www.nissan-global.com/jp/.
13. Japan automobile recycling promotion center, http://www.jarc.or.jp/documents/pdf/03_06.pdf, (accessed October 2016).
14. Jeongsoo Yu, Keinichi Togawa, "*Achievements and Problems of ELV Recycling Law in Korea*", Proceedings of 26th Annual Conference of the Japan Society of Material Cycles and Waste Management, 2015, pp. 185–186.
15. During ELV recycling processes, if the cost to recycle an ELV is larger than the benefit that a recycling factory can earn, the ELV will be defined as 'chargeable'. In fact, in Korea, no ELVs are chargeable at present. However, if the price of metal (especially iron) keeps falling in the future, ELVs in Korea may become chargeable, and if so, similar to in Europe, it will be the automobile manufacturers' responsibility to recycle ELVs.
16. Results from the interview survey of the Vice President of China National Resources Recycling Association, 2016.
17. Japan Economic Center Corporation, *The condition and foresight of Next-Generation Vehicle's market and technology*, smart energy group (ed.), 2015, chapter 1, p. 31.
18. Japan Automobile Manufacturers Association, *Evaluation and Investigation of ELV recycling regulation*, Japan Automobile Manufacturers Association, 2014, document 3-1, chapter 2, p. 19.
19. Same as above, Japan Economic Center Corporation, *The condition and foresight of Next-Generation Vehicle's market and technology*, smart energy group (ed.), 2015, chapter 1, p. 23.
20. Same as above, Japan Economic Center Corporation, *The condition and foresight of Next-Generation Vehicle's market and technology*, smart energy group (ed.), 2015, chapter 1, p. 31, p. 42.

21. Ministry of Economy-Manufacturing Industries Bureau-Automobile Division, *Correspondence to the Change in the Structure for Automobile industry*, Ministry of Economy, 2015, chapter 2, p. 20.
22. Nihon Keizai Shinbun, *Volkswagen is planning to develop Electronic vehicle instead of diesel-powered vehicle*, http://www.nikkei.com/article/DGXLASFZ21H4Y_R21C15A0K10100/, (accessed September 2016).
23. Nihon Keizai Shinbun, *Introduction of mid-priced EV-Tesla Model 3*, http://www.nikkei.com/article/DGXLASFK01H5G_R00C16A4000000/, (accessed August 2016).
24. State council of the People's Republic of China, http://www.gov.cn/zwgk/2012-07/09/content_2179032.htm, (accessed August 2016).
25. Figure 2 was created by reference to Japan Economic Center Corporation, *The condition and foresight of Next-Generation Vehicle's market and technology*, smart energy group (ed.), 2015, chapter 1, p. 14 and p. 31.
26. Figure 3 was created by reference to Japan Economic Center Corporation, *The condition and foresight of Next-Generation Vehicle's market and technology*, smart energy group (ed.), 2015, chapter 1, p. 31.
27. Same as above, Japan Automobile Manufacturers Association, *Evaluation and Investigation of ELV recycling regulation*, Japan Automobile Manufacturers Association, 2014, document 3-1, chapter 2, p. 19.
28. Figure 4 was created by reference to Japan Automobile Manufacturers Association, *Evaluation and Investigation of ELV recycling regulation*, Japan Automobile Manufacturers Association, 2014, document 3-1, chapter 2, p. 19.
29. Same as above, Japan Economic Center Corporation, *The condition and foresight of Next-Generation Vehicle's market and technology*, smart energy group (ed.), 2015, chapter 3, p. 124.
30. Table 2 was created by reference to Japan Economic Center Corporation, *The condition and foresight of Next-Generation Vehicle's market and technology*, smart energy group (ed.), 2015, p. 121.
31. Figure 5 was created by reference to Japan Economic Center Corporation, *The condition and foresight of Next-Generation Vehicle's market and technology*, smart energy group (ed.), 2015, p. 123 and p. 126.
32. Figure 6 was created by reference to Japan Economic Center Corporation, *The condition and foresight of Next-Generation Vehicle's market and technology*, smart energy group (ed.), 2015, p. 124.
33. IT media, *The reuse of EV's used batteries for power storage system*, http://www.itmedia.co.jp/smartjapan/articles/1507/13/news144.html, (accessed August 2016).
34. Next-generation Vehicle Promotion Center/Yano Research, *Investigation on the recycling of Lithium-ion battery in 2011*, 2012, chapter 1, p. 28 (10 years), and Mitsubishi Motors Corporation, http://www.mitsubishi-motors.co.jp/support/maintenance/service/warranty/miev.html (distance), (accessed September 2016).

35. Japan Environment Management Association for Industry, http://www.cjc.or.jp/raremetal/advanced-business-model/toyota-smm-toyochemi, (accessed September 2016).
36. TOYOTA Motors Corporation, https://www.toyota.co.jp/jpn/sustainability/environment/recycling_based/battery_recycle/, (accessed September 2016).
37. Figure 7 was created by reference to Japan Environment Management Association for Industry, http://www.cjc.or.jp/raremetal/advanced-business-model/toyota-smm-toyochemi.
38. Next-generation Vehicle Promotion Center/Yano Research, *Investigation on the recycling of Lithium-ion battery in 2011*, 2012, Chapter 6, p. 251.
39. 4R Energy Cooperation, https://www.4r-energy.com/company/about/, (accessed September 2016).
40. Figure 8 was created by reference to 4R Energy Cooperation, https://www.4r-energy.com/company/about.
41. Japan Automobile Manufacturers Association, *Efforts for the appropriate disposal and recycling of NEXT-Generation Vehicle*, associated session of Industrial Structure Council and Central Environment Council in 2014, 2014, document 5-3, Chapter 2, p. 5.
42. Table 3 was created by reference to Japan Automobile Manufacturers Association, *Efforts for the appropriate disposal and recycling of NEXT-Generation Vehicle*, associated session of Industrial Structure Council and Central Environment Council in 2014, 2014, document 5-3, Chapter 2, p. 5.
43. Automobile Inspection & Registration Information Association in Japan, https://www.airia.or.jp/publish/statistics/ao1lkc00000000z4-att/03_7.pdf, (accessed September 2016).
44. T. Kosuke, N. Hirotaka, Yu Jeongsoo, B. Erdenedarai, *The Current Status of Lead Pollution in Soils in Ulaanbaatar*, Mongolia, 8th Asian Automotive Environmental Forum Guidebook, 2015, p.48.
45. Abe Shin, *Comparison between Japan and Europe on the exportation of secondhand vehicles from the viewpoint of international resources circulation*, Discussion Paper Series No.530, 2010, p. 2.
46. Revenue Office of Ulaanbaatar City. 24th August 2016.
47. Based on the interview with the manager of an automobile inspection center in Mongolia, 25th August 2016.
48. Jeongsoo YU, *Scramble for waste automobile batteries between Japan and Korea*, Scrap watch Column, 19th November 2011.
49. Based on the interview with retail battery shops, waste battery collectors and nomads in Mongolia, 26th August 2016.
50. T. Batjargal, E. Otgonjargal, K. Baek and J. S. Yang, *J. Hazard. Mater.*, 2010, **184**, 872, DOI: 10.1016/j.jhazmat.2010.08.106.
51. J. Pareja-Carrera, R. Mateo and J. Rodriguez-Estival, *Ecotoxicol. Environ. Saf*, 2014, **108**, 210, DOI: 10.1016/j.ecoenv.2014.07.014.
52. Philippine Statistics Authority. Highlights of the Philippine Population 2015 Census of Population. https://www.psa.gov.ph/content/highlights-philippine-population-2015-census-population (accessed September 4, 2016).

53. Saira, Syed. End of the Rod for Jeepneys in the Philippines. BBC News. http://www.bbc.com/news/business-23352851 (accessed September 2, 2016).
54. Dancel, Raul. 2015. *End of the Road for Manila's Jeepneys*. The Strait Times. http://www.straitstimes.com/asia/se-asia/end-of-the-road-for-manilas-jeepneys#xtor=CS1-10 (accessed August 15, 2016).
55. Francisco, Katerina. 2016. *To improve traffic in 2016*, DOTC to overhaul bus, jeep systems. Rappler. http://www.rappler.com/nation/115175-dotc-plans-2016 (accessed August 20, 2016).
56. Francisco, Katerina. 2016. *To improve traffic in 2016*, DOTC to overhaul bus, jeep systems. Rappler. http://www.rappler.com/nation/115175-dotc-plans-2016 (accessed August 20, 2016).
57. Interview with a junkshop dealer, February 16, 2016, Quezon City, Philippines.
58. Japan Automotive Recyclers Association (JARA), Automotive Recycling, 2011, First Edition, Japan.
59. Interview with Prof. Jeong-soo Yu, March 2016.
60. Table 4 was referenced from Japan Automobile Manufacturers Association, Automotive Yearbook. 2015–2016, Nikkan Jidosha Shimbun and Automobile Business Association of Japan (ed.), Tokyo, 2015.
61. Table 5 was referenced from Global Note, http://www.globalnote.jp/, (accessed September 2016).
62. Table 6 was referenced from the tax office in Ulaanbaatar (accessed September 2016).
63. Figure 9 was referenced from Trade Statistics of Japan, http://www.customs.go.jp/toukei/srch/index.htm, and Ministry of Economy, Trade and Industry, http://www.meti.go.jp/ (accessed August 2016).
64. Figure 10 was referenced from Local Tax Bureau of Mongolia, http://en.mta.mn/ (accessed September 2016).

Life Cycle Assessment of Road Vehicles

MICHEL VEDRENNE,* JAVIER PÉREZ, MARÍA ENCARNACIÓN RODRÍGUEZ, JULIO LUMBRERAS AND RAFAEL BORGE

ABSTRACT

Life cycle assessment (LCA) is a methodology that supports decision making by quantifying the environmental impacts of products, processes or activities. Because of its holistic approach in the description of interactions, LCA is adequate for evaluating the sustainable character of different aspects of the road transport sector. This chapter provides an overview of LCA, its component stages and its application to road vehicles. Special emphasis is given to the processes and types of data that should be taken into consideration when describing the life cycle of vehicles. Finally, a discussion about the suitability of LCA as a complementary tool for addressing current policy issues, especially those related to air pollution, is provided.

1 Life Cycle Assessment: A General Concept

Life cycle assessment (LCA) is a methodology for supporting decision making that quantifies the potential environmental impacts of any system (*e.g.* product, process, service or activity) over its entire life cycle. Its main objective is to orientate environmental policies, to inform consumer choice, to design increasingly sustainable solutions and to mitigate the environmental impact of industrial activity.[1–3]

---

*Corresponding author.

In general terms, LCA describes and quantifies with as much detail as possible the various stages and operations that intervene in the manufacture, use and after-use of the system. These stages are defined as unit processes (UPs) and consist of the extraction, conditioning and transport operations of raw materials, their transformation into a specific product or service, this specific product or service's transport and distribution, its use and its end-of-life (*e.g.* reuse, recycling, valorisation or final disposal). The standard LCA methodology contemplates four main stages, which are: (i) the definition of the goal and scope of the assessment; (ii) the compilation of a life cycle inventory (LCI); (iii) the assessment of environmental impacts of the life cycle; and (iv) the interpretation of results. Each LCA stage is described in detail in the sections below.

1.1 Definition of the Goal and Scope of the Assessment

The first and perhaps most essential step in any LCA is the definition of the goal and scope of the assessment. This stage should outline in a clear and understandable manner the objective of the analysis, taking into consideration the intended applications of the results, the limitations of the assessment methods and the target audience of the deliverables. The goal and scope will also condition the narrative of the assessment and the number of appraisal iterations that are needed to find a sufficiently robust result. Examples of LCA goals can be improving a specific design, benchmarking goods and services, evaluating green procurement practices and supply chains, eco-labelling and informing policy development.[3]

A critical step in this phase is the definition of the functional unit, which is defined as a quantitative description of the service performance or the fulfilled needs of the system under consideration.[4] The definition of the functional unit needs to be harmonised with the objectives of the assessment, it should be representative of qualitative and quantitative aspects of the system and needs to provide insight into its environmental performance in a simple and concise manner (*i.e.* answering questions such as 'what', 'how much', 'how well' and 'for how long').[4] The goal definition phase also involves assigning geographic and temporal boundaries to the system under analysis. Along with the definition of the functional unit, the definition of the system boundaries plays a key role in any LCA, as it will condition the level of contribution of the different UPs in the entire life cycle, and therefore their overall impact (Figure 1).

The importance of the definition of the boundary stems from the fact that the life cycle of any system is usually composed of several UPs that are proportional to its level of sophistication. Consequently, practitioners may wish to simplify the LCA and leave out those UPs with a potentially marginal contribution by defining a cut-off threshold based on the defined boundaries and the functional unit. To determine the UPs that can be excluded from the analysis, general *a priori* knowledge of the system as well as its interactions is necessary. In every case, drawing a flow diagram and

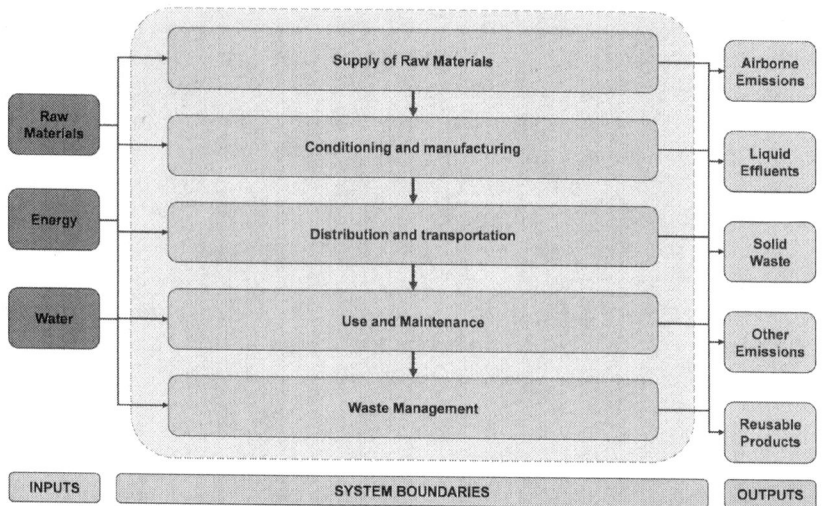

Figure 1 Schematic representation of a system in a LCA, its inputs and its outputs.

identifying all of the interactions within and outside the boundaries of the system can prove to be useful.

1.2 Life Cycle Inventory

The identification, compilation and quantification of the inputs and outputs of the system and its component UPs are carried out through a LCI. The inputs can be either materials or energy, while the outputs can take the form of airborne emissions, liquid effluents, solid waste, energy or reusable products. Fundamentally, a LCI consists of a mass and energy balance of the system and its composing UPs.

The compilation of the LCI requires that a substantial number of data are gathered, such as information on material and energy inputs, specific technologies involved, operation modes and parameters, geographic information, emission factors (EFs), *etc.* Moreover, it is an inherently complex activity that, apart from being data and time intensive, is critical for the correct estimation of the consumption of resources and emission/waste flows.[5,6] Implicit in this phase is the homogenisation of information regarding its provenance, data quality, reference year, reference technologies and geographic representativeness. The preparation of a LCI is not a one-off exercise; the iterative nature of a LCA may sometimes require methodological changes to improve the representation of the system or to incorporate better-quality data, which in turn calls for a certain degree of flexibility in the LCI.

The potentially large amount of data that are required for the compilation of a LCI, as well as the needs for transparency, traceability and flexibility, often require relying on specialised software that can handle information in a structured manner. An additional advantage of these types of software is that

Life Cycle Assessment of Road Vehicles

they incorporate inventory data of materials and characteristic processes such as power generation, raw material extractions or waste disposal and are routinely updated. Several software options, either proprietary or open source, are available in the market for practitioners, having varying degrees of complexity and coverage of processes. Perhaps the most common software examples are SimaPro, GaBi, EcoInvent (commercial) and OpenLCA (open source).

1.3 Life Cycle Impact Assessment

The environmental burden associated with the UPs of the system throughout its life cycle is estimated in the life cycle impact assessment (LCIA) stage. The characterisation of impacts is carried out by classifying the inputs and outputs of the UPs and multiplying them by a characterisation factor that determines the global attributed impact of the system on various environmental aspects, such as climate change, stratospheric ozone depletion, photochemical ozone formation, eutrophication, acidification, toxicological stress on human health and ecosystems, depletion of natural resources, noise, *etc.*[4,7] The attribution of impacts is often done through the application of particular methodologies, which can be classified as midpoint or endpoint depending on whether practitioners want to reflect the impact as cause–effect links (midpoints) or directly on sensitive receptors that require protection, such as human health, ecosystems or resources (endpoints).[8] Perhaps one of the best-known methodologies in Europe for the estimation of impacts is the International Reference Life Cycle Data (ILCD) system handbook, developed by the European Commission's Joint Research Centre; however, other methodologies exist, and practitioners are advised to choose the one that best describes the impacts based on the objective of the assessment. In general, and depending on the methodology of estimation, impacts on the following transport-relevant categories can be quantified:[7,9,10]

- *Climate change*, which represents the impact on global temperature due to the emissions of greenhouse gases.
- *Particulate matter formation*, representing the impacts caused by primary and secondary particulate matter on human health.
- *Human toxicity potential*, which describes the impact caused by the emissions of heavy metals and hydrocarbons on human health.
- *Photochemical ozone formation*, which represents the formation of urban haze or smog because of the chemical reactions between nitrogen oxides (NO_x), volatile organic compounds (VOCs) and aerosols, as well as its impact on human health and vegetation.
- *Acidification potential*, describing the potential impacts on the environment because of the emissions of nitrogen and sulphur dioxides, their chemical transformation in the atmosphere and their deposition on soils and ecosystems.
- *Eutrophication potential*, which represents the excessive growth of vegetal organisms in water due to the additional supply of plant

nutrients resulting from the emissions of nitrogen compounds in oxidised and reduced form. A distinction can be made between terrestrial, freshwater and marine eutrophication.
- *Land use changes*, which describe transformations in the structures of specific portions of the territory in terms of economic, social or natural variables (*e.g.* soil occupancy and loss of biodiversity).
- *Resource depletion*, which describes the decrease in the availability of raw materials due to human activity, compromising the security of future generations.
- *Other impacts*, such as noise, odour, ionisation damage, radiation, stratospheric ozone depletion, *etc.*

The obtained impacts can then be normalised using reference values with the objective of allowing 'like-for-like' comparisons between similar systems. These reference values are often used for benchmarking purposes and refer to a specific geography and time.[2,3] Normalising impacts is a common practice in transport-related LCAs, such as comparing different vehicle engine types or fleet configurations. Finally, impacts can be weighted based on value choices in order to facilitate comparison across different impact categories. This weighting or valuation process may be carried out attending to monetisation criteria, 'panels' or impact rankings derived through surveys on groups of people, distances of the impacts to the targets or people's preferences (*e.g.* willingness to pay).[7]

1.4 Interpretation of Results and Conclusions

This phase involves analysing the results of both the LCI and the LCIA stages based on the objectives and scope defined for the assessment; the final aim should be putting forward a series of conclusions and recommendations to the questions that motivated the assessment, based on the obtained results. An analysis of potential improvements aimed at reducing the environmental pressures of the system can be included, as well as the identification of the responsible processes. Finally, this phase can also accommodate sensitivity analyses in order to characterise the response of the system to different parameters such as feedstocks, operation conditions, transport options, *etc.*

It is worth highlighting that a LCA is an iterative process that requires continuous analysis and interpretation of all data inputs and outputs, as well as carefully monitoring the representativeness of the results by paying attention to the potential conclusions that can be reached after performing such an exercise. It is during this interpretation and scrutiny phase that the experience and analytic skills of the practitioner are essential in order to reach conclusions that have an informative value for policy making or strategic planning. A general schematic representation of the phases involved in a LCA and their interactions are shown in Figure 2.

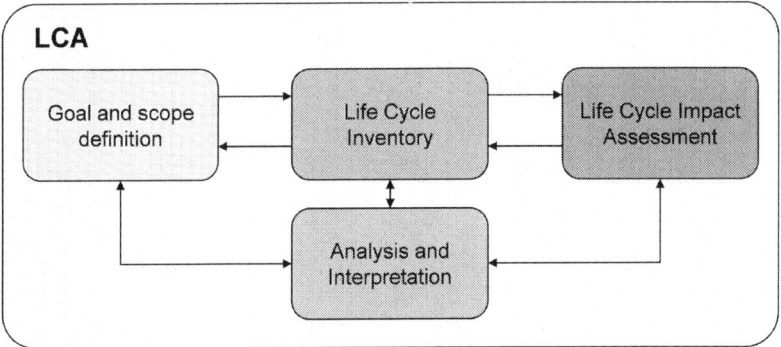

Figure 2 Schematic representation of the stages of a LCA and its interactions.

2 Life Cycle Analysis: Review of the State-of-the-art

LCA has been extensively used in the last two decades as an instrument for the identification of the sustainable character of systems, with a strong policy and industrial perspective. The overall objectives of most of these analyses are demonstrating the 'green character' of goods and services, assessing for competitive advantages and benchmarking, identifying the potential for greening processes, increasing consumer confidence, monitoring resource efficiency and waste generation or gaining a better understanding of supply chains.[10] Lately, the LCA field of knowledge has been prolific in terms of research and publications. Examples of recent publications in the field include the application of LCA to product packaging,[11] photovoltaic technology,[12] hydrogen fuel cells,[13] biodiesel,[14] oil and gas products,[15] meat and agricultural products,[16,17] wine,[18] universities and educational services,[19] airline services[20] and airport infrastructure.[21]

The road transport sector and the environmental evaluation of vehicles have been the areas of focus of several LCA-related studies. Most studies analyse the performance of different models, transportation options, engine types, fuels and vehicle technologies.[22] The environmental performance of conventional vehicles against hybrid or electric equivalents has been carried out for the USA,[23] as well as several European countries, such as Belgium,[24] Greece,[25] Italy[26] and Spain.[27] The conclusions reached by most of these publications are that environmentally friendly vehicles have higher efficiencies, which allow lower energy consumptions and emissions compared to conventional vehicles. In this line, substantial work has been carried out analysing the sustainability and cost effectiveness of electric, hydrogen and methanol-fuelled vehicles in the wider road transport sector.[28,29] LCA has also been used in numerous examples related to transport planning, such as assessing the global benefits of active forms of transport against private motorisation (*e.g.* cycling),[30] the possibility of reducing traffic in the urban environment,[31] the evaluation of an optimal distribution of alternative vehicles[32] or a comprehensive analysis of urban mobility.[33] The abovementioned areas highlight the wide applicability of LCA as an appraisal tool for environmental impacts, cost effectiveness and feasibility of transport options and policy actions.

As an example of an application of LCA in informing policies for the road transport sector, the findings of a study evaluating the effect of introducing hybrid and gas-fuelled cars in the taxi fleet of the city of Madrid, Spain, will be outlined. The greening of the taxi fleet of the city appeared as an important measure in the local air quality plan and aimed to reduce the high levels pollution originating from diesel vehicles. Four scenarios were evaluated for climate change, photochemical ozone formation, particulate matter, acidification and terrestrial eutrophication: 'business as usual', the scenario of the air quality plan, an all-diesel and an all-ecologic fleet. In every case, the all-ecologic fleet produced the lowest impacts across the board, followed by the scenario envisaged in the air quality plan. In the four scenarios, the use phase appeared to be very relevant for climate change and photochemical ozone formation. In the case of acidification and particulate matter, the fuel transportation phase contributed the most.[27] The results of this study provide a holistic view of the environmental benefits of a specific policy, as well as assessing the potential margins of improvement of the measure.

3 Life Cycle Analysis of Road Vehicles

The road transport sector is of great importance from a social and economic point of view. It is one of the largest sources of atmospheric pollution and greenhouse gases in several countries and is expected to continue to be so, as the world increasingly becomes an urban environment.[34] It is therefore essential that a correct characterisation of the environmental impacts of the road transport sector is made for any strategic planning purposes. This, however, is not an easy task, as the transport sector is an intricate combination of economic, technological and societal factors, whose interplay touches upon many aspects of modern human societies. In the urban environment, the road transport sector is a key factor to take account of when determining the level of development of a city, its quality of life, its needs for infrastructure or its safety.

Despite the complexities inherent to the road transport sector, one can simplify it by defining a group of vehicles with different characteristics that are driven for specific purposes during a specific period at a given geographic location. From this simple definition and from a LCA point of view, two main components can be identified as determinants of the environmental pressure exerted by the road transport sector: the vehicles and their use. With these components in mind, the entire LCA of the road transport sector is made of the sum of the individual life cycles of each of the comprised vehicles and their individual use. These two components are often described in the LCA literature as the material life cycle and the use life cycle. Since using a vehicle relies on the availability of energy from a source (most notably a fuel), the use life cycle of a vehicle can be further divided into a fuel life cycle and a vehicle use phase proper, which are described as follows in terms of their component UPs:[27,35]

- *Material life cycle.* Composed of: (i) the extraction and production of raw materials; (ii) the manufacturing of the vehicle; (iii) the manufacture of vehicle spare parts and auxiliary equipment; (iv) the transportation to the sale point; and (v) the scrappage and disposal of the vehicle after its use.
- *Fuel life cycle.* For conventional vehicles, this includes: (i) the recovery of crude oil and natural gas; (ii) its transport to the refinery; (iii) its transformation through refining; and (iv) its storage and transportation to the distribution centres. This group of phases is often referred to as the well-to-tank phase. For plug-in hybrid and battery electric vehicles (BEVs), these phases are complemented by: (i) the extraction of fuels for the generation of electric power; (ii) its transport to the power plant; (iii) the generation of electricity; and (iv) its distribution through the grid. A similar structure of phases can be proposed for hydrogen and fuel cell vehicles.
- *Use phase.* This phase involves the actual functioning of the vehicle through the consumption of fuel or electricity, and it may also include the consumption of spare parts, tyres, brakes or oil, as well as other inputs related to vehicle servicing or repairs; the tank-to-wheel phase is often used as a shorthand for the use phase.

Strictly speaking, a LCA should take account of the building and decommissioning processes of the extraction, distribution, production and sales facilities, the construction and deployment of distribution infrastructure (pipelines, vessels, pumps, *etc.*) and their related maintenance operations. In many cases, these stages are excluded from the LCA due to a lack of representative information.[36]

Having outlined the three basic parts of the life cycle of a vehicle, the following sections are dedicated to the detailed explanation of each of the material and fuel life cycles, as well as the use phase. This description provides information about the UPs that need to be considered, as well as the type of data that are required for the compilation of the LCI. It should be stressed that a full characterisation of the life cycle requires that emissions on all environmental receptors are quantified, such as air, land, water and waste generation. However, quantifying airborne emissions of both pollutants and greenhouse gases is comparatively simple because of the wealth of methods and considerations that are available for underpinning other policy applications (*e.g.* air quality modelling). Moreover, information is not always available for quantifying emissions to water or soil from all of the UPs that occur in the life cycle of a vehicle.[36,37] In this respect, the description of the material and fuel life cycles and use phase in the sections below pays special attention to airborne emissions. Finally, the information presented below aims to provide general guidance on the elements that should be considered in the LCA of road vehicles and does not intend to prescribe a specific analysis methodology.

3.1 Material Life Cycle of Vehicles

3.1.1 Extraction and Production of Raw Materials and Feedstocks.

The extraction and production of the raw materials used for the manufacture of vehicles depends on the specific material composition of the studied vehicles. Extraction processes may include activities such as mining operations of the mineral ores to produce metal parts and components, the exploration and extraction of crude oil to produce plastic parts and components, the exploitation of sand and limestone quarries to produce glass, *etc*. Production processes may include steel manufacturing, glass manufacturing, petroleum refining, petrochemical processes and plastics and fabrics manufacturing. Other more complex processes within this category involve the manufacturing of all those components that are not specifically produced at the vehicle assembly locations, such as high-tech hardware like on-board computers, control systems, brake mechanisms, engines, air conditioning equipment, *etc*. This stage also involves the logistics related to the supply chains between the different UPs, as well as the energy consumptions associated with the extraction, manufacturing and transport operations.

A proper characterisation of the extraction and production of raw materials and feedstocks requires a detailed knowledge of the material composition of a vehicle, which is related to a model and make, as well as the processes that originate the different material components of the vehicle. In many cases, this information will not be available to practitioners outside the concerned industries, as it may well be protected as a trade secret and not easily disclosed. In the absence of this information, the use of LCA databases is recommended (*e.g.* EcoInvent).[37] A review of the scientific and commercial literature enabled the identification of an average material composition of passenger cars for five different manufacturing companies, as presented in Table 1.[38–43]

Table 1 Average material composition of passenger cars by make (units: % weight).

Material	Daimler	Fiat Chrysler	Ford	Nissan	Volkswagen
Ferrous metals	46.9	63.0	76.0	59.4	58.99
Non-ferrous metals	24.3	10.0	—	14.1	11.96
Plastics	21.1	13.0	18.0	13.0	19.78
Fluids	—	5.0	0.8	—	4.69
Electronics	0.2	—	0.2	—	0.17
Others	7.5	9.0	5.0	13.5	4.41

3.1.2 Vehicle Assembly and Distribution. The assembly is perhaps the most intuitive process in the manufacturing of a vehicle, as it results in a finished product that is ready to be sold and used. The assembly of vehicles occurs in factories located in specific places around the world, depending on the model and make of the vehicle under consideration. The location and output of the assembly plants can be generally obtained from corporate documents and official communications from the vehicle manufacturers, vehicle associations or the press. A straightforward characterisation of the environmental pressures exerted by the vehicle assembly facilities can be made by using officially reported emissions or by consulting national and local emissions inventories or regulatory emission registers. In other cases, emissions can be estimated with the aid of standardised methodologies (*i.e.* European Monitoring and Evaluation Programme/European Environment Agency (EMEP/EEA) and United States Environmental Protection Agency (USEPA)) based on the characteristics of the assembly facility, or by relying on data compiled for similar installations elsewhere. It is also worth highlighting the tools and datasets that are openly available from the European Life Cycle Database for compiling LCIs. When compiling emission inventories for vehicle assembly facilities, practitioners should be aware that data such as outputs, operation conditions or consumption of raw materials may be confidential and thus not readily available.

As for the distribution of vehicles, practitioners should take into consideration the distance between the different assembly facilities of the vehicle models that comprise the studied fleet and the vehicle sale points within the geographic boundaries of the system. Based on these distances, vehicles can be distributed either by road only, by road and rail or by a combination of road, rail and maritime shipping. Distribution of vehicles by road is done almost entirely using vehicle carriers, whose airborne emissions can be characterised using standardised EFs available from emission models (*e.g.* Computer Programme to calculate Emissions from Road Transport (COPERT) for airborne emissions),[44] and as a function of the age of the vehicle (*e.g.* Euro standard), the travelled distance, the level of loading of the vehicle and the average circulation speed. The estimation of rail emissions follows a similar procedure, using EFs for representative locomotives with vehicle carrier racks and taking account of the length of the rail tracks between logistic centres. In the case of maritime transport, emissions will be based on specific EFs for vehicle carrier ships, as well as on the sailed distance between relevant ports of call. Further information on the details that need to be considered for estimating shipping emissions can be found in Section 3.2.2 of this chapter, as well as in the literature.[27] While a full characterisation of the complete distribution chain of vehicles is desirable, it is unlikely that it will be available to the desired level of granularity, which calls for carrying out assumptions where gaps are detected and, in many cases, relying on inventory databases (*e.g.* EcoInvent). The same applies when characterising the impacts originating in the manufacturing of spare parts and auxiliary equipment of the vehicles.

3.1.3 Vehicle End of Life. The end of life of a vehicle is considered to start immediately when the use phase (within the boundaries of the system) ceases to occur. Perhaps the best-known outcome of an end-of-life vehicle (ELV) is its reuse (*i.e.* sold and driven elsewhere). Depending on how the functional unit and the boundaries of the system have been defined, vehicle reuse can be considered as a part of the use phase or as a part of the end of life.

It has been made clear previously that vehicles are complex products that include many sorts of devices and materials with intrinsic value, so another common end-of-life outcome is their individual reuse or recycling. This necessarily involves dismantling the vehicle, either manually or mechanically, and attending to the potential uses of the specific materials to be recovered and to local environmental legislation. In many cases, specific parts of the ELV can be reused and sold at a residual value or entirely exported.[45] This is particularly true in developing countries, where the spare parts market relies on usable scrapped parts from local and foreign-owned vehicles.[46] The materials that comprise a vehicle (Table 1) can be recovered through different UPs such as oxygen cutting, shredding, smelting and compacting.[38] If the recycled materials are reintroduced into the life cycle as raw materials, the environmental impacts of the UPs that use them as inputs should logically decrease. Finally, other possible outcomes for ELVs are the disposal and landfilling of vehicle parts, the abandonment and uncontrolled degradation of the vehicle or its destruction (partial or total) by the action of fire or natural agents.

It is worth noting that more than one outcome is likely to affect any individual ELV. However, from a circular economy point of view, manufacturers should aim to produce vehicles that generate as little end-of-life waste as possible and to maximise the reuse or recycling potential of their components and materials.[47]

The characterisation of the end-of-life phase of a vehicle is a challenging task because of the uncertainties associated with modelling the recovery and disposal operations in terms of the quantity, time and composition of the returned products.[43] Although it has been found that the number of ELVs increases with the rise of vehicle sales, the fate of these vehicles remains an area of LCA that is known to carry uncertainties due to the inherent unpredictability of the fate of the vehicle or its parts once it stops being used.[48]

3.2 Fuel Life Cycle of Vehicles

At present, the transport sector is dominated by engine types consuming fossil fuels for motion, so a typical well-to-tank characterisation is essential for the appropriate description of baselines that will be used as references in the elaboration of scenarios and to inform potential environmental benefits of switching technologies or fuels. Even though the fuel life cycles of non-conventional vehicles (*e.g.* electric or hydrogen) may involve different UPs, they will not be described in detail in the following sections due to the relatively small proportions of these vehicles in the current vehicle fleets. However, when BEVs are being considered in the LCA, an appropriate description of end-of life UPs, such as battery disposal, is essential, as this is considered an important source of impacts such as human toxicity or resource depletion.[49]

3.2.1 Crude Oil and Natural Gas Recovery. The recovery of crude and natural gas is considered the first UP in the life cycle of the fuel and it is composed of several operations, with the final goal of making available these raw materials from their underground deposits. These operations include deploying exploration missions to those locations suspected of having rich deposits, drilling the oil well and extracting and processing the crude oil or natural gas with extraction agents (usually water and gases). The environmental impact of crude oil and natural gas recovery UPs is dominated by drilling operations, which are highly intensive in energy. The recovery of oil requires drilling a hole into the deposit, injecting a fluid to maintain its pressure and extracting the crude. The energy consumptions of the drilling operations increase exponentially with the depth of the reservoir,[50] while drilling rigs are usually powered by diesel engines. Occasionally, the lack of sufficient pressure to extract the crude, which becomes more common as the deposit ages, requires that pumps or other injection mechanisms are implemented. In other cases, pressurised steam may be injected to reduce the viscosity of the crude oil. In addition to this, there are emissions associated with the flaring of excess gases in the reservoirs, as well as the escape of gases trapped in the earth when drilling.[51] The use of water as an injection fluid also has an impact on the depletion of hydric resources and the potential emission of pollutants to water bodies.

As previously stated, the characterisation of the environmental impacts associated with the recovery of crude oil and natural gas requires a detailed knowledge of the equipment and the operation conditions, which are determined by the local conditions such as geology, depth, energy availability, *etc*. As a consequence, practitioners should have an idea of the countries of provenance of the oil and gas that ultimately originate the fuels used by the studied vehicles. As an example, Figure 3 provides information on the countries of origin of the crude oil imported into the UK and in Spain in 2014. In the case of the UK, almost half of the imported oil comes from Norway, and more than a third is from African countries; in the case of Spain, there is a wider variety in the origin of crude, with almost a quarter being imported from Latin America and another quarter from Africa.[52]

For the characterisation of the oil and gas recovery operations, including drilling, generic information is available for different regions of the world from the International Association of Oil and Gas Producers, as well as from LCA databases and software.

3.2.2 Transportation to the Refinery. Crude oil and natural gas are transported to the refining and distribution centres for consumption or

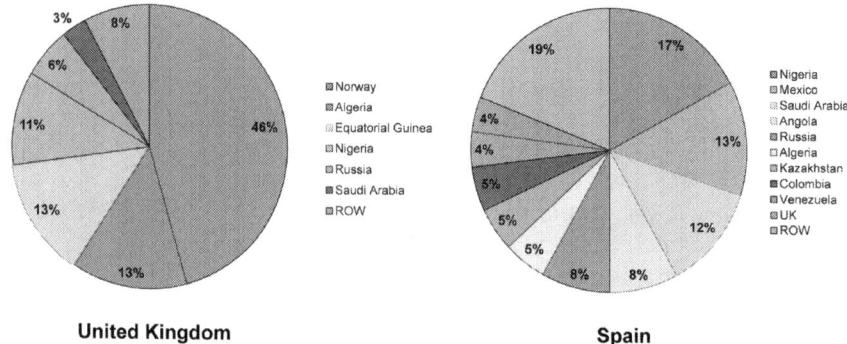

Figure 3 Countries of origin of the crude oil imported by the UK and Spain in 2014. ROW = rest of world.

processing using tanker ships in most cases, but also pipelines. The primary driving forces for crude oil and natural gas transportation are refinery requirements, which usually produce a specific set of refined products, adapting to changing crude grade availabilities and demand.[53] This requires that a heterogeneous oil tanker fleet transports several crude oil varieties from several loading ports to other discharging ports. Many of these loading ports only supply a single specific crude variety, while others supply more than one,[54] and as previously noted, the varieties consumed by the refineries that manufacture the fuels used by vehicles in the system will condition the loading and discharging ports and routes that need to be considered for the assessment. As an example, the origins of the crude consumed by a generic refinery in Spain in 2012, as well as their loading and discharging ports, are presented in Table 2.[27]

In every case, routes need to be calculated using the exact geographic locations of wells, pipelines and marine terminals. These routes should take into consideration the specific navigation routes of oil tankers based on ship movements, which are available from various sources (*e.g.* the Lloyd's List ship movements database).[55] This should also take account of mixed transportation routes involving intermediate pipelines (*e.g.* the Suez-Mediterranean (SUMED) or Yanbu pipelines).

Once the routes are defined, environmental impacts can be determined by the emissions of the vessels transporting crudes between the loading and discharging ports, as well as by the operation of pipelines. The estimation of emissions of ships requires combining activity data such as shipped tonnes, shipped miles or ship-equivalent traffic data (ship-km) with standardised EFs (*e.g.* EMEP/EEA). These EFs should ideally reflect the characteristics of the fleet that transports oil along a specific route, such as ship age, engine types (both main and auxiliary) and propulsion fuel.[56] EFs in many cases will depend on the gross tonnage, which is a measurement of the total volumetric capacity of the ship, and therefore of its size. Ideally, detailed information about the specific vessels that supplied raw materials to the refinery should be collected from the relevant port or customs authorities,

Table 2 Crude varieties, origins and transportation modes of crude oil imported to Spain in 2012.

Crude varieties	Origin	Loading port	Pipelines	Discharging port
Condensate	Algeria	Bejaia	In-Amenas-Bejaia	Cartagena
Sahara		Arzew	Hassi R'mel-Arzew	
Dalia	Angola	Platform	—	Rota
Girassol				
Albacora	Brazil	Platform		Rota
Castilla	Colombia	Coveñas	Caño Limón-Coveñas	Rota
Djeno	Congo-Brazzaville	Platform	—	Rota
Kirkuk	Iraq	Ceyhan (Turkey)	Kirkuk-Ceyhan	Cartagena
Foroozan	Islamic Republic of Iran	Kharg Island	Yanbu	Cartagena
Iranian Heavy		Yanbu (Saudi Arabia)	SUMED	
Iranian Light		Sidi Kerir (Egypt)		
Soroosh				
Bouri	Libya	Sirte	Amal-Ra's	Cartagena
El Sharara			Lanuf/Sirtica	
Es Sider			Waha	
Sirtica				
Mellitah				
Istmo	Mexico	Dos Bocas	Cantarell-Cayo Arcas	Rota
Maya		Cayo Arcas		
Oseberg	Norway	Sture	Oseberg-Sture	Rota
Siberia Light	Russian Federation	Novorossiysk	Omsk-Novorossiysk	Cartagena
Ural Light		Primorsk	Omsk-Primorsk	
Ural Novorossiysk		Tuapse	Irkutsk-Tuapse	Rota
Ural Primorsk				
Arabia Light	Saudi Arabia	Ras Tanura	Yanbu	Cartagena
		Yanbu	SUMED	
		Sidi Kerir (Egypt)		
Souedie	Syrian Arab Republic	Tartous	Karaichok-Tartous	Cartagena

from shipping companies or from the refinery operator. If this information is not available, generic vessel sizes can be selected based on the consumption of oil by the refinery, their maximum capacity and the given transportation route (*e.g.* Suezmax or smaller for vessels transiting the Suez Canal). Finally, and depending on the boundaries of the system, the quantification of the emissions of ships travelling in ballast back to the loading ports needs to be examined.

3.2.3 Refining and Distribution. Once the oil is received at the refinery, it will be transformed through a series of unitary operations and processes into a wide array of products and commodities, including transportation fuels. A refinery is a very complex part to characterise in LCA, due to the high number of processes that take place in the manufacture of transportation fuels and that require numerous feedstocks, some of which are produced in upstream processes within the same refinery. The quantification of the environmental impact of a refinery requires a realistic characterisation of the following processes:

- Crude oil distillation and vacuum distillation units;
- Catalytic reforming and naphtha hydro-treating units;
- Catalytic cracking units;
- Alkylation and isomerisation units;
- Coker units (delayed and fluid);
- Other operations: water cooling, process venting, flaring, storage and stationary combustion.

To robustly reproduce the material and energy interactions between those UPs in a refinery, a modelling system can be built using chemical process optimisation software and validated against independent data (*e.g.* actual energy consumptions by piece of equipment).[51] In the case of there not being enough information available about the feedstock and energy flows for each of the UPs of the refinery, practitioners may wish to rely on pre-compiled national or regional emission inventories for either the specific refinery or another one with similar a capacity and feedstocks.

The finished fuels of the refinery, most notably petrol and diesel, are transported to bulk terminals and to local filling stations. This transport can be achieved by different modes such as pipelines, rail, ships and barges from the refinery to the bulk terminal and to the filling stations by road using heavy goods vehicles (HGVs).[51] As already suggested in Sections 3.1.2 and 3.2.2 of this chapter, the emissions from these transportation modes follow specific methodological considerations in terms of the activity data and the estimation of EFs. Further details on the most common approaches for the estimations of road vehicles in LCA are given in Section 3.3 of this chapter.

Finally, the environmental impacts associated with the fugitive emissions of VOCs in the retail fuel supply chain need to be considered, especially when the quantification of impacts related to the formation of tropospheric ozone or human toxicity is envisaged. Filling stations are deemed to be the most relevant source of fugitive VOCs in the urban environment, so their appropriate consideration with regards to the intensity of fuel consumption by vehicles is essential.[57]

3.2.4 Electricity and Energy Mix. Many of the UPs in the aforementioned stages of the material and fuel life cycles require electricity as input. The generation of electricity is therefore an essential UP that influences the global environmental performance of the system, as it is one of the largest contributors to the emissions of greenhouse gases and airborne pollutants at the national level. In other cases, power generation is a substantial source of radiation, acidification, land use, freshwater and terrestrial ecotoxicity and resource depletion.[49]

Several electricity generation options are operative in the world, with different levels of penetration in the respective energy mixes. Examples of these are subcritical pulverised coal (with and without carbon capture and storage), natural gas (combined cycle gas turbine), nuclear (pressurised water reactor), solar photovoltaics, wind and biomass combustion.[58] The environmental pressure exerted by electricity consumption in the material and fuel life cycles or the use phase of the vehicle is in direct correlation with the fuel mix used by power plants connected to the grid at a given time and geographic location. It is well accepted that a higher reliance on fossil fuels and non-renewable energy sources is generally correlated with greater environmental impacts, particularly those related to climate change and resource depletion.[59] Figure 4 shows the energy mixes in 2012 of the UK and Spain, illustrating the level of penetration of renewable energies in both countries.[60,61] This difference will certainly condition the sustainable character of electricity-consuming UPs in the respective countries and is key in the overall performance of low-emission vehicles such as plug-in hybrids and electrics.

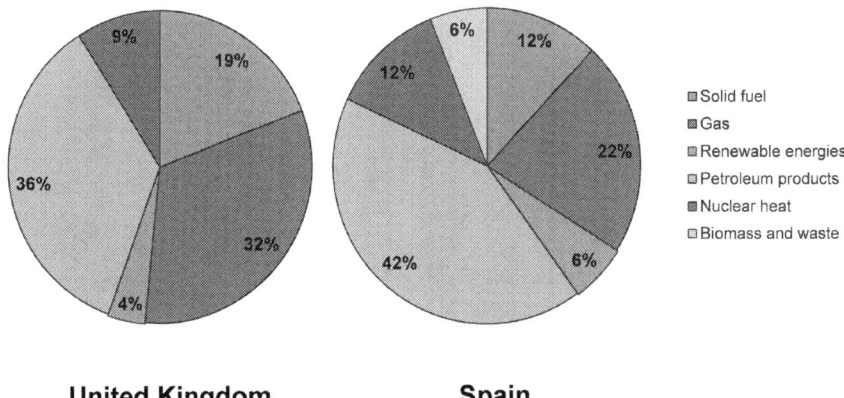

Figure 4 Energy mixes of the UK and Spain in 2012.

3.3 Vehicle Use Phase

The material and fuel life cycles converge at the vehicle use phase, also known as the 'tank-to-wheel' phase, and it characterises the environmental performance of the vehicle between the moment of purchase and the start of its end of life, or, in simpler terms, the impacts produced by the vehicle when driven. Due to its duration as well as its degree of energy intensiveness, the use phase of a vehicle is reputed to cause most the air pollution and climate change burden in the entire vehicle life cycle.[62] More specifically, the level of impact of the use phase is a function of several factors both intrinsic and external. Examples of intrinsic factors are the type of vehicle, its engine type, its age and its fuel efficiency, all of which already condition the vehicle's performance before its purchase.[63–65] In addition to this, external factors such as driving conditions, the configuration and state of the roads, the type of service the vehicle provides and the level of maintenance may exacerbate or attenuate such impacts to a significant extent.[66–68]

Depending on the scope and objective of the LCA, the characterisation of the use phase could involve describing an individual vehicle or an entire vehicle fleet under several assumptions with the goal of producing evaluation scenarios. Examples of these evaluations can be the comparison of environmental performance between equivalent fleets of conventional and non-conventional vehicles. Regardless of the objective of the evaluation and in line with what has been stated above, the description of the use phase of a vehicle minimally requires the definition of the vehicle attributes and the driving conditions, which are explained in the following sections.

3.3.1 Vehicle Attributes. A good understanding of the attributes of the vehicle is necessary for determining the EFs that are necessary for estimating their degree of environmental pressure. The EFs used in road transport are obtained from functional relations that predict the quantity of a pollutant that is emitted per distance driven, energy consumed or amount of fuel used. EFs are usually derived from vehicle categories and they depend on many parameters, such as vehicle characteristics, emission control technology and fuel specifications. There is also a component of the EFs that is determined by the ambient and operating conditions (*e.g.* cold start, cruising acceleration, *etc.*),[69] and is described in Section 3.3.2 of this chapter.

The emissions of several kinds of vehicles can be described through EFs. In the case of the UK, and perhaps most of Europe, information for the following vehicle classes and fuel types is usually available:[70]

- *Passenger cars.* Engine types: conventional (diesel or petrol), diesel hybrid, petrol hybrid, plug-in hybrid and battery electric. Engine sizes (conventional): <1400 cm^3, 1400–2000 cm^3 and >2000 cm^3.
- *Light goods vehicles (LGVs).* Engine types: conventional (diesel or petrol), diesel hybrid, petrol hybrid, plug-in hybrid and battery electric. Sizes: N1(I), N1(II), N1(III) and N2.
- *Urban buses.* Engine types: conventional (diesel), diesel hybrid and compressed natural gas (CNG). Sizes: ≤15 t and >15 t.

Life Cycle Assessment of Road Vehicles

- *Coaches.* Engine type: conventional (diesel). Sizes: ≤18 t and >18 t.
- *Rigid HGVs.* Engine types: conventional (diesel). Sizes: 3.5–7.5 t, 7.5–12 t, 12–14 t, 14–20 t, 20–26 t, 26–28 t, 28–32 t and >32 t.
- *Articulated HGVs.* Engine types: conventional (diesel). Sizes: 14–20 t, 20–28 t, 28–34 t, 34–40 t and 40–50 t.
- *Motorcycles and mopeds.* Engine types: conventional (petrol) and electric. Engine sizes (motorcycles): two-stroke >50 cm^3; four-stroke 50–150 cm^3, 150–250 cm^3, 250–750 cm^3 and >750 cm^3.

In addition to this, vehicle types are covered by emission limits imposed by national or international legislative frameworks and that have a direct impact in the magnitude of EFs. In Europe, vehicle technologies are classified through the staged application of tighter emission limits (the so-called Euro standards) with specific introduction dates for different vehicle types.[71] Euro standards are a convenient way of representing the degree of improvement of a vehicle or engine design, with the associated improvements in fuel economy and emissions, all of which are related to the age and composition profile of the fleet.[72,73]

The determination of EFs requires the use of a road traffic emission model, whose selection will depend on the application and objectives of the study (bearing in mind that the estimation of road transport EFs also informs other non-LCA applications such as air quality modelling). The most common traffic emission models are average-speed models (*e.g.* COPERT, USEPA Mobile Source Emission Factor Model (MOBILE) and California Environmental Protection Agency Emission Factors Database (EMFAC)), where EFs are a function of the mean travelling speed and rely on information from traffic models or field measurements. Other examples of emission models are traffic-situation models (*e.g.* Handbook Emission Factors for Road Transport (HBEFA) and Verkehr In Städten-Simulationsmodell (VISSIM)-VERSIT+), traffic-variable models, cycle-variable models and modal models.[72,74,75] The selection of the emission model is crucial as it will condition the representativeness of the EFs, which relies on the accuracy of the description of the actual emission level of a vehicle type and application conditions. For instance, EFs that are based on the average speeds of vehicles are assumed to be representative of the emissions at the national level, but less so of local situations, which tend to be influenced by highly changing speeds (stop-and-go situations).[69] Average-speed emission models are suitable for being used in LCA, as these are not bound by time constraints and do not usually incorporate data with a very high level of granularity.

The quantification of other impacts such as terrestrial ecotoxicity or stratospheric ozone depletion may require that specific characteristics of the vehicle, such as the chemical composition of the working fluids of air conditioning equipment (*e.g.* 1,1,1,2-tetrafluoroethane or HFC-134a), the formulation of refrigerant fluids and lubricants or the types of tyres and brakes.[49]

3.3.2 Driving Conditions. The conditions in which a vehicle is driven play a role in the magnitude of EFs, as well as in determining the level of activity. As previously indicated, EFs not only depend on the vehicle type, but also on the consumption of energy (*i.e.* level of use), and in average emission models, this is intimately related to the average circulation speed.[74] The average circulation speed can either be obtained from traffic models or be assumed based on the scientific literature. As an example, the average circulation speeds used in the National Emissions Inventory of Spain are shown in Table 3.[64] The more sophisticated the emission model is, the greater the amount of input information that will be required, which in turn calls for higher levels of data resolution (*e.g.* levels of service, speed limits by road, acceleration intervals, *etc.*).[73]

When the LCA has as its objective the assessment of vehicle fleets, an essential input that needs to be provided is the composition of such a fleet in terms of vehicle compliance with specific emission standards (*i.e.* Euro mix). Moreover, if the vehicle fleet is composed of more than one type of vehicle, the apportionment of the different vehicle classes needs to be considered as well. These compositions can either be provided based on vehicle registrations (*i.e.* the actual number of vehicles) or mileage (*i.e.* vehicle-kilometres), and they need to be consistent with the activity variable of choice. As an example, Figure 5 presents the total annual vehicle-kilometres by vehicle type in the UK and Spain for the years 2014 and 2012, respectively.[76,77]

Table 3 Average circulation speeds in Spain by vehicle type and road type.

Vehicle type	Motorways	Urban roads	Rural roads
Passenger cars	105 km hour^{-1}	25 km hour^{-1}	65 km hour^{-1}
LGVs	100 km hour^{-1}	25 km hour^{-1}	65 km hour^{-1}
Motorcycles	105 km hour^{-1}	25 km hour^{-1}	65 km hour^{-1}
Mopeds	Not applicable	25 km hour^{-1}	Not applicable

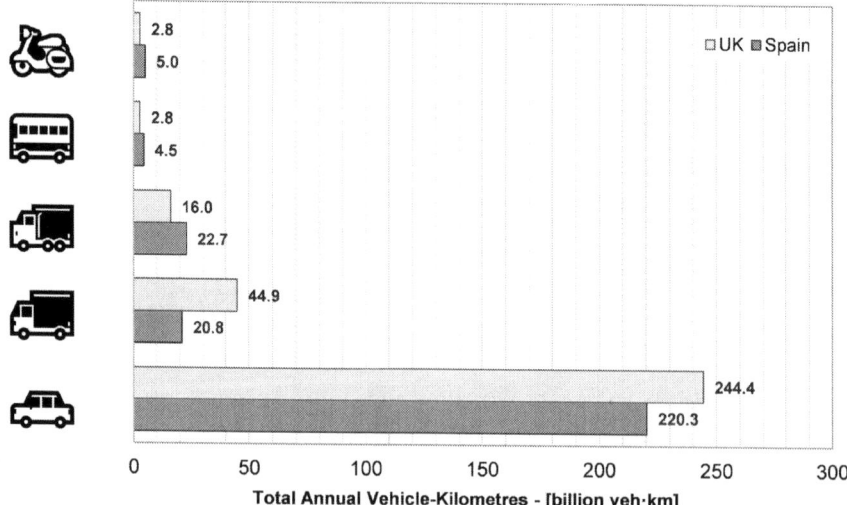

Figure 5 Total annual vehicle-kilometres by vehicle type in the UK (2014) and Spain (2012). Symbols in descending order: motorcycles and mopeds, buses and coaches, HGVs, LGVs and passenger cars.

Life Cycle Assessment of Road Vehicles

A total characterisation of the use phase of a vehicle or fleet requires an estimation of the duration of the vehicle's life. The estimation of the vehicle's life is based on survivability and vehicle mileage schedules, based on statistics obtained from road transport and energy surveys. These studies generally indicate that the probabilities of a vehicle being scrapped increase with its age, whose elasticities depend on the vehicle type and its predominant use.[78,79] This information can also be complemented with the results of fleet composition observation campaigns. In the absence of data, assumptions about the duration of the vehicle life's can be made. Based on estimations made by vehicle manufacturers in the UK, the average age of scrappage of a car is 13.2 years.[80]

4 Uncertainties and Limitations

Despite the comprehensive character of LCA, there are associated uncertainties and limitations that may substantially condition the results of the assessment and therefore the conclusions that can be drawn from it. The main source of uncertainties in LCA is the accuracy in the description of the UPs, which is directly correlated with the amount and quality of input data. In every case, data should normally be taken from the most reliable and representative sources of information. Ideally, these data should have been compiled using bottom-up methodologies that rely on detailed information being gathered at the local level. Data uncertainties in the compilation of a LCI can come from: (i) EFs; (ii) activity data; (iii) data resolution; and (iv) data quality. Examples of the methodological considerations that can be taken in order to mitigate the aforementioned uncertainties are as follows:[81]

- *EFs.* Operation/running conditions and technological specifications of the UPs within the temporal and geographic boundaries of the system.
- *Activity data.* Consistency with statistics within the temporal and geographic boundaries of the system (*e.g.* fuel sales, electricity consumption, traffic counts, *etc.*). Consistency with design, feedstock and maintenance levels of the UPs under consideration.
- *Data resolution.* Spatial resolution that is fine enough to represent the UPs within the temporal and geographic boundaries of the system. Data extrapolations that are based on defensible proxies or assumptions.
- *Data quality.* Reliability of the data sources (official institutions, research bodies, *etc.*). Data QA/QC procedures and robustness checks.

Other important sources of uncertainties are related to the allocation of burdens in the description of the UPs, the selected methodology for the estimation of impacts in the LCIA phase, normalisation, grouping and weighting criteria and inconsistencies in referencing data to the functional unit, all of which can introduce unintentional bias in the assessment.[7]

In general, data should be consistent and comparable, transparent, accessible, actual, valid and constantly reviewed, while the underlying assumptions should be always specified. Regardless of whether information is available or not, practitioners may encounter the problem of failing to achieve accurate representativeness of a UP, despite having relied on the best methodological resources available. If this appears to be the case, the experience of practitioners should be able to combine scientific and technical knowledge and apply expert judgement where possible. As an alternative, many LCA practitioners curb the uncertainties related to the choice of data and methodologies by relying on standardised databases, approaches and methodologies.

5 Practical Example of Life Cycle Assessment: Comparison of Fuel Types for Cars

A comparison of the environmental performance of different fuels used by passenger cars is presented in this section for illustrative purposes, applying the LCA principles described in the previous sections. The objective of this comparison is to understand the levels of environmental impacts that are produced by a specific passenger car using five different fuels: conventional diesel, conventional petrol, liquefied petroleum gas (LPG), CNG and battery electric. To complete the evaluation, the following methodological choices and assumptions were made:

- All passenger cars were assumed to be of a medium size (1400–2000 cm^3) with an average age of 5 years (Euro 5).
- The geographic reference of the analysis is Spain in the year 2016.
- The average use phase of a passenger car is assumed to be 160 000 km.
- The LCI was built relying on the EcoInvent 3 database.
- The LCIA was performed using the ILCD 2011 Midpoint+ V1.08/EU27 2010, equal weighting methodology. The selected impacts were climate change, particulate matter formation, human toxicity (cancer and non-cancer effects), photochemical ozone formation, acidification, terrestrial eutrophication, land use and resource depletion.
- All impacts were weighted relative to those of the conventional diesel car.
- The electricity mix of Spain in 2012 was considered, as reported in Section 3.2.3 of this chapter.[61]
- All data were obtained and estimations carried out using SimaPro version 8.2.0.0.

The impacts of the different fuel types are presented in Figure 6 in reference to those of the diesel passenger car, which are presented before normalising in Table 4. The decision of normalising all impacts to those of a diesel car is useful for elucidating potential environmental benefits that may be brought about by the substitution of these vehicles. In general, the fuel type that delivers the fewest environmental impacts compared to a diesel car

Life Cycle Assessment of Road Vehicles

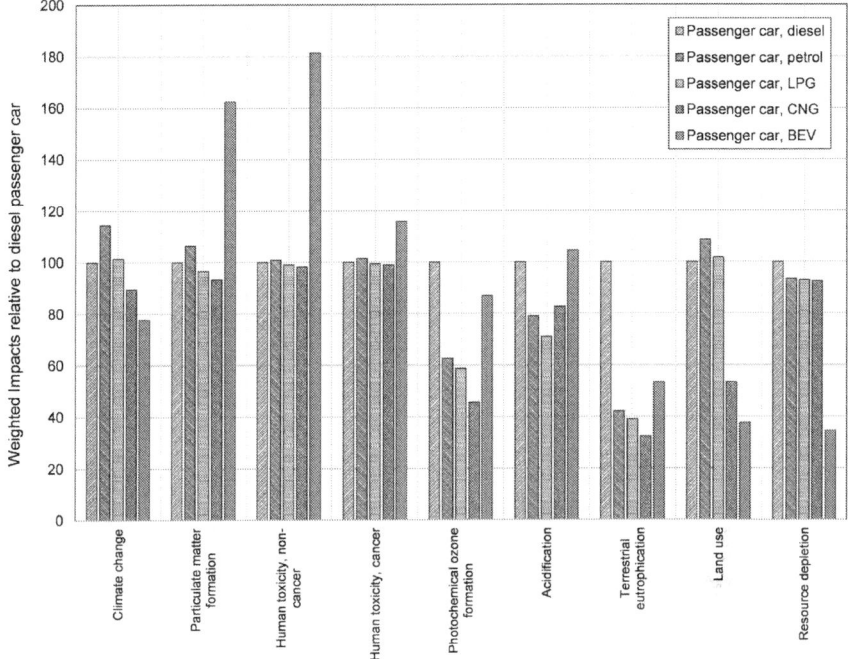

Figure 6 Comparison of the total life cycle impacts produced by different fuel types of a medium-sized Euro 5 passenger car in Spain. Impacts are weighted relative to the diesel passenger car.

Table 4 LCA results for a representative medium-sized Euro 5 diesel passenger car in Spain (ILCD midpoint level methodology). CTUh = comparative toxic unit for humans; NMVOC = non-methane VOC.

Impact		Units
Climate change	48 027	kg CO_{2eq}
Particulate matter formation	29.3	kg $PM_{2.5eq}$
Human toxicity, non-cancer	0.016	CTUh
Human toxicity, cancer	0.003	CTUh
Photochemical ozone formation	213	kg $NMVOC_{eq}$
Acidification	249	$mol_C\ H^+_{eq}$
Terrestrial eutrophication	713	$mol_C\ N_{eq}$
Land use	139 833	kg $C_{deficit}$
Resource depletion	15.1	kg Sb_{eq}

is a CNG alternative, as it performs better for each of the evaluated impacts. The CNG passenger car is also the one that produces the least photochemical ozone formation and terrestrial eutrophication impacts of the five evaluated options (32% and 45% of the diesel car, respectively). A similar performance is observed for the LPG car, except for climate and land use change impacts, which are in both cases 101% of those of the diesel car. Petrol cars seem to have a very similar performance to that of diesel cars for impacts such as human toxicity (cancer and non-cancer effects), particulate matter formation and resource depletion. However, its photochemical ozone formation, acidification and terrestrial eutrophication impacts are significantly lower (62%, 79% and 42%, respectively). Finally, the performance of a BEV is significantly better than that of a diesel car for climate change impacts (77%), terrestrial eutrophication (53%), land use change (37%) and resource depletion (34%). By contrast, BEVs have a worse performance than diesel cars in terms of human toxicity (181% and 115% for non-cancer and cancer effects, respectively), particulate matter formation (162%) and acidification (104%). The increased impacts of BEVs on particulate matter formation compared to the other technologies are a consequence of both the electric mix and the increased weight of these vehicles. In the case of the Spanish electric mix, increased particulate matter formation is correlated with the use of solid fuels (12%) and biomass (6%). Additionally, BEVs are on average 24% heavier than conventional internal combustion vehicles. This increased weight is directly related to higher emissions of non-exhaust particulate matter from brakes, tyre and road abrasion and road resuspension.[82]

The findings above indicate that none of the alternative fuel options that are commercially available for passenger cars is unquestionably better than conventional diesel vehicles. In terms of urban air quality, gas-fuelled vehicles (LPG and CNG) appear to perform better than diesel vehicles under the conditions of Spain. However, in terms of climate change, land use change and resource depletion, BEVs seem to produce significantly less impacts than conventional diesel cars. While this simple comparison focused on individual vehicles with a similar age and starting condition, the philosophy behind this exercise can be extended to evaluating the effects of complex policy actions that act on complete fleets of vehicles with specific compositions and ages.[27]

6 Life Cycle Assessment and the Role of the Road Transport Sector in Urban Air Quality

Road transport is one of the main sources of pollution in urban centres, and it also plays a very important role in their economy and degree of quality of life. The latest European Air Quality Report published by the European Environment Agency (2015) shows that exceedances of the limit values of Directive 2008/50/EC are being consistently detected in several locations of Europe. In the specific case of NO_2, which is perhaps the most-associated pollutant with road traffic, 19 of the 28 Member States measured annual mean concentrations above the allowed maximum (40 $\mu g\,m^{-3}$).[83,84] An analysis of the air quality plans of 18 of the cities that registered exceedances revealed that road traffic was one of the emissions sources responsible for high pollution.[85] In consequence, cities are effectively putting forward measures aimed at substituting conventional vehicles with low-emission options (*e.g.* hybrids, plug-in hybrids and BEVs). As an example, the recently published air quality plans for the UK contemplate the introduction and

Life Cycle Assessment of Road Vehicles

incentivisation of low-emission vehicles as key components of the package of measures envisaged to achieve compliance with European legislation.[86] The main motivation behind these measures is the lower EFs that these vehicles exhibit compared to their conventional counterparts under the same driving conditions.

In line with that is recommended by Directive 2008/50/EC, modelling vehicle substitution scenarios and assessing their impacts on the air quality of the agglomeration using dispersion and chemical transport models are routinely pursued in many air quality plans. While dispersion modelling is valid for regulatory applications, it only gives a partial picture of the total impacts, benefits and synergies of adopting specific policy actions. This is especially relevant as, despite the current understanding that air quality and climate change are intimately connected, their view as two separate issues in science and policy is still widespread. Furthermore, it has been evidenced that policy actions aimed at mitigating either issue must necessarily consider the feedbacks with the other in order to avoid benefitting one sector while worsening the situation in the other.[87] Perhaps the most common example of this disconnection is the substantial dieselisation of European vehicle fleets due to the constant fiscal incentives that were given to replace petrol vehicles with diesel equivalents, with the objective of increasing fuel efficiency and reducing greenhouse gas emissions. This eventually translated into higher NO_x emissions and a consistent worsening of air quality throughout Europe,[88] which is in part responsible for the exceedances of the NO_2 limit values of Directive 2008/50/EC as already mentioned.

The aforementioned information strongly highlights that robust policy making requires that the full impact of these vehicle transitions is characterised not only in terms of local air pollution, but also in relation to other environmental pressures, and not only for the limited period of the use phase, but also throughout the vehicle life cycle. Beyond the climate change–air quality conundrum, in many cases policy makers could be interested in assessing within a single and consistent analysis platform the effects of policy trade-offs, such as the effect in terms of pollution that an increase in the consumption of electricity from electric vehicles will have on the activities of power generation plants. In the same way, if a vehicle is analysed attending not only to its intrinsic polluting potential (conditioned by its fuel and technology type), but also to broader factors such as the energy mix of the country or the relative position of the city in which it will be driven, a more comprehensive picture of its friendliness to the environment can be achieved. For example, an electric vehicle being used in a country with a renewable energy-orientated energy mix will have less of an environmental impact than the same vehicle being used in a country with a coal-reliant mix. The same argument can be applied to any vehicle that is driven far away from the facility where it was originally assembled, as the there is an additional impact associated with its transportation, usually by ship, train or carrier. In this respect, LCA appears to be an adequate tool for obtaining answers regarding these specific aspects, being an ideal complement to diagnosis/prognosis models (*e.g.* air quality modelling) for a robust appraisal of the effects of transport-related measures, as it allows for the quantification of the simultaneous impacts of any policy action on many relevant aspects of the environment.[89,90]

7 Concluding Remarks

As a decision-support tool, LCA allows for characterising the environmental impacts of many aspects of the road transport sector in a consistent and homogeneous way. This is a particularly important feature, as the road transport sector is inherently very complex due to it being composed of sophisticated processes and components (*e.g.* a road vehicle or a refinery). The results that can be obtained from a LCA may shed light on impacts to several relevant aspects of the environment simultaneously, such as air pollution, climate change or human toxicity. The overarching character of LCA allows for these impacts to take account of processes that occur upstream of the use of road vehicles proper, such as the extraction of raw materials, transportation and logistics operations, refining or end of life. This diagnostic advantage requires that a substantial amount of data is gathered and built into a LCI, which will underpin the quality of the impact characterisation. The construction of the LCI requires a good understanding of each of the unitary processes under consideration, as well as an adequate definition of the system boundaries and a functional unit of reference. The compilation of the inventory also requires that representative data of the best quality are used every time, aiming to realistically characterise the environmental pressures exerted by each of the unitary processes. In many cases, data absences will occur, so expert judgement is essential to filling in any gaps and to carrying out modelling assumptions. This calls for endowing LCA with enough transparency and clarity to allow for comprehensive scrutiny of the results and conclusions, always aiming towards better policy making. Finally, the undisputable relevance of road transport in determining the environmental quality of urban environments makes LCA a uniquely positioned assessment framework for characterising the effect and co-benefits of policies.

References

1. A. Levasseur, O. Cavalett, J. S. Fuglestvedt, T. Gasser *et al.*, *Ecol. Indic.*, 2016, **71**, 163.
2. *ISO 14040*, International Organisation for Standardisation, Lausanne.
3. *ISO 14044*, International Organisation for Standardisation, Lausanne.
4. Joint Research Centre, *International Reference Life Cycle Data System (ILCD) Handbook*, European Commission, Ispra, 1st edn, 2010.
5. G. Rebitzer, T. Ekvall, R. Frischknecht, D. Hunkeler *et al.*, *Environ. Int.*, 2004, **30**, 701.
6. S. Suh and G. Huppes, *J. Cleaner Prod.*, 2005, **13**, 687.

7. D. W. Pennington, J. Potting, G. Finnveden, E. Lindeijer, *et al.*, *Environ. Int.*, 2004, **30**, 721.
8. J. C. Bare, P. Hofstetter, D. W. Pennington and H. A. U. de Haes, *Int. J. Life Cycle Assess.*, 2000, **5**, 319.
9. C. Bauer, J. Hofer, H. J. Althaus, A. del Duce and A. Simons, *Appl. Energy*, 2015, **157**, 871.
10. Joint Research Centre, *The International Reference Life Cycle Data System (ILCD) Handbook. Towards More Sustainable Production and Consumption for a Resource-efficient Europe*, European Commission, Ispra, 1st edn, 2012.
11. G. Martinho, A. Pires, G. Portela and M. Fonseca, *Resour., Conserv. Recycl.*, 2015, **103**, 58.
12. M. D. Chatzisideris, N. Espinosa, A. Laurent and F. C. Krebs, *Sol. Energy Mater. Sol. Cells*, 2016, **156**, 2.
13. A. G. Venetsanos, P. Adams, I. Azkarate, A. Bengaouer, *et al.*, *Int. J. Hydrogen*, 2011, **36**, 2693.
14. F. Fernández-Tirado, C. Parra-López and M. Romero-Gámez, *Energy Sustainable Dev.*, 2016, **33**, 36.
15. V. Rajović, F. Kiss, N. Maravić and O. Bera, *Energy Convers. Manage.*, 2016, **118**, 96.
16. C. Lamnatou, X. Ezcurra-Ciaurriz, D. Chemisana and L. M. Plà-Aragonés, *J. Cleaner Prod.*, 2016, **137**, 105.
17. T. Winkler, K. Schopf, R. Aschemann and W. Winiwarter, *J. Cleaner Prod.*, 2016, **116**, 80.
18. M. Meneses, C. M. Torres and F. Castells, *Sci. Total Environ.*, 2016, **562**, 571.
19. V. G. Lo-Iacono-Ferreira, J. I. Torregrosa-López and S. F. Capuz-Rizo, *J. Cleaner Prod.*, 2016, **133**, 43.
20. R. Mayer, T. Ryley and D. Gillingwater, *J. Air Transp. Manage.*, 2015, **44–45**, 82.
21. H. Wang, C. Thakkar, X. Chen and S. Murrel, *J. Cleaner Prod.*, 2016, **133**, 163.
22. M. Granovskii, I. Dincer and M. A. Rosen, *J. Power Sources*, 2006, **159**, 1186.
23. L. Gao and Z. C. Winfield, *Energies*, 2012, **5**, 605.
24. M. Messagie, K. Lebeau, T. Coosemans, C. Macharis and J. van Mierlo, *Sustainability*, 2013, **5**, 5020.
25. E. A. Nanaki and C. J. Koroneos, *J. Cleaner Prod.*, 2013, **53**, 261.
26. I. Bartolozzi, F. Rizzi and M. Frey, *Appl. Energy*, 2013, **101**, 103.
27. M. Vedrenne, J. Pérez, J. Lumbreras and M. E. Rodríguez, *Energy Policy*, 2014, **66**, 185.
28. R. Faria, P. Moura, J. Delgado and A. T. de Almeida, *Energy Convers. Manage.*, 2012, **61**, 19.
29. Y. Bicer and I. Dincer, *Int. J. Hydrogen*, 2017, **42**, 3767–3777.
30. S. Gössling and A. S. Choi, *Ecol. Indic.*, 2015, **113**, 106.
31. M. V. Chester and A. Cano, *Transp. Res., Part D*, 2016, **43**, 49.

32. N. C. Onat, M. Kucukvar, O. Tatari and Q. P. Zheng, *J. Cleaner Prod.*, 2016, **112**, 291.
33. C. François, N. Gondran, J. P. Nicolas and D. Parsons, *Ecol. Indic.*, 2017, **72**, 597.
34. C. Chavez-Baeza and C. Sheinbaum-Pardo, *Energy*, 2014, **66**, 624.
35. H. Ma, F. Balthasar, N. Tait, X. Riera-Palou and A. Harrison, *Energy Policy*, 2012, **44**, 160.
36. A. Garg, S. Vishwanathan and V. Avashia, *Energy Policy*, 2013, **58**, 38.
37. R. Frischknecht, H. J. Althaus, G. Doka, R. Dones *et al.*, *Overview and Methodology. Final Report EcoInvent 2000 No.1*, Swiss Centre for Life Cycle Inventories, Dübendorf, 2004.
38. J. Tian and M. Chen, *J. Waste Manage.*, 2016, **56**, 384.
39. A. G. Daimler, *Sustainability Report 2014*. 2015. Available online at: http://ddd.uab.cat/pub/infsos/146256/isDAIMLERa2014ieng.pdf. Accessed: 15/10/2016.
40. Fiat-Chrysler Automobiles (FCA), *2014 Sustainability Report*. 2015. Available online at: http://2014sustainabilityreport.fcagroup.com/sites/fca14csr/files/allegati/2014_sustainability_report_2.pdf. Accessed: 15/10/2016.
41. Ford Motor Company, Sustainability Report 2014/15. 2015. Available online at: http://corporate.ford.com/microsites/sustainability-report-2014-15/doc/sr14.pdf. Accessed: 15/10/2016.
42. Nissan, Sustainability Report 2014. 2015. Available online at: http://www.nissan-global.com/EN/DOCUMENT/PDF/SR/2014/SR14_E_All.pdf. Accessed: 15/10/2016.
43. Volkswagen AG, Sustainability Report 2014. 2015. Available online at: http://www.volkswagenag.com/content/vwcorp/info_center/en/publications/2015/04/group-sustainability-report2014.bin.html/binarystorageitem/file/Volkswagen_Sustainability_Report_2014.pdf. Accessed: 15/10/2016.
44. L. Ntziachristos, D. Gkatzoflias, C. Kouridis and Z. Samaras, *COPERT: A European Road Transport Emission Inventory Model*, ed. I. N. Athanasiadis, P. A. Mitkas, A. E. Rizzoli and J. Marx Gómez, Springer, 2009, pp. 491–504.
45. V. Simic, *Resour., Conserv. Recycl.*, 2016, **114**, 1.
46. M. Nieto, *La basura es una mina. Reportaje*, Diario El País, Madrid, 2014.
47. European Environment Agency (EEA), *Circular economy in Europe. Developing the knowledge base.* EEA Report No. 2/2016. Copenhagen, 2016.
48. S. Ene and N. Öztürk, *Technol. Forecast. Soc.*, 2017, **115**, 155–166.
49. T. R. Hawkins, B. Singh, G. Majeau-Bettez and A. H. Strømman, *J. Ind. Ecol.*, 2012, **17**, 53.
50. A. R. Brandt, *Sustainability*, 2011, **3**, 1833.
51. M. M. Rahman, C. Canter and A. Kumar, *Appl. Energy*, 2015, **156**, 159.
52. R. Hausmann, C. A. Hidalgo, S. Bustos, M. Coscia *et al.*, *The Atlas of Economic Complexity. Mapping Paths to Prosperity*, Centre for International Development at Harvard University, Cambridge, 2011.

53. F. Henning, B. Nygreen, M. Christiansen, K. Fagerholt et al., *Eur. J. Oper. Res.*, 2012, **218**, 764.
54. F. Henning, B. Nygreen, K. C. Furman and J. Song, *Eur. J. Oper. Res.*, 2015, **243**, 41.
55. P. Kaluza, A. Kölzsch, M. T. Gastner and B. Blasius, *J. R. Soc., Interface*, 2009, **10**, 1.
56. P. Campling, L. Janssen, K. Vanherle, J. Cofala, C. Heyes and R. Sander, *Specific evaluation of emissions from shipping including assessment for the establishment of possible new emission control areas in European Seas*. Final Report. Flemish Institute for Technological Research NV. Mol, 2013.
57. S. P. Karakitsios, V. K. Delis, P. A. Kassomenos and G. A. Pilidis, *Atmos. Environ.*, 2007, **41**, 1889.
58. L. Stamford and A. Azapagic, *Energy Sustainable Dev.*, 2014, **23**, 194.
59. R. Faria, P. Marques, P. Moura, F. Freire et al., *Renewable Sustainable Energy Rev.*, 2013, **24**, 271-287.
60. Deloitte, *European energy market reform*. Country profile: UK. Zurich, 2015.
61. Deloitte, *European energy market reform*. Country profile: Spain. Zurich, 2015.
62. K. Danilecki, M. Mrozik and P. Smurawski, *J. Cleaner Prod.*, 2017, **141**, 208.
63. O. V. Lozhkina and V. N. Lozhkin, *Transp. Res., Part D*, 2016, **47**, 251.
64. C. Pastorello, P. Dilara and G. Martini, *Transp. Res., Part D*, 2011, **16**, 121.
65. J. A. García, J. M. López, J. Lumbreras and M. N. Flores, *Energy*, 2012, **47**, 174.
66. A. Alam and M. Hatzopoulou, *Transp. Res., Part D*, 2014, **29**, 12.
67. H. Boogard, N. A. H. Janssen, P. H. Fischer, G. Kos et al., *Sci. Total Environ.*, 2012, **435-436**, 132-140.
68. G. Bel, C. Bolancé, M. Guillén and J. Rosell, *Transp. Res., Part D*, 2015, **36**, 76.
69. V. Franco, M. Kousoulidou, M. Muntean, L. Ntziachristos et al., *Atmos. Environ.*, 2013, **70**, 84.
70. M. Vedrenne, S. Cooke, J. R. Stedman and A. J. Kent, *Streamlined PCM Technical Report. Report for Department for Environment, Food & Rural Affairs (project AQ0959)*. London, 2015.
71. M. Kousoulidou, L. Ntziachristos, G. Mellios and Z. Samaras, *Atmos. Environ.*, 2008, **42**, 7465.
72. R. Smit, L. Ntziachristos and P. Boulter, *Atmos. Environ.*, 2010, **44**, 2943.
73. Ayuntamiento de Madrid, *Estudio del parque circulante de la ciudad de Madrid. Año 2013*. Dirección General de Sostenibilidad y Planificación de la Movilidad. Madrid, 2014.
74. R. Borge, I. de Miguel, D. de la Paz, J. Lumbreras et al., *Atmos. Environ.*, 2012, **62**, 461.
75. C. Quaassdorff, R. Borge, J. Pérez, J. Lumbreras et al., *Sci. Total Environ.*, 2016, **566-567**, 416-427.

76. Ministerio de Agricultura, Alimentación y Medio Ambiente (MAGRAMA), *Inventarios Nacionales de Emisiones a la Atmósfera 1990-2012. Volumen 2: Análisis por Actividades SNAP. Capítulo 7: Transporte por carretera*. Madrid, 2013.
77. Department for Transport (DfT), *Transport Statistics Great Britain*. London, 2014.
78. National Centre for Statistics and Analysis (NCSA), *Vehicle Survivability and Travel Mileage Schedules*. Springfield, 2006.
79. J. Lumbreras, R. Borge, A. Guijarro and J. M. López, *Technol. Forecast Soc.*, 2014, **81**, 165.
80. Society of Motor Manufacturers and Traders (SMMT), *SMMT Annual CO_2 report*. 2006 market. London, 2006.
81. M. Vedrenne, R. Borge, J. Lumbreras, M. E. Rodríguez *et al.*, *Atmos. Environ.*, 2016, **145**, 29.
82. V. R. J. J. Timmers and P. A. J. Achten, *Atmos. Environ.*, 2016, **134**, 10.
83. C. B. B. Guerreiro, V. Foltescu and F. de Leeuw, *Atmos. Environ.*, 2014, **98**, 376.
84. European Environment Agency (EEA), *Air quality in Europe – 2015 report. EEA Report No. 5/2015*. Copenhagen, 2016.
85. C. Nagl, L. Moosmann and J. Schneider, *Assessment of plans and programmes under the Air Quality Framework Directive*. Umweltbundesamt Österreich, Vienna, 2010.
86. Department for Environment, Food & Rural Affairs (Defra), *Improving air quality in the UK. Tackling nitrogen dioxide in our towns and cities. UK overview document*. London, 2015.
87. M. Maione, D. Fowler, P. S. Monks, S. Reis *et al.*, *Environ. Sci. Policy*, 2016, **65**, 48.
88. R. M. González and G. A. Marrero, *Energy Policy*, 2012, **51**, 213.
89. E. C. Gentil, A. Damgaard, M. Hauschild, G. Finnveden *et al.*, *Waste Manage.*, 2010, **30**, 2636.
90. A. Laurent, I. Bakas, J. Clavreul, A. Bernstad *et al.*, *Waste Manage.*, 2014, **34**, 573.

Subject Index

Abrasion wear 88
Adaptation 157, 163–172
Adiposity 123
Adsorption 96
Advanced biofuels 36, 37
　　combustion 8, 10, 16
After-treatment 1, 4, 5, 6, 18, 25, 32
Agricultural 26, 31, 32, 33, 36, 37
　　N$_2$O 33
Air pollution 46, 47, 48, 55, 56, 60, 62, 63, 64, 65, 68, 71, 74, 116
Air quality 236
Air-guided systems 7
Al 92
Alcohols 3, 9, 10
Ammonia 32
Annoyance 125
Antecedent dry period 95
ASR disposal fees 178
Assessment 215
Atherosclerosis 122
Atrial fibrillation 122
Automobile manufacturers 179
Autonomic nervous system 108
A-weighted equivalent value (L_{Aeq}) 111

Backstop technologies 157, 158, 162–164, 171
Batteries 138, 195
Battery 137
　　lifetime and degradation 149
　　market 182
　　recycling 153, 184
　　utilisation 147
BC 55, 58, 59, 62, 72, 75
BEVs 134
Bioalcohols 1

Bioavailability 93
Biobutanol 10
Biodiesel 3, 10, 11, 12
Bioethanol 10
Biological mechanisms 108
Black carbon 55
Blood lipids 124
　　pressure 115
　　samples 203
　　samples of livestock 204
Body mass index (BMI) 123
Brake 91
Build-up 95
Burden of Disease from Environmental Noise 108
Butanol 10, 17

CadnaA 111
Car 235
Carbon dioxide 3, 26, 25, 27
　　intensity 40
Carbonates 99
Carburettors 6
Carcinogens 94
Cardiovascular disease 115
　　mortality 118
Carotid artery intimal media thickness (CIMT) 122
Catalysts (DOCs) 5
Catalytic converters 93
Cd 92
Cerebrovascular disease 118
Cetane 8–10, 13, 17
CH$_4$ 26, 34
Charging infrastructure 137
China 179, 195
Circulation 231

Climate change 97
CMB 64, 68, 70
CNOSSOS-EU 111
CO 48–51, 55, 57–59, 63, 65, 71, 72, 75
CO_2 4–6, 9–16, 20, 26–35, 39–42, 48–59
    -eq. 33–36
    ROI 147
CO_2/GHG 28
Coarser particles 98
Common rail 4, 5
Component wear 88
Composition 232
    ratios for vehicles 182
Compression ignition 3, 4, 47
    ratio 6–8, 9, 13, 16, 19, 20
Computerization 181
Conclusions 218
Confounding 125
Cr 92
Cross-border environmental problems 186, 193, 209
Crude varieties 226, 227
Cu 92
Cylinder deactivation 16, 18, 19

Day-time and night-time average sound levels 110
Decision-support tool 238
Definition 215
Degree of hybridisation 142
Developing countries 174
Diesel engine 47–52, 70, 75
    oxidation 5
    oxidation catalysts (DOCs) 4–5
    particle filter (DPF) 5
DOC 5
Double glazing 125
Down-sizing 4, 7, 16
DPF 5
Drive cycles 2, 142
Driving conditions 231
Driving cycle 2, 142
Dry deposition 89
Dual-fuel 13

Eastern Europe 177
EC 56, 59, 61
Eco-friendly vehicles 175, 191
Effect-modifying factors 125
EFs 230
EGR 4, 5, 8, 12, 16, 17
Electric vehicles 3, 14, 15
Electricity 229
    production 146
Elemental carbon 56
ELV Directive 176
ELV recycling 174, 175, 206–208
    rates 177
    regulation 208
Emission limits 1, 2
    model 232
End-of-life 224
    HVs 183
    NGV battery 183
    Recycling Law 178
    vehicles (ELVs) 174
Energy demand 2
    intensity 40
    mix 229
Energy recovery 40, 41, 177
    return on investment 145
    security 157, 158, 163–165, 172
Environmental damage 197
    health 88
    impact 146
    noise 107
    problems 174
Epidemiological studies 114
EPR principle 209
    -based ELV recycling laws 176
Equivalent 28
    carbon dioxide 25, 33
EROI 147
Ethanol 3, 10, 14, 17
EU 177
Euro 5 235
European Noise Directive 111

Subject Index

EV 181, 184
Exhaust emissions 88
    gas recirculation (EGR) 4
Extended producer responsibility (EPR) 174, 175
Extraction 222

Facade noise level 114
FC system 141
Fe 92
    -Mn oxides 99
Fertilisers 32, 33, 35
Finer particles 98
First flush 96
Flywheels 138
Fossil fuels 1–3, 8, 9, 11
Free trading 177
Fuel cell vehicles 134
    consumption 1, 6, 7, 11–20
    delivery system 6
Fuel 88
    -air mixture 4, 17
    economy 25, 28, 30, 40, 41, 153
    properties 8

Gasoline engine 47, 50
Gas-online-range organics 94
GCI 17
GDI 6–8, 14, 15, 17
GHG emissions 208
GHG 26–40
Global ELV recycling system 209
    environmental problems 209
Glucose 124
Glycated haemoglobin (HbA1c) 124
Greenhouse gas 25, 26, 31
    gas (GHG) emission 134
Gross damage 157, 165
Ground deposition 88
Guidelines 179

HCCI 16, 17
Heart failure 121
    rate 122
High-octane 8

Housing type 125
Human and to ecosystem health 92
HV 181
HVO 11
HVO's cetane number 11
HWFET 144
Hybrid 3, 14, 15, 16, 20
    electric vehicles 14, 15
    EVs 134
    vehicles 141, 152
Hydrocarbons 92
Hydrogen 10–13
Hydrotreated vegetable oil 11
Hydrous metal oxides 99
Hypertension 115

Illegal dumping 187
Impervious surfaces 98
Imported used cars 174
International 208
    resource circulation 186
    resource recycling 209
Ischaemic heart disease (IHD) 118

Japan 178, 190
Jeepneys 205, 207

Knock 6–8, 19
Knocking 17
Korea 178, 190

L_{den} 108
Land use regression (LUR) modelling 114
Lead acid batteries 139, 140
    batteries 193
    battery recycling 197
    battery recycling factories 201
    concentration 199, 200, 202, 204
    concentrations in the blood 203
    exposure 205
    refining operator 195
    toxicity 198

Subject Index

Leaded fuel 92
Lean NO_x traps 5, 8
LIBs 184
Life cycle analysis 220
    cycle assessment 145, 214
    cycle inventory 216
    impact assessment 217
Limitations 233
Lithium-air 140
Lithium-ion 140
    -ion battery 146, 152, 180
Local junkshop 207
London, UK 112
Low carbon 40
    -temperature combustion 5, 16
LP 11
LPG 3, 13, 14
Lubricants 89
Lubrication system leakages 88

Manual dismantling 179
Meta-analysis 115
Metals 92
Methane 26–29, 31, 41
Methanol 10
Metro Manila 205
Mileage 232
Misclassification of individual noise exposures 114
Mn 92
Modelling 237
Mongolia 187, 188, 190, 197
Monitoring 205
    system 179
Monitors 114
Motorization 175, 191
Myocardial infarction(MI) 118

N_2O 26, 32–34
Natural gas 1, 3, 9, 12, 13
    gas transportation 226
New European Drive Cycle 144
    Regulation on the disposal of End-of-Life Vehicles 179
    -generation vehicles (NGVs) 174, 175

Next-generation vehicles 180
NGV recycling 209
NH_3 32
Ni 92
Nickel 184
    -cadmium batteries 139, 140
    -hydrogen batteries 195
    -hydrogen battery (NiMH) battery 180
    -metal hydride 139, 140
Night Noise Guidelines for Europe 109
Nissan 184
Nitrate 59
Nitric oxide 3
Nitrous oxide 25–27, 32
NMVOC 72
NO 47
NO_3^- 55, 58, 59
Noise impacts on sleep 108
    level errors 114
    modelling 110
    sensitivity 126
Nomadic people 198, 204
Non-authorized recycling facilities 178
    -tailpipe emissions 47, 53, 54, 57, 60
NO_x 5, 8, 16, 17, 20, 26, 32, 41, 47–59, 61, 65, 69, 71, 72,75
    -particulate emissions trade-off 4
NYCC 144

O_3 47, 48, 52, 55, 57–59, 61–63, 65, 66, 68–70
Octane 8–10, 12, 13, 17
Oil 225
Organic carbon 92
    matter 99
Ozone 47, 66

PAHs 56, 57, 59, 71, 94
Particle 25, 26, 37, 38, 39
    mineralogical composition 99
    -specific surface area 99

Subject Index

Particulate matter (PM) 1, 3, 26
    solids 95
Pavement 91
Pb 92
PCCI 17
PFI 6, 7
Philippines 205
Photochemical smog 47, 48, 58
Plastic 182
PM 4–20, 26, 41, 47–58, 60, 62, 63, 67, 70, 74, 75
PM_1 69
PM_{10} 47, 53, 54, 57, 58, 61
$PM_{2.5}$ 47, 53–65, 68–72, 75
PMF 64, 69
PM-NO_x trade-off 4
    -NO_x-thermal efficiency 41
Policy actions 237, 238
Pollutant build-up 95
Pollutant characteristics 88
    wash-off 96
Polycyclic aromatic hydrocarbons 47, 90
Power generation 152
Powertrain 8, 9, 135
Precious metals 182

Rainfall patterns 97
Range extender 137
Raw materials 222
Recovery 225
Recyclable resources 208
Recycling 151, 184
    fee 178
    fund 178
    industry 205
    process 193
    rate 176
    schemes 186
Refinery 199, 204, 225, 228
Regenerative braking 137, 144
Renewable energy 185
Residual fractions 99
Resource circulation 209
    recycling 192
Reuse 184
    market 192

Review 219
Rightsizing 7
Road dust 88
    layout 91
    paint 92
    slope 92
    surface wear 88
    surfaces 87
    traffic noise 108
Roadside soil 92
Runoff volume 96

Sampling 200
Scope 215
SCR 5, 32
Scrap metal and parts 208
Second life 151
    life batteries 153
Secondary organic aerosols 47
Second-hand 183
    Japanese vehicles 188
    vehicle 185, 187, 190, 195
    vehicle exportation 186
    vehicle imports 188
Selective catalytic reduction (SCR) 5, 32
Smart meters 148
SO_2 47– 49, 54, 55, 66, 67, 71, 72, 74, 75
SO_4^{2-} 55, 58, 59
SOA 47, 58, 59
Social benefits 207
Socio-technical 157, 161–163
Soil environment 199
Sound pressure level (SPL) 110
SoundPLAN 111
Source apportionment 53, 64, 69
Spark ignition 1, 3, 5, 47
    ignition powertrains 1
Specific energy 135
Steel content 182
Stop-start 16, 20
Storage batteries 195, 197
    device 185
Stormwater runoff 88
Street canyons 114

Sulfate 48, 59, 68, 75
Supercapacitors 138
Surface coatings 99
    complexation 99
    electrochemical
        properties 99
    texture 91
Synthetic fuels 3, 11
Systematic review 115

Tailpipe 57
    emissions 47, 50, 53, 57
Tank-to-wheel 25, 41, 42
Texture depths 92
Thermal efficiency 40, 41
Thoracic aortic calcification 122
Three-way catalyst 32
Topsoil 202
Toxicity 94
Trace metals 92
Traffic characteristics 89
    congestion 89
    density 101
    -generated emissions 87
    mix 89
    pollutants 87
    volume 89
TRANEX: TRAffic Noise
  Exposure 112
Transnational environment
  problems 187
Transport 236
Transportation 157, 159, 162–167, 172, 225
    modes 227
Turbochargers 4, 5
TWC 5, 6, 8
Type 2 diabetes mellitus 124
Tyre 91

UK Calculation of Road Traffic Noise (CoRTN) method 111
Ulaanbaatar 198, 199
Unauthorized routes 195

Unburned fuel 3
Urban agriculture 101
    environment 89
    surfaces 98
Urbanisation 87
Used parts 192
    parts market 193
    vehicles 208

Variable valve 5, 7, 19
Vehicle ages 191
    assembly 223
    attributes 230
Vehicle efficiency 35, 41, 145
    emissions 46–49, 52, 55, 60, 62, 66, 74, 76
    -to-grid 148
Vehicles 220
Vehicular traffic 87
VOCs 47–59, 65, 69–71
Volatile organic compounds 47
Volatility 8, 9, 16, 17

Waist circumference (WC) 123
    /hip ratio (WHR) 123
Wall guided system 7
Wash-off 95
Waste lead batteries 200
    management 206
    materials 192
Water environment 93
Weight reduction 137, 181
Well-to-tank efficiency 136
    -to-wheel (WtW)
        efficiency 135, 145
Wet deposition 89
Window opening 125
World Health Organisation (WHO) 108

X-ray fluorescence 202

Zinc-air 140
Zn 92